REALITY
CHECK

Donald R. Prothero

REALITY CHECK

How Science Deniers Threaten Our Future

Foreword by Michael Shermer

Illustrations by Pat Linse

INDIANA UNIVERSITY PRESS

Bloomington & Indianapolis

This book is a publication of

Indiana University Press
Office of Scholarly Publishing
Herman B Wells Library 350
1320 East 10th Street
Bloomington, Indiana 47405 USA

iupress.indiana.edu

Telephone orders   800-842-6796
Fax orders   812-855-7931

© 2013 by Donald R. Prothero

⊖ The paper used in this publication
meets the minimum requirements of
the American National Standard for
Information Sciences—Permanence
of Paper for Printed Library
Materials, ANSI Z39.48-1992.

Manufactured in the
United States of America

Cataloging information is available
from the Library of Congress.

ISBN 978-0-253-01029-2 (cloth)
ISBN 978-0-253-01036-0 (eb)

1 2 3 4 5   17 16 15 14 13

THIS BOOK IS DEDICATED TO MY SONS,
ERIK, ZACHARY, AND GABRIEL PROTHERO

*May their future be brighter than ours,
and governed by more rationality than is our current world.
May they not curse the previous generations for the
problems we left behind.*

Facts do not cease to exist because they ignored.

ALDOUS HUXLEY

To treat your facts with imagination is one thing,
but to imagine your facts is another.

JOHN BURROUGHS

Reality is that which, when you stop believing in it,
doesn't go away.

PHILIP K. DICK

You are entitled to your own opinion, but you
are not entitled to your own facts.

FORMER SENATOR DANIEL PATRICK MOYNIHAN

Reality must take precedence over public relations,
for Nature cannot be fooled.

RICHARD FEYNMAN

CONTENTS

FOREWORD

*Denialism vs. Skepticism: How to Think
about Controversial Issues*

MICHAEL SHERMER

Was 9/11 a conspiracy? Yes, it was. By definition, a group of nineteen al-Qaeda members secretly plotting to fly planes into buildings constitutes a conspiracy. But that is not what the so-called 9/11 Truthers believe. They think that 9/11 was an inside job orchestrated by the Bush administration in order to implement its plan for global domination and a New World Order launched by a Pearl Harbor–like attack (which was also an inside job by Roosevelt and Churchill) on the World Trade Center, the Capitol, and the Pentagon, thereby providing the justification for war.

What is the evidence for this conspiratorial claim? There is no *positive* evidence whatsoever—no security camera videotape of people planting explosive devices, no explosive device debris in the World Trade Center ruins, no letters, e-mails, memos, or documents of any kind, no confessions by conspirators or their friends, family, or colleagues who might have overheard a clandestine conversation, and no one coming forward to tell all in a book or on a television talk show about what they saw or heard. Nothing. Instead, Truthers rely on alleged anomalies in the government's explanation for what happened, such as how the World Trade Center buildings collapsed, or why WTC building 7 fell, or the damage to the Pentagon, or cell phone peculiarities, or . . .

The belief that a handful of unexplained anomalies can undermine a well-established theory lies at the heart of all conspiratorial thinking, and is easily refuted by noting that beliefs and theories are not built on single facts alone, but on a convergence of evidence from multiple lines of inquiry. This principle of converging evidence lies at the heart of determining the difference between skepticism and denial. There is nothing wrong with being skeptical of one's government, for example, because we know that governments lie to their citizens and that politicians can be bought off by special interest groups. But when ideology trumps facts—when commitment to a political, economic, or religious belief takes precedence over evidence—skepticism merges into denial. Never is this more evidence than in politics, particularly regarding such questions as these: Should gay marriage be legal? Should marijuana be decriminalized? Should health care be universal? Science has little to say on these matters except on specific points within the larger questions: For example, does the legalization of gay marriage lead to a decline in traditional marriage? (No, it does not.) On such questions, people typically line up according to their religious, political, or social beliefs and corresponding cohorts, and listen to their opponent's arguments only in order to shoot them down in a public debate.

The adversarial structure of modern politics invites liberals and conservatives to deny the other side's position a priori. In this sense, denialism is part and parcel with politics—you are supposed to deny your political opponent's position, otherwise you are not a good party member. Not so in science . . . at least in principle.

Donald Prothero has emerged as one of America's foremost experts on and debunkers of pseudoscience of various stripes. As a world-class paleontologist and geologist he diverted precious research time to the cause of taking on the evolution deniers—creationists and their intelligent design brethren—because of the threat they pose to good science education in America. Prothero noticed that global warming skeptics and climate deniers employed the same tactics as creationists: focusing on minor anomalies in the data, interpreting normal scientific debates as indications that mainstream science is flawed, and quote mining experts to make it sound as if they were saying something in support of their denialist cause. *Reality Check: How Science Deniers Threaten Our Future* is Prothero's magnum opus on all things pseudoscience, covering not only creationism and climate denial, but also other threats to a rational and sane society, including the anti-vaxxers (those who believe vaccinations cause autism and other problems and should be abandoned), the AIDS deniers (yes, believe it or not, there are still people who do not believe that HIV causes AIDS), alternative medical practitioners who deny the benefits of modern science–based medicine, the tobacco deniers (primary smoking deniers have morphed into secondhand smoking deniers), the peak oil deniers (those who hold that the supply of oil is nearly endless), and many others who employ tried-and-true strategies of selling doubt as a product. As Prothero demonstrates, it is almost as if all these deniers went to the same school of denial, employing parallel methods to sow seeds of doubt into the mind of the public, who as non-experts often have a difficult time distinguishing the difference between denial and skepticism.

Denial or denialism is the automatic gainsaying of a claim regardless of the evidence for it—and sometimes even in the face of evidence. Denialism is typically driven by ideology, politics, or religious beliefs, in which the commitment to the belief takes precedence over the evidence for or against it. Belief comes first, reasons for belief follow, and those reasons are winnowed to assure that the belief is always supported.

Prothero is a skeptic. So am I. When we call ourselves skeptics we mean simply that we take a scientific approach to the evaluation of claims. Science is skepticism and scientists are naturally skeptical because most claims turn out to be false. Weeding out the few kernels of wheat from the substantial pile of chaff requires extensive observation, careful experimentation, and cautious inference to the best conclusion. Donald Prothero is a scientist's scientist in this regard, and as the editor of *Skeptic* magazine I have leaned on him many times over the years not only for his expertise in a particular field but for his overall agility in thinking critically about any controversy in mainstream and borderlands science. In this volume you will indeed get a reality check on some of the most important issues of our time.

Michael Shermer is publisher of *Skeptic* magazine, a monthly columnist for *Scientific American,* adjunct professor at Claremont Graduate University, host of the Skeptics Society Distinguished Science Lecture series at Caltech, and author of *Why People Believe Weird Things, Why Darwin Matters,* and *The Believing Brain*

PREFACE AND
ACKNOWLEDGMENTS

This book was originally written during summer 2010, and extensively rewritten in spring 2012 as events changed many of the points made in the first draft. It represents over thirty years of research and teaching on my part, from my firsthand familiarity with the issues of creationism and global warming science to the issues of growth, population, and resources—which have been sadly overlooked in recent years. It reflects my professional expertise in geology, biology, medicine, astronomy, and many other subjects that were the foundation for the chapters. Of course, I have learned much from people on the cutting edge of many of these fields, and their work is cited in the appropriate places.

Many recent books have dealt with the issues of creationism (including my 2007 book on the topic) and climate change (including my 2009 book on the topic), but few have tried to write a book that connects the common threads among science denial movements, from creationism to climate change deniers to anti-vaxxers and AIDS deniers, to medical quackery, to astrology, to the issues of resource and population denial. In particular, it is striking how many of these deniers use exactly the same tactics pioneered by the Holocaust deniers, and refined by tobacco companies seeking to cloud the scientific issues. Yet, as I emphasize throughout, science is one of our most precious discoveries and assets, and our only hope for the future. Whether we take the path of science and rationality, or superstition and denial, will determine whether we survive another century on this planet.

For a project that encompasses so many topics, and so many different fields of expertise, it is essential that there be sufficient input from experts in various fields to catch errors and inconsistencies. I thank the following individuals for reviewing all or part of the manuscript in various drafts: Gilbert Klapper, James Lippard, Michael Shermer, Steve Novella, and several anonymous reviewers. I thank Pat Linse for her masterful job preparing the art. I thank my editor, Robert Sloan, and the Indiana University Press staff for help with production. Finally, I thank my family—especially my patient and understanding wife, Teresa LeVelle, who allowed me the time to work on this project and took the kids on vacation for three weeks in summer 2010 so I could finish the book without distractions.

REALITY CHECK

1

Reality Check

To treat your facts with imagination is one thing,
but to imagine your facts is another.

John Burroughs

What's real is what's real, and, like it or not, no one can change
the nature of reality. Except, of course, with mushrooms.

Bill Maher

Let us imagine a scenario:

- This scientific consensus on this idea is accepted by 95–99% of all
  the scientists who work in the relevant fields;
- This scientific topic threatens the viewpoints of certain groups
  in the United States, so it is strongly opposed by them and those
  they influence;
- Their antiscientific viewpoint is extensively promoted by websites
  and publications of right-wing fundamentalist institutes such as
  the Discovery Institute in Seattle, and is often plugged by Fox
  News;
- Opponents of this consensus cannot find legitimate scientists
  with expertise in the field who oppose the accepted science, so
  they beat the bushes for so-called scientists (none of whom have
  relevant training or research credentials) to compose a phony list
  of scientists who disagree on the topic;

- Deniers of the scientific consensus resort to taking quotations out of context to make legitimate scientists sound as though they question the consensus;
- Deniers of the scientific consensus often look for small disagreements among scholars within the field to argue that not everyone in the field supports their major conclusions;
- Deniers often nitpick small errors on the part of individuals to argue that the entire field is unsound;
- Deniers of the scientific consensus often focus on small examples or side issues that do not seem to support the consensus to argue that the consensus is false;
- Deniers of the scientific consensus spend most of their energies disputing the scientific evidence, rather than doing original research themselves;
- By loudly proclaiming their "alternate theories" and getting their paid hacks to question the scientific consensus in the media, they manage to make the American public confused and doubtful, so only half of U.S. citizens accept what 99% of legitimate scientists consider to be true;
- By contrast, most modern industrialized nations (Canada, nearly all European countries, China, Japan, Singapore, and many others) have no problems with the scientific consensus, and treat it as a matter of fact in both their education and in their economic and political decisions;
- The deniers are part of the right-wing Fox News echo chamber, and repeat the same lies and discredited arguments to themselves over and over again;
- Powerful Republican politicians have used the controversy over this issue to force changes in the teaching of this topic in schools.[1]

Most people reading through that list would immediately assume that it describes the creationists and their attempts to target the overwhelming scientific consensus on evolution. Indeed, the list could describe creationists, or evolution deniers—but it also describes the actions of the climate deniers (who deny global warming is real and caused by humans) as well. Indeed, the membership lists of creationists and climate change deniers have a great deal of overlap, and both causes are promoted equally by right-wing political candidates, news media (especially Fox News), and religious organizations such as the Discovery Institute.

Even more revealing is how these denier movements get the money to make such a fuss. As Deep Throat said in *All the President's Men,* "Follow the money." Both kinds of denialism are heavily funded by wealthy entities with vested interests that further their causes while characterizing them as populist grassroots movements in opposition to unpopular scientific topics. The creationists are funded not only by many rich fundamentalist churches, but also by powerful right-wing businessmen or institutes—such as Howard Ahmanson, Jr., the Coors family, the McClellan Institute, and the Stewardship Foundation. The climate deniers receive massive funding and support from the oil, coal, and other energy industries—especially ExxonMobil and Koch Industries—that are threatened by the possibility of our reduced dependence on oil and coal.

Let us make an important distinction here: these deniers are not just "skeptics" about climate change or any other scientific idea that they do not like. A skeptic is someone who does not believe things just because someone proclaims them, but tests them against evidence. Sooner or later, if the evidence is solid, then the skeptic must acknowledge that the claim is real. A denier, by contrast, is ideologically committed to attacking an opposing viewpoint, and no amount of evidence will change their minds. In the words of astronomer Phil Plait,

> I have used the phrase "global warming denialists" in the past and gotten some people upset. A lot of them complain because they say the word *denial* puts them in the same bin as holocaust deniers.
>
> That's too bad. But the thing is, they do have something in common: a denial of evidence and of scientific consensus.
>
> Moon hoax believers put themselves in this basket as well; they call themselves skeptics, but they are far from it. Skepticism is a method that includes the demanding of evidence and critical analysis of it. That's not what Moon hoax believers do; they make stuff up, they don't look at all the evidence, they ignore evidence that goes against their claims. So they are not Moon landing *skeptics,* they are Moon landing *deniers.* They may start off as skeptics, but *real* skeptics understand the overwhelming evidence supporting the reality of the Moon landings. If, after examining that evidence, you still think Apollo was faked, then congratulations. You're a denier.

Really, it's this difference that biases people against skeptics like me. I am always accused of having a closed mind—of being a denier. But that's not only not true—I can be convinced I am wrong by evidence or a logical argument—but it's usually the person accusing me that has a mind closed against reality. No matter how much evidence you put in front of them showing them clearly and obviously that they are wrong, they refuse to see it.[2]

BELIEF VS. REALITY

Reality is that which, when you stop believing in it, doesn't go away.

Philip K. Dick

Climate denialism and creationism have a lot in common with many other kinds of denialism. In each case, a well-entrenched belief system comes in conflict with scientific or historic reality, and the believers in this system decide to ignore or attack the facts that they do not want to accept. Holocaust deniers are a classic example of this. Despite the fact that we have hundreds of survivors who were victims and witnesses of the Holocaust (sadly, fewer and fewer of them remain) and accounts written by the Nazis themselves, the deniers keep on pushing their propaganda to a younger generation that has no memory of the Holocaust and does not get to hear about it in school. When you dig deep enough, the Holocaust deniers are nearly all hard-core antisemites and neo-Nazis who want to see the return of the Third Reich, but for public appearances they attempt a façade of legitimate scholarship. Most people regard the Holocaust deniers as a minor nuisance, but to the Jewish community they represent the threat that the Holocaust might happen again. In Germany and in several other European countries, it is a crime to deny that the Holocaust happened, and prominent deniers (such as David Irving) have been convicted and gone to prison. Yet in the Muslim world, Holocaust denial is commonly used to incite Muslims against Israel. Just in the past few years, we have heard numerous Muslim leaders (such as Mahmoud Ahmadinejad of Iran) make statements of Holocaust denial with the full approval of his government and many other Muslims.

Human beings have many ideas that conflict with reality. Most of the time we regard them as just harmless cranks and curiosities. Just Google the term "Flat Earth Society" and you will find websites describing small but sincere groups of believers who are convinced that the earth is not a sphere but a flat disk. When confronted with photographs of the earth from space, they always claim that these images are fraudulent or doctored in some way. When the topic of the moon landings is raised, they claim it was all a NASA hoax filmed in a soundstage. Their insistence on a flat earth and a geocentric view of the world (with the earth, not the sun, at the center of the solar system) is based on biblical literalism. There are many verses in the Bible (e.g., Isaiah 11:12, 40:22, 44:24) that say so, and they believe the Bible must be literally true. Most people find them amusing and silly, but their belief system is just as strongly held as the beliefs of many of their audience. When these same people who laugh at the flat-earthers are confronted with aspects of their own belief systems that conflict with science and reality, they do not find the issue so amusing after all.

Likewise, there is an entire group of religious fanatics who still believe that Galileo (and Copernicus and every astronomer since then) was wrong and the Church was right in insisting the earth was the center of the universe. They held a conference in November 2010 that featured many speakers with impressive-sounding credentials (but none with any true scientific training in astronomy).[3] The list of talk titles reveals a mix of weird science, paranoid conspiracy theories that claim the shots of earth from space are hoaxes, and apologias for the literal interpretation of the Bible that does indeed claim the earth is the center of the universe (as all ancient cultures believed). Ironically, the Catholic Church has long ago apologized for its persecution of Galileo and for its long rejection of the heliocentric solar system, so clearly they do not endorse these views by Catholics who do not follow their own church's teachings.

Sometimes, however, these crazy ideas have negative consequences and cannot just be dismissed as the human propensity to believe "weird things." In Appalachia, there are churches that routinely handle rattlesnakes, copperheads, and other poisonous serpents out of a conviction that their connection to God will protect them against snakebite and death. After all, the Bible says so (Mark 16:17–18; Luke 10:19). Most individuals in our society regard these people as a deluded cult, and find them amusing or appalling. Nevertheless, more than seventy people in these small churches have died of snakebite over the past eighty years—proof that their faith does not stop Mother Nature. The latest such person was Pastor Mack Wolford, who died in May 2012 from snakebite, just as his pastor father did in 1983.[4]

Sometimes the belief systems are so dangerous that the cult followers lose their lives, as in the case of Jim Jones's People's Temple in Guyana in 1978 (913 people died drinking cyanide-laced Flavor Aid) or the Heaven's Gate cult, whose 39 members committed suicide in 1997 in the belief that aliens were aboard a comet and about to take them to heaven. Other belief systems demand that their followers physically abuse themselves, or stare into the sun until they are blind, or starve themselves.

These examples are indeed extreme, but most humans practice many behaviors that are in denial of reality. These include superstitions that a certain activity or item of clothing will bring luck to their favorite team, or gambling in the lottery or casinos. We like to think of ourselves as rational beings, but we fall back on irrational thinking and behavior time and time again. For example, we often make the mistake of assuming that if two events happen together, one must have caused the other. An example is the urban myth of "earthquake weather," the idea that earthquakes happen when a particularly hot day occurs. There is no link, of course, and it makes no sense, since earthquakes are generated many kilometers down in the earth's crust and cannot feel the daily changes in weather, which only penetrate a few centimeters down into the ground. This, and many other examples, of two phenomena that seem to be connected but are not, is commonly known as the post hoc fallacy (from the Latin *Post hoc ergo propter hoc*, "After this, therefore because of this"). Scottish philosopher David Hume put it in a more modern context by pointing out that correlation does not equal causation. Just because two events occurred together does not mean they are related or causally connected.

Another example is the old myth that sleeping while wearing your shoes causes a headache. A closer examination of the facts shows that the shoes did not cause a headache; drunks who fall asleep without taking off their shoes almost always have a hangover headache the next day.

This is similar to the situation with the anti-vaccine deniers discussed in chapter 7. They claim to have noticed signs of autistic behavior in their children about the time that the child received certain shots (such as the MMR vaccine), and assume that the shots caused autism. But as chapter 7 shows, this is a coincidence, not cause and effect. It just happens that autistic symptoms show up in most developing children at about eighteen months, the same age when these shots are given. Correlation does not necessarily prove causation.

Why do people fall for this type of thinking? Much of it is hardwired in our brains because it conferred survival value, and the ability to see patterns and connections was a particularly important skill. When we were small and helpless and hunted by a wide variety of large, terrible predators during the Ice Ages, a fast reaction on our part to a sound or to a movement might have meant escape and survival. Making links and connections between various events is how people navigate complex environments. In the past, it helped us to hunt, find food, and avoid death; now it helps us deal with people and keep track of large amounts of information. The curious hominid who stopped to discern whether a threat was real might end up as lunch for a saber-toothed cat. So, like a skittish deer or bird, we correlate any unusual sounds or movements with threats, even though these things rarely threaten us any more.

Another common fallacy hardwired into our brains is *confirmation bias*. Humans tend to see what they expect to see, and forget when things do not match expectations. We hear evidence that appears to support our existing belief systems, and ignore or try to discount evidence that suggests we might be wrong. We correlate events with a wide variety of belief systems, and whenever the events seem to respond to our prayers, we claim that our beliefs made it so. A typical fortune-teller or medium or psychic or faith healer works with confirmation bias when they conduct a session, using the principle of cold reading. They start by describing very general things that commonly trouble most people and that may or may not be true about you. If you give them any positive feedback (by speech or body language), they then zero in on your cues and make more and more specific predictions. Yet if you sit down with a transcript of a psychic reading, you will find that the psychic or fortune-teller was more often wrong than right—but every time they get one thing right by random guessing and following our cues, we forget the ten times they were wrong.

In short, humans are very easily fooled, and believe all sorts of "weird things" that are manifestly not true. We are very easily deceived and duped, especially by our own instincts and training, and often make disastrous decisions based on these false beliefs. As the Nobel Prize–winning physicist Richard Feynman said, "The first principle is that you must not fool yourself—and you are the easiest person to fool."

So how do we avoid fooling ourselves? How do we avoid getting caught up in weird beliefs and find out what is real? Many people have their own ideas about this—from religious beliefs to political dogmas—but the one method that has worked time and again is the *scientific method*. That is the subject of our next chapter.

2

Science,
Our Candle in the Darkness

Science is nothing but developed perception, interpreted intent, common sense rounded out and minutely articulated.

George Santayana

There are in fact two things, science and opinion; the former begets knowledge, the latter ignorance

Hippocrates

A WORLD TRANSFORMED

Foreign travel is an extremely valuable experience. Not only do you get to see amazing sights that occur nowhere else and receive exposure to languages and foods and cultures very different from our own, but if you travel to the underdeveloped world, you also begin to appreciate how lucky we are to live in industrialized nations. Seeing the huge numbers of poor, diseased, starving people in Africa—or much of Asia or Latin America—usually comes to a shock to pampered Westerners. Once in a while, a hit foreign movie like *Slumdog Millionaire* disturbs us with scenes of a young boy jumping into a latrine full of feces, or children being blinded by scoundrels who exploit them so they will get more sympathy when they beg. Such a movie breaks through our self-absorption and isolation, and reminds us sheltered Americans and Europeans that there are still wretched masses living on the verge of death in slums, or subsistence farmers, or hunter-gatherers starving in the wilds of Africa, Asia, or Latin America.

Just a century or two ago, these conditions applied to humanity in general, even in the richest, most developed countries in the world, such as England. One need only turn to the novels of Dickens to read about the misery and wretchedness of the lower classes. Most people suffered from a wide spectrum of diseases and malnutrition and had a typical life expectancy of only twenty-five to thirty-five years, as had been true since human prehistory. The lower classes were typically illiterate and had little opportunity for social advancement. If diseases or starvation did not kill, then the high crime rates in the slums ensured that most poor people never lived very long. The infant mortality rates, in particular, were very high; about three in every ten children died young.

Even the wealthiest and most powerful were still subject to a host of diseases that medicine could do nothing about. In many cases, the practices of doctors using leeches for bleeding, or touching uninfected patients with infected hands made the cure worse than the disease. Charles Darwin attended medical school in Edinburgh, and soon found he had no stomach for medicine as practiced in the 1820s. He could not tolerate watching surgery on patients screaming in agony as their limbs were sawn off without anesthetic. Many patients died of secondary infections, because there were no antibiotics or antiseptics (the germ theory of disease had not yet been established). Doctors did not know to wash their hands after treating a patient, so they transferred germs from a sick patient to a relatively healthy one. About all that doctors could do back then was comfort and reassure patients (often with all sorts of quack cures not too different from modern quack medicine), and hope that the patients' own natural immunity kept them alive.

Consider how much things have changed in the past century or two:

· *Life expectancy:* For most of human history, people typically lived only 25–35 years. Currently average life expectancy is 67.2 years worldwide, and it is 82–83 years in developed countries with good health care and dietary habits, such as Japan, Iceland, Switzerland, Sweden, France, and Canada (the United States, with its lack of universal health care and many bad habits, is ranked only thirty-eighth on that list; its average life expectancy of 78.2 years).
· *Infant mortality:* Before modern medicine and health care, about 30% of all children born died in infancy or childhood. By

contrast, nations with good health care now lose only 0.5–0.6% of all children. If you visit an old cemetery with graves dating prior to 1900, you will be struck by the large number of headstones of infants and young children.

· *Diseases:* Most diseases that wiped out humans by the thousands only a century ago (smallpox, measles, malaria, cholera, tuberculosis, whooping cough, tetanus, meningitis, syphilis) are virtually extinct in the developed world (although still a problem in underdeveloped countries). This is entirely thanks to modern medicine, which has allowed diseases typical of older age (e.g., cancer, heart attacks, strokes, diabetes, and Alzheimer's) to emerge as the leading killers in the more developed countries.

· *Transportation:* A trip through a history museum shows a wide variety of slow, mostly horse-drawn carriages and wagons, the fastest known form of transportation for thousands of years once the horse was domesticated. If you did not have a horse, your only way to get around was to walk. As French historian Fernand Braudel put it, "Napoleon moved no faster than Julius Caesar." Since the invention of the steam locomotive, then the internal combustion engine for automobiles, and finally aircraft, our maximum travel speeds have jumped dramatically—from about 10 mph in a stagecoach or early steam locomotive to 400 mph in a modern jetliner to over 800 mph in a supersonic jet, and 225,000 mph in modern spacecraft. Today we routinely fly 3,000 miles across the United States without spending more than 5–6 hours in the air, but in the mid-nineteenth century, a trip of only 2,000 miles along the Oregon Trail from Missouri to the Pacific Coast involved almost six months of arduous travel and many dangers.

· *Communication and technology:* One of the paradoxes of American history was Andrew Jackson's victory at the Battle of New Orleans, fought on January 8, 1815—*after* the treaty ending the War of 1812 had been signed on December 24, 1814, in Ghent, Belgium. Thanks to slow communications, however, news of the war's end didn't reach Jackson until February 1815. Just two centuries ago, the fastest form of communication was handwritten letters carried by mounted couriers. Messages that traveled across oceans often took months or years to reach their destination, as they had for centuries. But over the course of the nineteenth and twentieth centuries, communications became faster and more reliable with the invention of the telegraph, the transatlantic cable connecting Europe and North America, and then the telephone. Today communication is virtually instantaneous: satellites beam signals around the world in milliseconds, and computers allow us to communicate with people anywhere on the planet in an instant for free. The average home computer can do more than a giant room-sized mainframe computer could accomplish in the 1960s. I remember using carbon paper and typewriters and a slide rule right through college, and calculators were forbidden on exams because they were still too expensive for most students to own. In 1982 I was the last doctoral student in my program to type my own dissertation on a typewriter; this was just before students had access to word processing and eventually to early personal computers. I learned to do multivariate statistical analysis in the late 1970s on a huge, slow mainframe computer with FORTRAN punch cards; now the same data can be analyzed on a small personal computer with just a few clicks on a software routine.

· *Education and information:* Less than two centuries ago, most people were illiterate or barely literate. College education was a rare opportunity, and was primarily for the children of the wealthy and powerful. As far as news of the world, people knew what was happening only from word of mouth, or possibly from a small local newspaper. News of national events often took a long time to reach the smaller towns in rural areas, whether in Europe, Japan, or North America. Today, most people in developed countries receive education for ten years or more, and a high percentage of people in many countries are college educated as well. We can find out the news of the world by clicking on a few websites or listening to the news on radio or television.

All of these trends have been observed and discussed many times before, notably by futurists such as Alvin Toffler, author of *Future Shock*. There are numerous laws of computer science, which all point to the fact that computers continually become faster, smaller, and cheaper in a matter of months and each piece of hardware or software is obsolete in a few years due to the rapid pace of improvement. We denizens of the modern Information Age take these changes for granted, but they seem magical to people in underdeveloped countries who have not been exposed to them, or even to the people of our grandparents' generation. As science fiction writer Arthur C. Clarke put it, "Any sufficiently advanced technology is indistinguishable from magic." This progress is almost entirely due to one thing: *science.*

We citizens of the developed world are now the longest-lived, healthiest, best educated and informed, fastest traveling, and most technologically advanced humans that have ever lived. Cartoonists and comedians and pundits like to poke fun at the younger generation—plugged into their iPods and iPhones and iPads and oblivious to the world around them—but they are the future. The way that they think and learn and act are already radically different than my Baby Boomer generation, raised on the early days of T V, movies that appeared only in theaters, and music that required a phonograph or tape player. These changes even affect things like politics. During the 2008 presidential campaign, Obama and his staff transformed the political landscape with their modern computerized and web-based campaign and outreach methods. As the polls kept coming in, it was clear that those pollsters who sampled voters with landlines only (and did not sample cell phones) were skewing Republican—because many of Obama's young supporters had no landlines, only cell phones, and were thus underrepresented in the polling. This was even more obvious in the 2012 election, when pollsters who called landlines only did much worse than those who called cell phones. Already the number of people without landlines has risen dramatically, and soon a plug-in landline phone will be just as obsolete as an 8-track tape player.

The next time you hear a modern Luddite—from a creationist who rejects all of modern astronomy, biology, and geology, to a faith healer or homeopath or other quack who rejects modern medicine—just ask yourself one thing: Would you want to go back to the world of the late eighteenth century and its high death rates and short life expectancy, suffer exposure to many deadly diseases, and live in an isolated world with limited education and widespread poverty? That is the choice they are offering you—even as those same creationists and other Luddites benefit from modern medicine, and even exploit modern technologies like the internet to push their antiscientific causes. As Michael Shermer put it, science and critical thinking are "the most precious things we have."

WHAT IS SCIENCE?

There are many hypotheses in science which are wrong. That's perfectly all right; they're the aperture to finding out what's right. Science is a self-correcting process. To be accepted, new ideas must survive the most rigorous standards of evidence and scrutiny.

Carl Sagan

So what is science, and why do we consider it so useful and important? Despite the Hollywood stereotypes, science is not about white lab coats and bubbling beakers or sparking apparatuses. Science is *a way of looking at the world* using a specific tool—the *scientific method.* There are many definitions of the scientific method, but the simplest is a method by which we generate explanations about how the natural world works (*hypotheses*), and then try to test or shoot down those ideas using evidence of the real world (*testability* or *falsifiability*). As philosopher of science Karl Popper pointed out, most scientific ideas have to be tested by proving them wrong (*falsified*), since no number of favorable observations will prove a statement true.[1] Thus, we don't speak of proving something true; instead, if a scientific hypothesis has survived numerous tests and attempts to falsify it, it is considered to be well corroborated or well supported—but never the "final truth." Scientific hypotheses must always be tentative and subject to revision, or they are no longer scientific—they are dogma. Strictly speaking, science is not about final truth, or about certainty, but about constructing the best models of the world that our data allow, and always being willing to change those models when the data demand it.

Since Popper's time, not all philosophers of science have agreed with the strict criterion of falsifiability, because there are good ideas in science that don't fit this criterion, yet are clearly scientific. Pigliucci proposed a broader definition of science that encompasses scientific topics that might not fit the strict criterion of falsifiability. All science is characterized by *naturalism*—we can only examine phenomena that happen in the natural world, because we cannot test supernatural hypotheses scientifically. We might want to say that the statement "God did it" explains something about the world, but there is no way to make a test of that hypothesis; *empiricism*—science studies only things that can be observed by our senses, things that are objectively real not only to ourselves but also to any other observer. Science does not deal with internal feelings, mystic experiences, or anything else that is in the mind of one person and no one else can experience; and *theory*—science works with a set of theories which are well-established ideas about the universe that have survived many tests. Gravitation is just a theory, as much as evolution is a theory. This is very different from the popular use of the word "theory" to mean wild speculation, such as reasons for JFK's assassination. From well-established, highly explanatory theories such as gravity, evolution, or plate tectonics, scientists then make predictions as to what nature should be like, and go out and test those predictions.

In this way, science is very different from dogmatic belief systems such as religion and Marxism, which take certain absolute statements to be true and then try to twist the world to fit their preconceptions. None of these other belief systems are willing to critically test their claims and discover that they might be false, because their core beliefs are sacrosanct and unchanging. By contrast, science is constantly changing not only the small details of what it has discovered, but occasionally even fundamental ideas.

In 1962, philosopher Thomas Kuhn described this phenomenon as a *scientific revolution*. For example, the evidence gathered by Copernicus and Galileo showed the earth was *not* the center of the universe, and by the eighteenth century, Copernican astronomy overthrew the old geocentric system of Ptolemy and many other ancient and medieval scholars. Yet a dogmatic religious organization, the Catholic Church, condemned Galileo of heresy and placed him under house arrest for the remainder of his life. Only recently (1992 and again in 2000) did the Church and the pope formally recant their false position on astronomy and admit that it was wrong in condemning Galileo.[2]

In 1859 Charles Darwin revolutionized biology with his theory of evolution, and biology has never looked back. Once again, dogmatic religious organizations, the creationist fundamentalists (both Protestant and Muslim), continue to fight against this scientific breakthrough over 154 years later, despite overwhelming evidence that has accumulated in Darwin's favor since then (see chapter 4). Other revolutions, such as the Einsteinian revolution in physics, or the plate tectonics revolution in geology, have had equally dramatic effects on their respective fields, although they have not generated the same antiscientific opposition since they do not threaten as many religious or political dogmas.

Thus, the scientific viewpoint of the natural world is in many ways a humble one: we do not have absolute truths, but we are trying to understand nature as best we can. As scientists, we must be ready to abandon any cherished hypothesis when the evidence demands it. As Thomas Henry Huxley put it, it is "the great tragedy of science—the slaying of a beautiful hypothesis by an ugly fact."[3] As scientists, we must be careful when we use words such as "truth" and "belief," because science is not about *believing* final truths, but *accepting* extremely well corroborated hypotheses about nature that approach truth in the everyday sense. In the vernacular, scientists are comfortable using the words "real" or "true" to describe phenomena that are so well established that it would be perverse not to admit they exist. We all agree that gravity is real, but we still do not understand how it works in detail. Despite this, objects fall through the sky no matter whether we fully understand why. Likewise, evolution happens all the time around us (see chapter 4), whether we fully understand every detailed mechanism.

Science does not give us comforting certainties or higher truths about morals and ethics that we crave, but that's because science is only capable of examining testable explanations of the natural world. Yet many people are uncomfortable about this, and turn to nonscientific belief systems, like religion and Marxism, for these answers. As the lyrics to David Bowie's "Law" put it, "I don't want knowledge, I want certainty!" As scientists, we cannot evaluate claims of the supernatural, or the Marxist view of humanity, other than to point out that many of the predictions these systems make about the world have been proven false. Many people turn away from science because it does not give them the answers they want to hear.

Science is also a human enterprise, subject to the same fallacies and foibles of any other human enterprise. Scientists are humans and can make mistakes, or be fooled into believing something that is false or mis-led by their biases and ideologies into erroneous ideas. There are whole books written about the topic, such as Martin Gardner's 1957 classic *Fads and Fallacies in the Name of Science,* or Robert Park's 2001 *Voodoo Science: The Road from Foolishness to Fraud.* In some cases, scientists not only are fooled by their biases, but consciously fudge the data, or cheat in other ways, as outlined in William Broad and Nicholas Wade's 1985 exposé *Betrayers of the Truth: Fraud and Deceit in the Halls of Science,* Horace F. Judson's 2004 *The Great Betrayal: Fraud in Science,* and David Goodstein's 2010 *Of Fact and Fraud: Cautionary Tales from the Front Lines of Science.* Certain fields, such as anthropology, are often hugely influenced by the cultural biases of their time, as Roger Lewin showed in his 1997 *Bones of Contention: Controversies in the Search for Human Origins,* or Stephen Jay Gould documented in his 1981 *The Mismeasure of Man.* Reading the titles of these books, one might come away with the perception that all scientists are biased and incompetent, but as the books themselves point out, these topics are newsworthy because they are unusual, rare, and exceptional. Unlike any other kind of academic endeavor, science is checked against an external reality that other scientists can access and do experiments on in many cases. Unlike many belief systems, science is self-correcting and self-policing through the process of peer review. Individual scientists may be able to deceive themselves or their peers for some duration, but sooner or later, if their work is important enough, someone will recheck it and correct it if it is faulty or fraudulent. Really important ideas are checked immediately in the process of peer review, and if they do not pass muster, they never make it to print.

In many cases, a scientific discovery may be premature and may have fooled the experimenter, but the rest of the scientific community will quickly try to replicate the results. If they cannot be replicated after enough attempts, then the research is refuted. The most famous recent example of this was cold fusion, claimed by two researchers, Stanley Pons and Martin Fleischmann, in 1989.[4] The claim was publicized by Pons's institution, the University of Utah, and made worldwide news. Had they achieved nuclear fusion without the extremely high temperatures and pressures previously required to produce it (such as occurs inside the sun), it would have been a revolutionary discovery and a solution to our energy problems. Scientists around the world dropped their own research projects and tried to replicate Pons and Fleischmann's results, and within a month, it was clear that cold fusion was impossible. We may never know *what* the two scientists were seeing when they ran their experiment, but the scientific community was quick to check their results, and the mistake was corrected immediately.

But if scientists are human and can make mistakes or try to commit fraud, then why do we think of science as a better descriptor of nature and the natural world that religion or philosophy? The answer is simple: *because it works.* Science and critical thinking have proven over and over again to be the primary method of screening out nonsense and determining which ideas pass the rigorous scrutiny and which ones do not. Science and technology have produced the practical benefits of our modern society that we discussed above, which were held back for the entire Dark Ages while religious dogma held thrall over the human mind. Science may not provide all the answers we want, but the answers it does provide are well tested and produce practical results for the real world.

As British comedian and actor Ricky Gervais put it,

> Science seeks the truth. And it does not discriminate. For better or worse it finds things out. Science is humble. It knows what it knows and it knows what it doesn't know. It bases its conclusions and beliefs on hard evidence—evidence that is constantly updated and upgraded. It doesn't get offended when new facts come along. It embraces the body of knowledge. It doesn't hold on to medieval practices because they are tradition. If it did, you wouldn't get a shot of penicillin, you'd pop a leech down your trousers and pray.[5]

BALONEY DETECTION

> Skeptical scrutiny is the means, in both science and religion, by which deep thoughts can be winnowed from deep nonsense.
>
> *Carl Sagan*

As mature adults, we have learned not to be naïve about the world. By hard experience, we are all equipped with a certain degree of healthy skepticism. We have learned that politicians and salesmen are often dishonest, deceptive, and untruthful, and that most advertisements are misleading, lying, exaggerating, or distorting the truth. We tune them out and don't even listen to most of them. We are always cautious when buying something, worried that the seller might cheat us. We follow the maxim *Caveat emptor,* "Let the buyer beware." Such a view may seem cynical, but we learn about human nature the hard way, and it is essential to our survival to be skeptical and not too trusting.

It is a paradox, then, that people who are hard-boiled and skeptical when they are listening to politicians or salesmen are so gullible when they fall prey to quack medical claims or the lies of religious extremists. Humans clearly have a deep need for answers to questions that science cannot answer, but the pseudoscientist or quack is happy to sell you an answer you want to hear. Some kinds of pseudoscience are harmless, but more often than not pseudoscience robs people of their time or money or other resources that they really need in moments of stress and hardship, and sells them phony answers and snake-oil cures, just for some temporary reassurance. In some cases, the victims of the con artist realize after it is too late that they haveve been swindled, but others keep on believing in the magical cure or religious nostrum without ever noticing they have been hoodwinked.

If we want to avoid being conned by pseudoscientists, we need to have what Carl Sagan called "a baloney detection kit," or what Penn and Teller (in their cable TV show *Bullshit!*) describe bluntly as "bullshit detection." Sagan, Shermer, and my 2007 book give lists of skeptical rules to follow, particularly when pseudoscientists and ideologues are trying to muddy the waters of a scientific consensus.[6] The list that follows is by no means as long or comprehensive as the lists in the books above, but there are certain basic "baloney filters" we always need to keep in mind.

Much of the debate about issues such as global warming, evolution, and medical topics revolves around determining who has the expertise to be taken seriously. We often find global warming or evolution deniers quoting some supposed authority to say what they claim. How do we decide whether the authority is really credible? I discuss this in detail in the next section, but below are several red flags that often come up.

Quote Mining

First, if the source uses quotations by real scientists, one should check to see if the quotation of an authority is not taken out of context and used to mean the opposite of what its author intended. Creationists are notorious for mining works by real scientists and using these quoted words dishonestly. Whenever someone engages in quote mining out of context, it is a sure indicator that they either do not understand what they have read, or they do understand but are deliberately trying to mislead people who lack the time or inclination to go back and check what the quotation really says or means. Climate deniers do the same thing—for example, pulling small pieces of prose out of stolen e-mails from the Climate Research Unit of the University of East Anglia and using the prose in ways contrary to their authors' intentions (as is clear from reading the entire text).

Credential Mongering

It is another red flag when authors makes claims and wave their credentials in front of us in an attempt to intimidate. This is particularly common among creationist books that flaunt their authors' Ph.D.s on their covers, but it occurs even in science when a fringe scientist-author wants to be taken seriously. But as those of us who have earned a Ph.D. know, you prove yourself on the strength of your data and arguments, not by flaunting your degree. When you see "Ph.D." on the cover of the book, it is often an indicator that the book cannot stand on the strength of its own arguments and evidence.

Expertise in the Appropriate Topic

A related problem is that the general public is often impressed when someone has a Ph.D., and assumes that it makes the degree holder somehow smarter and more expert than the average person. The only thing that getting the Ph.D. demonstrates is that the degree holder was able to struggle through five to seven years of graduate school, jump through all the hoops, and produce writing and research in a very narrow topic. As most of us with doctorates know, all this focus on a narrow topic actually makes you *less* broadly trained that you were before you started. More importantly, the Ph.D. degree *qualifies you to critique only topics in which you were trained.* When you hear a creationist with a Ph.D. in hydraulic engineering or biochemistry trying to attack evolutionary biology, you know it is a clear case of baloney. That creationist has no relevant training in biology or paleontology, and is no more qualified to critique evolutionary biology than to write a symphony or build a skyscraper. You would not entrust your car repair to a person with a Ph.D. in biochemistry; instead you would send it to a licensed auto mechanic. Why would you trust the arguments of someone with a Ph.D. in hydraulics when they are talking about evolutionary biology or paleontology, and yet they have never published in those fields or can not tell one fossil from another? Another example is evaluating the arguments of the deniers who trot out purported lists of scientists who disagree with evolution, global climate change, or what have you. First of all, a thorough analysis of these lists show that very few of the so-called scientists on it have any relevant training, whether it be in climate science research or evolutionary biology. Yet the deniers know that pointing to this phony list of dissenting scientists will impress the layperson unaware of the difference.

Conflict of Interest

We are always suspicious of politicians who push a particular cause, especially when we find out that their largest campaign donors stand to benefit from their proposed legislation. We tend to be suspicious of the hired-gun expert witnesses in a trial who have a track record of saying whatever they are paid to say by lawyers trying to win cases. Likewise, we should be suspicious of so-called experts who claim to know a topic, but have something to gain from their advocacy. We have seen this before when we find that medical studies that deny the problems with a drug, or deny the dangers of smoking, were funded directly by the companies who benefited from this research. Likewise, most of the experts who deny global warming are either from the oil or coal or mining industries (such as mining geologist Ian Plimer, who has no training in climate science), or work in conservative think tanks that are heavily supported by industries that stand to gain from their advocacy. As Upton Sinclair put it, "It is difficult to get a man to understand something, when his salary depends upon his not understanding it!"

Burden of Proof

When scientific consensus has been reached, and a large amount of evidence supports a particular explanation of a phenomenon, then the burden of proof falls on the dissenter who seeks to overthrow the consensus. It is not enough for a climate denier to point to one small piece of contrary data, or a creationist to nitpick one little inconsistency in the huge data set supporting evolution. Like a lawyer in a civil court case, they must show that the *preponderance of the evidence* supports their view before scientists and others will take their arguments seriously. Likewise, the Holocaust denier cannot simply point to a few inconsistencies in the documentation of certain events in 1942 and claim that the Holocaust never occurred. If there is overwhelming evidence showing that events like the Holocaust occurred or supporting evolution, we expect dissenters to prove otherwise with even stronger evidence before we give them much credence.

Correlation Is Not Causation

It is human nature to try to link two events that coincide in time and assume that one causes the other. But for a scientist, a much more rigorous standard is required. Before we say that two events are related, we need to examine many cases of these events occurring over and over again. We must perform statistical tests to determine whether the connection between the two events is real, or just due to random chance. In the case of large-scale medical studies, many different individuals have to be examined, not just a few individual cases. Rigorous statistical tests must be performed to convince scientists that the connection could be real before we take it seriously.

Anecdotes Do Not Make Science

As humans, we are always impressed by the personal testimony of friends, family, and other witnesses. Telemarketers get celebrities to tout their products, and people are swayed to buy it. We often consider the advice of our back neighbor to be convincing, but for the scientific community, a handful of cases are not solid evidence. As Frank Sulloway said, "Anecdotes do not make a science. Ten anecdotes are no better than one and a hundred anecdotes are no better than ten."[7] Scientists and researchers must look at dozens to hundreds of cases, and also look at a control group that was not, for example, exposed to a particular treatment. If the control group is not significantly different from the exposed group, then we cannot establish that event A (such as inoculation) causes result B (autism), as discussed in chapter 7. If people who take the placebo recover just as often as those who get the medicine being tested, then we have not established that this medicine has any real effect. This is particularly a problem in the alternative medicine community, which promotes their snake oil and can always find one or two people who got better and will provide testimony. None of these purported medicines has been met the strict FDA standards for an approved drug or medication. Thus, when they advertise, they will not overtly claim their product cures something, but instead they make vague statements about "supports thyroid health" or "promotes healthy bladder function," because those are not medicinal claims and are not subject to the strict standards of truth in advertising. What they do not mention is that nearly all these alternative medicines *have* been scientifically examined and proven not only to have no beneficial effect, but some actually have harmful side effects and can be very dangerous.

Special Pleading and Ad Hoc Hypothesis

When a scientific hypothesis comes up against a number of well-established observations that truly shoot it down, it is falsified . . . dead . . . kaput! The advocates of that hypothesis must admit that their original idea is dead, and it is time to start over with a new hypothesis. But pseudoscientists are emotionally attached to their cherished ideas, and will not give them up no matter what evidence goes against them. They resort to bending and twisting the evidence, or special pleading to make things work for them. These are known as ad hoc (for this purpose) explanations, and such maneuvers are a clear sign of an idea in trouble. Special pleading may provide comfort to the believer in pseudoscience, but it is not permissible scientifically. If a snake-oil cure fails to work, the quack might say, "You didn't use it right," or "It doesn't work when the moon is full." If the séance fails, the medium might say, "You didn't believe in it sufficiently," or "The spirits didn't feel like contacting us tonight." If you point out the absurdities of the idea of fitting millions of species into Noah's ark, the creationist says, "Only the created kinds were aboard," or "Insects don't count," or "Fish were not on board. " If the scientific data shows no connection between the MMR vaccine and autism, then the anti-vaxxers move the goalposts and say, "It must be due to some other vaccine." Such maneuvers may reassure the true believer, but they reveal a belief system that is unscientific and untestable.

WHOM CAN WE TRUST?

I maintain there is much more wonder in science than in pseudoscience. And in addition, to whatever measure this term has any meaning, science has the additional virtue, and it is not an inconsiderable one, of being true.

Carl Sagan

For every expert there is an equal and opposite expert; but for every fact there is not necessarily and equal and opposite fact.

Thomas Sowell

This discussion about authority and experts raises a good question: Whom can we trust on issues of science and pseudoscience? How do we tell who is the truly qualified, unbiased expert and who are the paid hacks or phony con artists? How can we judge the back-and-forth arguments among creationists and scientists, or between climate scientists and their critics? The media, unfortunately, make the situation worse in their attempts at fairness and equal time. They feel obligated to let both sides have an equal hearing, no matter what the merits of each case. When the evidence is inconclusive, this might make sense, but on most scientific issues consensus has been reached and it is foolish to give pseudoscientists equal time to spout their nonsense. No responsible journalist thinks that the flat-earth creationists, or the religious fanatics from Galileowaswrong.com who think that the earth is the center of the universe, merit equal time with real scientists on TV, yet they will give the equally discredited young-earth creationists or intelligent design creationists equal time. Likewise, the global climate change deniers try to get an equal hearing every time global warming is mentioned, no matter how fallacious their arguments or data are. The anti-vaxxers get their celebrity spokespeople Jenny McCarthy and Jim Carrey on TV all the time, just because they are celebrities, but the relatively unglamorous medical researchers who have the data on their side do not seem to have the same media impact. This is an inherent problem with the modern media, which are driven by an "If it bleeds, it leads" mentality, and feature sensationalism over sound science. The scientific community does not have much chance to change this as long as the general population is willing to pay attention and give good ratings to shows that feature UFOs, Bigfoot, quack medicine, and creationist nonsense.

Goldman (2006) and Pigliucci (2010) provide an interesting set of criteria by which we can evaluate expertise. In some cases it is sufficient for a layperson to listen to the arguments from both sides and judge for themselves. If we are critical listeners and use some of the baloney filters mentioned above, we can often spot whose arguments are full of garbage. Lately, political lies and hypocrisy are daily staples of most news. We can listen closely to politicians and often find that their pronouncements flagrantly contradict what they might have said to a different audience, or said in the past.

But the evaluation of many arguments is beyond the education and training of most people, so then the next best criterion is the evidence of *agreement or consensus by experts in the field.* This is where the scientific method works best. Scientists are a very hard-boiled, skeptical group, always criticizing each other and spotting the flaws in the work of other people. This is the process of *peer review,* and it is one of the best guarantees that a specific body of science has passed quality control. Peer review is a very rigorous and often nasty process whereby all scientific ideas must undergo intense scrutiny.

Many nonscientists do not even know what peer review is, or how it makes a big difference in the reliability of scientific statements. In a nutshell, nearly all scientific research must be evaluated by a panel of experts in a given field of science ("peers"), who critique and vote on it before a grant can be funded or a paper can be published. The process is very strenuous and challenging for the scientist, since for some journals, fewer than 5% of the articles that are submitted are accepted for publication. In my branch of science funding, the National Science Foundation (NSF), fewer than 10% of the applicants (many of whom are the among the world's top scientists) receive funding in any given grant cycle. As a publishing scientist, as well as a frequent scientific reviewer and NSF panelist, I have been on both sides of this process. It can be painful and sometimes nasty, but it works—and very little gets past it. Scientists have to have thick skins to survive the harsh anonymous critiques of other scientists, and persist in trying to get funded and published despite this highly critical environment. It takes a lot of careful review and a huge amount of good data to make an idea acceptable to the majority of these skeptical scientists. Occasionally, the process fails to spot bad data or otherwise breaks down, but even when this happens, it is exposed sooner or later *by scientists,* so the entire process is very rigorous and self-correcting. Thus, when scientists do speak in a common consensus (as in the case of the 95–99% of climate scientists who agree that anthropogenic global climate change is happening, or the 99.99% of biologists and geologists who agree that life is evolving), this is the most convincing voice of authority we have.

Of course, pseudoscientists can beat the bushes and find a handful of scientific dissenters who dispute the consensus. Then Goldman's third criterion comes into play: *look at independent evidence of the expertise.* We quickly see that very few of the creation scientists actually have a Ph.D. in a field relevant to evolutionary biology, so they have no credibility when they criticize evolution. Almost none of the climate skeptics has a degree in climate science or is actively working in climate science research. They are all outsiders who do not know the data firsthand.

Goldman's fourth criterion is to *look for bias or conflict of interest;* this, too, is very revealing when it comes to evaluating the critics of mainstream science. Almost without exception, every one of the scientific experts who disagree with evolution reached their positions due to religious conviction. I know of no antievolutionist who looked at the evidence fairly and came to doubt it, because they always start with their religious biases distorting their perception. The ones who claim they were atheists first and then rejected evolution did so because of a religious conversion experience, not because they found the evidence lacking. Likewise, nearly all the global climate change critics come from right-wing think tanks. Their opinions were influenced by their libertarian or conservative anti-government laissez faire attitudes, not from a dispassionate study of the climate data. Those employed by the oil, coal, and mining industries clearly have a conflict of interest, since their livelihood depends on denying that their product is causing global climate change. The leading anti-vax doctor, Andrew Wakefield, did his research after he was retained by lawyers trying to win a class-action suit against the MMR vaccine; he was also trying to develop his own vaccine to replace it. In contrast, most research scientists who support evolutionary biology or climate science are paid (usually much less) by nonprofit universities and nonprofit government organizations, none of whom tell their scientists what to believe or what conclusions they want them to reach.

Finally, Goldman says if all else fails, *look at the track record of the expert*. A classic case is creationist minister Kent Hovind, who calls himself "Dr. Dino" even though he has no training in paleontology, and his phony doctorate is from a diploma mill that sends you your degree if you pay the fee. He has been repeatedly caught in lies about evolution in his debates and books, but the pattern of deception and cheating runs even deeper: he is now serving a ten-year prison sentence in the Federal Correctional Institution in Florence, Colorado, for tax evasion. Creationists such as Duane Gish have a long track record of their distortions and lies and double-talk, all documented in books and in the blogosphere, and naturally you would not trust what they wrote (see chapter 6 for documentation). Creationists almost never publish in peer-reviewed journals or present at professional scientific meetings, but avoid the scrutiny of the real scientific community and publish only in their own sympathetic house journals and websites. Anti-vax doctors like Andrew Wakefield have a track record of bad research that has been repudiated by the medical community. Due to his many transgressions, he has been banned from practicing medicine.

With all the screaming and shouting and name-calling in the media and blogosphere, it is often hard to tell who is telling the truth, and who is just a shill for a powerful industry or political faction or religious group. But with the baloney filters we just mentioned, and with a careful examination of who has real expertise, it is possible to find out where scientific reality lies. In the chapters that follow, we will be using critical thinking and the scientific method to evaluate a wide variety of claims and determine whether they meet the standards of acceptable science.

FOR FURTHER READING

Goldman, A. I. 2006. "Experts: Which Ones Should You Trust?" in *The Philosophy of Expertise,* ed. E. Selinger and R. P. Crease, 14–38. New York: Columbia University Press.

Gore, A. 2007. *The Assault on Reason.* New York: Penguin.

Pigliucci, M. 2010. *Nonsense on Stilts: How to Tell Science from Bunk.* Chicago: University of Chicago Press.

Sagan, C. 1996. *The Demon-Haunted World: Science as a Candle in the Dark.* New York: Ballantine.

Shermer, M. 2002. *Why People Believe Weird Things: Pseudoscience, Superstitions and Other Confusions of Our Times.* New York: Holt.

3

Betrayers of the Truth:
Selling Out Science

Doubt is our product since it is the best means of competing
with the "body of fact" that exists in the minds of the general
public. It is also the means of establishing a controversy.

Tobacco company memo

"CANCER BY THE CARTON"

As citizens of the early twenty-first century, we have long ago come to
terms with the fact that tobacco and smoking are dangerous and addictive. Hundreds of studies have shown that smoking and other tobacco
use causes a wide variety of diseases (especially cancer, heart disease,
emphysema, and lung disease). We have been paying a huge price in
terms of both health care costs and loss of the lives of many generations
of smokers. The World Health Organization estimates that smoking-
related diseases killed 5.4 million people in 2004 alone,[1] and caused 100
million deaths during the twentieth century.[2] Given all these realities,
you would think that tobacco is so obviously harmful that the entire
foolish habit would be gone by now, like the old habit of snorting snuff,
or wiping out bird species in order to wear silly hats with plumes.

Yet tobacco companies are doing just fine, and continue to be profitable; they are even expanding in underdeveloped countries. All of this is
in spite of the fact that decades have passed since the 1964 surgeon general's report and increased government effort (from additional taxes to
warning labels to huge ad campaigns) and social ostracism (smaller and
smaller areas where smoking is permitted) to discourage smoking. There

are some positive signs. In the United States, smoking rates have fallen by half from 1965 to 2006, dropping from 42% to 20.8% among adults.[3] There have been small declines in the smoking rates among younger Americans, but the habit has not vanished like other destructive social fads.

The simple reason for this persistence is that tobacco contains nicotine, a highly addictive substance that creates a physiological-psychological need that is almost impossible to ignore. Those who try to stop smoking often have to go through extraordinary measures to placate their nicotine addiction, from gum to patches to injections and many other remedies. Many smokers try to kick the addiction, only to fall off the wagon as soon as stressful events cause them to resort to their old habits. In regions where smoking is banned nearly everywhere, you see the signs of this addiction in people rushing outside buildings to shiver in the cold while they desperately finish a cigarette, or smokers getting fidgety during long movies or meetings because of so-called nic-fits.

In this regard, smoking is an addictive, deadly habit not much different from alcoholism or the consumption of illegal addictive substances such as cocaine or heroin or crystal meth. But not all deadly addictions are treated equally. Smoking is a multibillion-dollar industry and powerful lawmakers are paid to defend it, so it is tough to legislate against it, despite the fact that it costs the nation millions each year in additional health costs for people who are dying from the habit. As we learned from Prohibition (and the continuing drug wars), it is very hard to legislate morality, or use laws to restrict dangerous addictive substances if people get a buzz from them or become addicted. The greatest successes have been the gradual reduction of smoking and excessive drinking due to social pressure and ostracism, as well as fees and taxes that help defray at least some of the societal costs of these deadly habits. I feel very fortunate to live in a part of the world in which smoking is truly frowned upon and smokers are restricted to small areas, hiding, or quitting, so I seldom breathe secondhand smoke. A 2012 study showed that the longest life expectancies in the United States are in New York City,[4] because their laws make smoking very rare and discourage many other unhealthy habits. Yet it is a shock to travel to Las Vegas, or Europe, or Asia, and find lots of smokers still puffing away as they did in the United States before the 1970s.

All of these lessons learned about smoking and its deadly nature have been known since the 1950s, both by independent researchers and even by scientists funded by tobacco interests. In a perfect world, the discovery that smoking was dangerous would have led to all sorts of immediate efforts to restrict or ban it, as the FDA does right away when a medication is found to be even remotely risky. But this did not happen. Smoking remained a common habit for at least fifty years after its dangerous nature was discovered in the 1930s and confirmed over and over again; reduced tobacco usage in the United States has only occurred over the past decade or so. In most other countries, smoking is actually on the rise as people become wealthier and able to afford cigarettes, even though most governments around the world warn their people of the dangers of smoking. Why has the conclusive evidence of the dangers of smoking not immediately curtailed tobacco use?

The answer to this curious dilemma is well known: *tobacco companies actively fought scientific research and spread lies and disinformation to protect themselves.*[5] In the 1930s, German scientists had studied the link, and the Nazi government was one of the first to push an active anti-smoking campaign,[6] but this research was considered tainted because of its Nazi origins and had to be rediscovered in the 1940s and 1950s. By 1953, the studies of the link between smoking and cancer were conclusive. The news of this link was publicized in the *New York Times, Life,* and even *Reader's Digest,* one of the most widely read publications in the world at that time; the article was entitled "Cancer by the Carton." As Oreskes and Conway document, the tobacco companies were thrown into a panic, and on December 15, 1953, they met with Hill and Knowlton, one of the biggest public relations firms in the country, to do whatever they could to prevent science from changing people's deadly habits.[7] They began a PR strategy that is the blueprint for nearly every effort powerful vested interests have undertaken since then to deny or corrupt science they do not want to hear.

"No Proof"

It is a great PR tactic to claim that science has not proven smoking causes cancer with 100% certainty, or that anthropogenic climate change has not been proven to be occurring, or whatever inconvenient scientific reality we try to deny. But *nothing in real science is 100% proven*. Science is always about tentative hypotheses that are tested and retested, and only after hundreds of such studies are conducted and all agree on a common conclusion do scientists regard something as *very likely*, or *very well established*, or *well corroborated*—but never proven. "Proved" is a word appropriate only to the world of mathematics, where proofs are possible within the limited artificial scope of the discipline. The real world is too complex and messy to allow absolute proof. Based on statistical analysis, we can show that if something has a 99% likelihood of occurring, or being true, then this level of confidence is so overwhelming that it would be foolish to ignore it. We can tell a person about to jump off a building that the odds are 99% that he will be seriously injured or killed, and this should be a sufficient level of confidence to prevent a non-suicidal person from jumping. We cannot *prove* that he will not be hurt, but most people would not like the odds on that bet. Sure, there is a tiny possibility that by some miracle someone will suddenly put a safety net or giant airbag below him after he leaps, but that is extremely improbable. Likewise, the link between cancer and smoking is about 99%, and only a fool ignores this level of confidence—but it is not proven.

"Other Causes"

Another tactic used by denialists is to claim that there may be other causes that contribute to the problem, and therefore we should not try to assign blame or restrict discussion to a single cause. In some cases, a phenomenon does indeed have multiple complex causes (such as cancers not related to smoking or the multiple causes of autism-spectrum disorders), and it is hard to tell if one in particular can be isolated. But in the case of smoking-related cancers, the link is overwhelming, and it would be a lie to deny the reality that smoking is the main cause of these forms of cancer. Like a magician onstage who is trying to fool you, denialists of all stripes have long used this form of misdirection to get people to look away from the ugly reality on center stage. And suppose it were true that there are additional causes? In cases such as smoking and cancer, does it not make sense to eliminate this primary, well-established cause (smoking) and see whether the other potential causes are really important?

"Both Sides of the Question"

Purveyors of junk science who are trying to protect their favored idea or product always appeal to journalistic fairness and argue that we need to listen to both sides of a controversial question. Most journalists, not knowing the facts of the case, play along. In some cases, in which the arguments for each side are inconclusive or equally balanced, this is appropriate—but not in the case of topics where the scientific evidence is overwhelming and conclusive. Journalists don't give equal time or present both sides of the question of whether the earth is flat, even though the flat-earthers mentioned in chapter 1 sincerely believe they are right and demand equal time. No journalist runs a story in which the Holocaust deniers get equal time for spouting their antisemitic bile. Likewise, the junk science presented by medical quacks, anti-vaxxers, creationists, and global warming deniers does not deserve the same credibility, because it has already failed the crucial tests of science and critical scrutiny, and because the overwhelming consensus of the scientific community has found the ideas wanting and rejected them.

"Fund and Promote 'Alternate Research'"

One of the main strategies used by the tobacco companies and their PR firms to fight the scientific evidence was to pay for their own research, in hopes these scientists might find something that helped their cause. As several of the books cited below document (based on many different sources, especially the famous 1996 study by Glantz et al.[8]), the tobacco companies paid a lot of money for such research, and certainly did fund studies that found other causes of cancer. They then publicized these studies to the hilt based on the theory that if other causes were mentioned, the strong link between smoking and cancer would be ignored. However, not everything went as planned. By the 1960s most of the scientists funded by the tobacco industry had bad news for their patrons:

the research clearly showed the link between cancer and smoking was real. So what did the tobacco companies do? Did they publicize these results, as any honest scientific organization would do? No. They actively suppressed and buried the inconvenient truth. The scientists could not prevent this, since the tobacco companies had funded the research and had made them sign agreements controlling its release and publication. Meanwhile, Big Tobacco's PR machinery continued to crank out denials of the link well into the 1990s, more than thirty years after their own scientists (and outside scientists) had proven they were liars. The same strategy is employed by the energy companies who fund research that promotes their ends (especially research that might question the reality of anthropogenic global warming), or the creationists who fund the strange studies by fringe scientists that support their cause.

"Hire 'Experts' to Promote Your Cause"

As Oreskes and Conway (2010) document, one of the key strategies of tobacco companies and other organizations trying to deny an inconvenient scientific reality is to find anyone with credentials who will serve as a front man for their cause and give them scientific credibility. Ironically, these so-called experts often turn out to be scientists with no formal training in the field they are defending, yet because of their prior (irrelevant) scientific laurels, they are taken seriously by the press and public. The shocking thing that Oreskes and Conway document is that just a few individuals (Fred Seitz, Fred Singer, William Nierenberg, Robert Jastrow, Edward Teller, and a handful more) were at the front of every single one of these attempts to deny scientific reality. Most of these men gained their reputations as nuclear physicists, and some actually built the atomic bomb. After the Cold War ended and there was no more Commie bogeyman to fear, they retained the Cold Warrior mentality that anything threatening capitalism and free enterprise is bad—even if the scientific case for it is overwhelming. Thus, these nuclear scientists headed panels and commissions (often secretly and lavishly funded by special interest money) defending tobacco companies, energy companies, and the like against the evidence for smoking-related cancer, secondhand smoke, anthropogenic global warming, the ozone hole, acid rain, and the

nuclear winter scenario. Never mind that a background in nuclear physics gives one absolutely no qualifications whatsoever to evaluate studies
in medicine or climate science. These few men have done more to harm
the country and stunt the dissemination of scientific research than any
Soviet threat ever could have accomplished.

"Doubt Is Our Product"

When all else fails, use a smokescreen and confuse the public. In the 1953
PR report that Hill and Knowlton prepared for the tobacco companies,
they made it clear that their primary strategy was to muddy the waters
of public opinion and confuse people so that "scientific doubts must
remain."[9] For decades afterward, the tobacco companies pursued this
strategy, spending huge amounts of money to trumpet "scientific" studies which cast doubt on the smoking-cancer link (not revealing that they
had paid for it, so the impartiality was compromised). Even in the late
1970s and early 1980s, the tobacco company denials were uncompromis-
ing, and the research by their paid hacks continued to be publicized. But
the lawsuits kept mounting, one after another.

Although none of the initial suits had been successful, the evidence
kept piling up. In 1964, the surgeon general first made the link official
in the eyes of the United States government, and as the years went on,
the surgeon general's warnings got more and more scary. Finally, the
tobacco companies had to be prosecuted under the RICO (Racketeer Influenced and Corrupt Organizations) Act. This law was originally passed
to give prosecutors and congressional committees the power to break
up organized crime and racketeering. During these hearings in 1999,
the tobacco executives repeatedly lied under oath in front of Congress,
denying their product was harmful, and denying that they knew this
years ago. But their own internal company documents showed these
statements were lies, and that companies had long suppressed research
that they themselves had funded and that demonstrated the tobacco-
cancer link (the verdict against them was upheld by the Supreme Court
in 2006). The most revealing document of all was the internal memo that
came to light during the investigation, revealing the tobacco executives'
full knowledge of what they were doing. It read, "Doubt is our product

since it is the best means of competing with the 'body of fact' that exists in the minds of the general public. It is also the means of establishing a controversy."[10] Today, we have the energy companies funding research to create doubt about anthropogenic climate change, and many other organizations who are not interested in scientific truth, but only in protecting the status quo—and their bottom lines.

"THE TRUTH WILL SET YOU FREE"—IF YOU CAN FIND IT

In the remaining chapters, we will look at specific cases of how powerful organizations thwarted scientific research and its open dissemination, or created doubt through PR campaigns and paid scientific hacks. But it is worth briefly mentioning a number of other examples to show this pattern is much more widespread than just the examples that we can discuss at length. Oreskes and Conway (2010), who did most of the hard digging into public records for their research, documented each of these cases in detail. In each example, the patterns are the same: first there is an inconvenient scientific discovery, then powerful interests react, then there is great controversy as they try to discredit the discovery and create a smokescreen to prevent the public understanding of the truth, and then—in a lot of cases—the truth wins out.

Secondhand Smoke Kills, Too

It took years of fighting on the part of the government, health, and medical organizations; hundreds of warning labels; bans on cigarette ads; a congressional investigation; and a Supreme Court decision to finally force the truth out of the tobacco companies that smoking can kill you—more than fifty years after the original scientific evidence had shown this to be true. One might think after these public and humiliating losses, tobacco companies would stop stonewalling and lying, and instead be more honest about their product liabilities—but you would be wrong.

In 1986, another surgeon general's report pointed out an additional health risk of tobacco—secondhand smoke. Once again, the tobacco companies' own research had already revealed this risk back in the 1970s. It found that secondhand smoke actually spreads *more toxic* chemicals to the nearby nonsmokers than to the smoker. But the tobacco companies buried the results or used them to develop products that gave off less obvious secondhand smoke.[11] In the 1980s, independent research began to document in study after study that those who are exposed to cigarette smoke (but never smoke themselves) are at a high risk of contracting all the known smoking-related diseases.[12] The most conclusive studies were those comparing the nonsmoking spouses of smokers, which found that even though the spouses never put a cigarette to their lips they died of smoking-related diseases at very high rates.[13] This kind of study is conclusive, because it rules out all the other possible variables: different work environments, different cultures, genetics (spouses are normally not closely related by blood), or even differences in life habits. None of these factors showed any correlation with disease; the only positive correlation was smoking. In addition, the spouses of heavy smokers were more at risk than those of light smokers, exactly as predicted.

Despite the efforts of the tobacco companies to deny this evidence, as more and more independent studies were conducted, the results were overwhelming: nonsmokers who are exposed to smokers are at higher risk of dying of smoking-related diseases. The United States Department of Health and Human Services wrote, "There is no risk-free level of exposure to second-hand smoke: even small amounts can be harmful to people's health."[14] Soon a number of states began passing laws pertaining to smoking in public, and in other environments where nonsmokers could not avoid the smokers' poisonous fumes. Naturally, the tobacco companies fought back, attacking the results of the research any way they could, and commissioning their own research in attempt to nullify the evidence. They spent huge amounts of money fighting efforts by local governments to pass smoking bans and to preserve smoking areas for their clients. They pushed the idea of accommodation—that smokers had rights (as did handicapped people) for their own facilities (except that smoking is a voluntary act, whereas handicapped people have no choice in the matter). And most of all, they spent money to maintain the controversy, pouring $16 million into ads and "research" that kept the public confused about the deadliness of secondhand smoke.[15]

Last but not least, they cast the issue as one of individual liberty. What right does the so-called nanny state have to infringe upon the liberties of

those who want to exercise their right to suicide by smoking? This argument appeals to conservatives, libertarians, and free-market types, but
there is an important difference: the smokers choose their own deadly
habit and suffer the consequences of later diseases and death. But the
nonsmoking bystander is innocent, and had no choice whether to not
to breathe the smoke (especially a spouse who must live with a smoker
or let the marriage end). They are involuntarily forced to inhale deadly
fumes until they, too, get sick and die. By law, the smoker has no right to
force other people to suffer from his or her bad habits any more he has
the right to force others to endure other types of obnoxious or dangerous behavior—such as dangerous driving, dangerously loud music, or
shouting, "Fire!" in a crowded theater.

Then, in 1992, as the Environmental Protection Agency released its
own research and reports on the topic, the tobacco industry brought out
its usual hired guns: Fred Seitz and Fred Singer. These nuclear physicists
(*not* trained in any area relevant to smoking, medicine, or epidemiology) found their own creative ways to fight the mountain of evidence
stacked against the tobacco companies. They blamed the diseases on
other environmental factors; they said that the link was not 100% proof
of secondhand smoke being dangerous; they quibbled about the statistical confidence levels used by the EPA (95% rather than 99%, but still
significant)—all the familiar denialist dodges. Singer created his own
Science and Environmental Policy Project (SEPP) to "promote 'sound
science' in environmental policy."[16] This all sounds legitimate until you
look closely at the actions of this organization (entirely junk science that
favored tobacco companies) and the money trail (entirely funded by
tobacco and other industry money). Ironically, it was these people who
first coined the term "junk science" to criticize the good science that they
disputed. Because of this, you have to read the context closely when the
term "junk science" is used. It is often in the employ of people who are
enemies of reality and good science, are using junk science themselves,
and want to cloud the issue and confuse people.

The 1999 RICO hearings brought out all this dishonesty and conniving
by the tobacco companies and their minions. It gave much more force to
antismoking laws as well as final validation of all the decades of research
that tobacco tried to discredit. Today, the tobacco companies are still

fighting a rearguard action against restrictions on smoking in public and private places, but most of the time local governments win because more people want to ban smoking—there are now fewer and fewer smokers. But if you travel outside the United States, you will find that most other countries are smokers' paradises, and secondhand smoke is not even an issue. We still have a long way to go.

Star Wars vs. Nuclear Winter, or How to Crucify Carl Sagan

Another example described by Oreskes and Conway is the battle during the early 1980s over missile defense systems and the nuclear buildup.[17] I remember this one vividly, since I was actively involved in political protests against the efforts of the Reagan administration to challenge the Soviet Union by spending billions of our tax dollars on more nukes. Most Americans regarded this as folly, since we already had enough nukes to wipe out the Soviet Union many times over. As Oreskes and Conway show, this movement began in the 1970s with the old Cold Warriors' worry that the Soviets were catching up in the arms race to build more and more nuclear missiles, and the fear that there was a missile gap. When Reagan's election brought them to power in 1980, they soon pushed a hard-core militarist agenda while starting a new Red Scare and claiming that the Soviet Union was becoming too strong. Ironically, once-secret documents from that era show that the Soviet Union was a hollow shell. It never had the capabilities of being the threat that it was supposed to be, and the Soviets had never even started building a missile defense system.[18] Nor did they want to start a nuclear war or believe such a war was winnable; they wanted to avoid using nukes at any cost.[19]

Reagan's 1980 election also had the effect of mobilizing thousands of antiwar and antinuke activists—who were worried about his bellicose rhetoric and his excessive spending on weapons—worldwide. In March 1983, the Reagan administration countered with the promotion of the Strategic Defense Initiative (SDI, nicknamed "Star Wars" after the popular movie series). As Reagan presented it to the American people, it would be a magical umbrella in space that could shoot down any Soviet nuclear missile headed our way, and provide us more protection than the old policy of building so many nukes that we were guaranteed mutually assured destruction (appropriately abbreviated MAD). It immediately drew a storm of controversy. Not only did it violate many previous international treaties about weapons in space and antiballistic missiles (signed by Nixon and other Republican presidents), but it also provoked an escalation in the arms race (the Soviets might build even more nukes, or release many at once, to overwhelm it). Most surprising of all, however, was that nearly all the scientists who worked on these kinds of missile and laser projects said it was entirely impractical! Not only would it be hugely expensive and virtually impossible to engineer, but—as Carl Sagan pointed out—all it takes is one missile with multiple warheads to get through and consequences would be just as bad as if SDI did not exist. Most serious in this regard is that no field tests could reassure us that the umbrella was leakproof. In the space business, you get only one chance, and if the risky venture failed, we would have spent huge amounts of money for a worthless leaky umbrella.

In May 1986, 6,500 scientists signed a pledge not to accept funds from the missile defense research program. Most of the leading scientists in the country, including such giants as Hans Bethe, who helped develop the first atomic bomb, were against SDI. Many of the scientists in Reagan's own administration were against the idea. As Oreskes and Conway wrote, "Historically, it was unprecedented. Scientists had never before refused to build a weapons system when the government had asked."[20]

The Reagan administration and the Cold Warriors (especially Edward Teller, the hawkish developer of the H-bomb and a hard-core anticommunist) struck back with everything they had, attacking the peaceniks and so-called traitor scientists, and falling back on the Red Scare tactics of the 1950s. They pointed to satellite photos they claimed showed a Soviet beam weapon facility. Ironically, after the Soviet Union fell and one of Teller's protégés toured the facility, it turned out to be a conventional rocket engine facility and had nothing to do with nukes (shades of the recent experience with the falsely promoted "weapons of mass destruction" in Iraq!).[21]

As the battle escalated between hawks and doves, another surprise threw the issue into a totally different direction. At the NASA Ames Research Center, some scientists were using computer models to try to understand the atmospheric effects of the recently proposed idea that an asteroid impact 65 million years ago had wiped out the dinosaurs.[22] As the atmospheric models ran on the supercomputers, they realized that any large explosion on the earth's surface would generate huge dust clouds that could spread across the planet in a matter of days, block out the light, and bring about instant winter. Thus, the same model could be used to simulate the effects of a nuclear war. From this came the nuclear winter scenario: any nuclear exchange would generate worldwide clouds of dust and ash that would not only spread deadly radioactive particles in the fallout, but chill the entire planet so that the land surface would freeze for months, most plants would die, and people would die of mass starvation. In 1964, Stanley Kubrick filmed the black comedy *Dr. Strangelove,* which postulated a doomsday machine that would destroy the entire planet if anyone started a nuclear war. Without realizing it, the nuclear winter models showed that the world's atomic powers had already built a doomsday machine by the mid-1950s, but we didn't know that there would no winners if a nuclear exchange took place.

The original "nuclear winter" paper was published in the December 23, 1984, issue of *Science* by Richard Turco, O. Brian Toon, Thomas Ackerman, James Pollack, and Carl Sagan.[23] Ironically, it came to be known by the initials of the authors ("TTAPS"), yet theirs was not a deliberate attempt to create an acronym evoking the bugle call signaling the end of the day, or the end of life.

The original TTAPS paper soon received the typical criticism from within the scientific community about various assumptions and calculations, and underwent the stringent peer review it deserved and required. Another group at the National Center of Atmospheric Research (NCAR) in Boulder, Colorado, ran its own models and found results that were not quite as extreme as the TTAPS study, but still serious. Nevertheless, the nuclear winter hypothesis stood the test of unbiased scientific peer review and seemed to be well established, at least in some form (as it still is today, almost thirty years later).

But this was not good enough for those whose political or economic interests were served by SDI. Edward Teller, astronomer Robert Jastrow (whom I knew when I was a student at Columbia University), William

Nierenberg (retired director of Scripps Institute of Oceanography), and the ubiquitous Fred Seitz (whom we shall encounter shortly) launched a campaign of attacks on the various nuclear winter models. These retired physicists, hawkish anticommunists, and former Cold Warriors decided to fight back by setting up their own institute—the George Marshall Institute—to further their goals and "raise the level of scientific literacy of the American people in fields of science with an impact on national security and other areas of public concern."[24] Their mission statement sounds high minded and innocuous enough, but their actions spoke louder than words. All of the efforts of the Marshall Institute went to disputing scientific claims that did not fit their right-wing agenda, and releasing their own counter-propaganda to boost SDI and other Cold War causes. With Jastrow and Nierenberg leading the charge, the Marshall Institute got right-leaning media entities such as *Commentary* and the *Wall Street Journal* to publish many critiques of the nuclear winter scenario or support the feasibility of SDI. When public television stations were about to air a program on SDI, the Marshall Institute demanded equal time under the Fairness Doctrine for its propaganda, and the SDI show was aired in only a few places. The Marshall Institute threatened journalists who did not balance their reporting about SDI with its point of view.

In 1986, as Jastrow raised money from the Coors Foundation, the Olin Foundation, the Scaife Foundation, and other wealthy organizations that funded right-wing causes, they set on a full assault on the nuclear winter scenario. The fact that later studies had modified the original harsh TTAPS scenario to a milder (but still catastrophic) nuclear autumn was evidence to Jastrow that scientists had been dishonest and deceptive and were willing dupes of the Communists. (As a working scientist, Jastrow should have known that this is normal in the scientific process—new ideas get modified as further testing and peer review occurs, but scientists should not have political goals in this process.) Jastrow claimed that the TTAPS and NCAR studies had not taken into account the oceans or the effects of the rainout of the dust, but this was a lie: they are both accounted for in every model for nuclear winter or nuclear autumn. These and other outlandish and exaggerated claims were repeated again and again by the Cold Warriors, who accused the

TTAPS group of doing bad science and trying to push its Communist agenda. They even began to claim the entire scientific establishment was corrupted by left-wing politics. (Never mentioned in these articles and editorials was that Jastrow, Nierenberg, and Seitz were clearly motivated by right-wing politics, and had funding from conservative institutions.)

Oreskes and Conway suggest that this is where the right-wing turn against all science began.[25] Today it is manifested by the antiscience attitudes on right-wing media (especially Fox News) attacking global warming, evolution, and other ideas accepted by the vast majority of scientists. Rush Limbaugh and Michael Crichton continued to attack Carl Sagan even in the present decade, long after he died and could no longer defend himself. The attitude goes beyond the simple Cold War politics that started it to the idea that anything in science that interferes with the free markets and capitalism must be attacked—so global warming, acid rain, the ozone hole, and many other environmental issues are routinely attacked by right-wingers as well. As Chris Mooney documented in *The Republican War on Science* (2006), the George W. Bush administration was one of the most hostile toward science—oil company lobbyists wrote energy policy and censored environmentalists, science policies stopped stem cell research and many other critical ventures in medicine, and there was even a sop to creationism. Obama's promise in his 2009 inaugural address to "restore science to its rightful place, and wield technology's wonders to raise health care's quality and lower its cost,"[26] was heard with great relief in the beleaguered scientific community, especially when he appointed top-notch Nobel Prize–winning scientists to head his adminstration's key science posts. But as the current battles over global warming and creationism (see chapters 4 and 5) show, the war to restore scientific integrity to the government is not over.

Meanwhile, whatever happened to SDI? The Soviet Union and Communist Bloc vanished in 1990, and with it the threat that SDI was supposed to neutralize. With the end of the Reagan-Bush years and the election of a Democrat, Bill Clinton, one would assume that it had died quietly by congressional defunding. Yet in the Clinton years, SDI was changed to the Ballistic Missile Defense Organization (BMDO), and its scope changed to defending local-theater missiles. In 2002 the George W. Bush administration renamed it again (it is now the Missile Defense

Agency). There are still millions of federal dollars flowing into different missile defense projects, and by some estimates over $100 billion has been spent on SDI and its successors. Yet there are no longer any superpowers threatening us with a huge array of nukes pointed at the United States There *are* worries about terrorists getting bomb material from old unguarded Soviet stockpiles, or hostile countries such as Iran or North Korea launching a stray nuke at a neighboring country, but the last North Korean test launch exploded ineffectually, and SDI would never have protected us against these kinds of threats. Nevertheless, the spending goes on, entrenched in the budget of our military-industrial complex and almost impossible to defund. Republicans constantly complain about large federal social programs that, once started, cannot be cut or eliminated, but both sides have loaded the federal budget with projects that are obsolete and will never die.

CONCLUSION

From the efforts of corporations (primarily tobacco) to defend their deadly product by corrupting science, we have seen the attack on the science of inconvenient truths grow until the strategies of the tobacco companies (deny, fund alternate research, create doubt) are now routinely employed by people wishing to prevent scientific reality from hampering their agenda. The scientific reality is that smoke kills (both the smoker and the nonsmoker), SDI is now irrelevant and would never have worked in the first place, and nuclear autumn is still a realistic scenario. But corporate and political interests who want to deny reality are so powerful that none of these scientific ideas was accepted by the public as soon or as completely as it should have been. As we shall see in the next few chapters, the scenario is played out again and again on almost any scientific consensus that might hamper big business or go against the interests of the right wing. These efforts are so effective that most Americans think these issues are controversial, even though the scientific consensus has been conclusive for a long time. This is a total corruption and sellout of the integrity of the scientific process. Chapter 13 discusses this issue further.

FOR FURTHER READING

Hoggan, J., and R. Littlemore. 2009. *Climate Cover-Up: The Crusade to Deny Global Warming*. Vancouver: Greystone.

Michaels, D. 2008. *Doubt Is Their Product: How Industry's Assault on Science Threatens Your Health*. Oxford: Oxford University Press.

McGarity, T. O. 2010. *Bending Science: How Special Interests Corrupt Public Health*. Cambridge, Mass.: Harvard University Press.

Oreskes, N., and E. M. Conway. 2010. *Merchants of Doubt: How a Handful of Scientists Obscured the Truth on Issues from Tobacco Smoke to Global Warming*. New York: Bloomsbury.

4

Making the Environment the Enemy: Acid Rain, the Ozone Hole, and the Demonization of Rachel Carson

A comprehensive history of great business fortunes would show a disconcertingly large number that were made [in which] the enterpriser devised a silent way to commonize costs while continuing to privatize the profits.

Garrett Hardin

THE TRAGEDY OF THE COMMONS

Before the 1960s, there was no significant "environmental movement" in the United States. Pioneering writers and thinkers such as Aldo Leopold (*The Sand County Almanac,* 1949) and Rachel Carson (who published *Silent Spring* in 1962) demonstrated the dangers of our impact on nature to a mostly apathetic public. But during the turbulent 1960s, more and more warning signs began to appear. Environmental disasters seemed to occur over and over: water pollution so bad that the Cuyahoga River in Cleveland caught fire;[1] air pollution so bad that over two thousand people died of smog in London in 1952;[2] attempts to put a dam in the Grand Canyon and flood it; oil spills fouling the beaches and killing the sea life near Santa Barbara; plus the effects of urban sprawl, excessive logging and mining, and oil drilling on natural lands and waters. Added to this was the antiestablishment attitude of young people alienated from mainstream society by the Vietnam War and the civil rights movement (especially hippies and a "back to nature" movement). By April 1970, when Senator Gaylord Nelson of Wisconsin helped establish the first Earth Day, environmentalism was truly a movement, and a large portion of society had swung against the idea that big corporations should be allowed to do anything they wanted.

Despite its popularity among the hippies and young people, the environmental movement was popular with nearly every part of the political spectrum back in the 1970s. Few politicians, Democrat or Republican, wanted to be associated with polluters or openly serve as the mouthpiece for environmental exploiters. The Wilderness Act of 1964, which designated over nine million acres as wilderness free from any human impact, passed the House of Representatives by a vote of 373–1, and the Senate by a vote of 73–12. We do not usually think of Richard Nixon as a friend of the environment or as an enemy of big corporations, but his administration set up the EPA. Congress passed and Nixon signed into law the Clean Air Act, the Clean Water Act, the Endangered Species Act, and the National Environmental Policy Act. Going even further back, it was Republican Theodore Roosevelt who first led the nation into conservation and established the national parks and national forests. John D. Rockefeller, who made his millions from the Standard Oil Corporation, nevertheless donated millions to preserving and enriching public lands. Conservation and environmentalism were once broadly popular among both Democrats and Republicans. Not until Reagan came along with statements like "If you've seen one redwood tree, you've seen 'em all," with the weird idea that trees cause more air pollution than automobiles do, and with an administration that was actively and openly against environmental protection and regulation did the Republican Party became the party of environmental exploiters and hostile to environmental protection and regulation.

One of the most influential and original thinkers of the environmental movement was Garrett Hardin. Originally trained in microbiology, he began teaching as a professor of human ecology at the University of California, Santa Barbara, in 1946 and spent his entire career in this idyllic campus setting, stimulating generations of students to think differently about ecology. One of his most famous concepts was "the tragedy of the commons," on which he first published in 1968.[3] He describes a parable (recycled from William Foster Lloyd, who first wrote it in 1833) about the old practice of everyone grazing their livestock on the commons, a grassy area or parkland that is shared by all residents. The commons can only support so many cows or sheep, so if an individual farmer puts too many animals on the land, it will be ruined. However, a farmer does not own the commons and does not have to pay for its restoration, so there is a dilemma: individuals acting in their own self-interest in a free market without regulation will have no incentive to preserve a common resource. Instead, they will profit from exploiting and exhausting it as quickly as possible, since they do not have to pay the costs of its restoration. Thus, most shared resources will be depleted or destroyed, even though it is in no one's long-term interests to do so.

In economic terms, this is what is known as an "externalized cost": the cost of depleting a common resource is not calculated in the original price of the product, and there is no incentive for the capitalist to spend money conserving a resource they use unless some other kind of outside entity (such as the government) puts a tax or cap-and-trade program or other restriction into effect that forces businesses to internalize these costs.

Much of environmental economics revolves around this central principle. In most parts of the world, exploiters of a common resource (the air, the water, the mineral resources, the forests, the fisheries) have little incentive to preserve this resource but instead tend to grab as much profit as they can in a short period of time. They leave when the resource is exhausted—and someone else (typically the taxpayer) gets to clean up their mess. If you travel, you will see the scars of this practice: huge areas of forest that have been cut down and then allowed to rot with no effort to clear and replant them; scars of old mines and their tailings piles, many leaking toxic chemicals; old oilfields where the machinery remains, rusty and broken, and the ground is saturated with spilled toxic chemicals; abandoned factories where the groundwater is full of waste products; and coral reefs that are dead and bleached with no life around them.

Thanks to modern environmental laws, however, these same polluters must factor in the cost of the cleanup. I used to teach at Knox College in Galesburg, Illinois, and on geology field trips we would visit the remnants of the old coal mines. These regions were literally turned upside down by strip mining, and all that remains are piles of dirt with stagnant water in between, only good for duck hunters in the fall, and for my classes to hunt fossils in the spoil piles of limestones and shales that were once overburden above the coal layer. Afterward my students and I would visit a modern strip mine, such as Rapatee Mine. Their dragline bucket was the size of a house and could remove many tons of topsoil and overburden of rock just to reach a 2-foot-thick seam of coal over 100 feet below the surface. Except for the active mine excavation, there was no trace of mining across the rest of the landscape. Thanks to strict environmental mitigation laws, the land had been restored and replanted as a productive farm. During the mine tour, the company representative joked that the farming the topsoil was more profitable than digging down to retrieve the coal seam in the first place. This coal deposit was small and expensive to reach, and the coal had a lot of sulfur in it that causes acid rain, so it was not worth as much on the market anymore.

ACID RAIN: DEATH FROM THE SKIES

Saying sulfates do not cause acid rain is the same as saying that smoking does not cause lung cancer.

Drew Lewis, Secretary of Transportation, 1981

The idea that air pollution might cause damage on the ground and in the biosphere goes back to the Industrial Revolution in the nineteenth century, when it was clear that soot from factories were damaging the forests of the British Midlands and the Black Forest of Bavaria. But the modern concept of acid rain was pioneered by a study done in 1963 at the Hubbard Brook Experimental Forest in New Hampshire. A group of Dartmouth College faculty (biologists Herbert Bormann and Gene Likens and geologist Noye Johnson, a friend of mine) and U.S. Forest Service biologist Robert S. Pierce studied the patterns of gradual plant death, documenting that the rain and snow in the region was indeed more acidic than natural. Rainwater has a slight natural acidity (pH = 5.6; 7.0 is neutral) due to formation of carbonic acid when water mixes with carbon dioxide in the atmosphere. By 1974, Likens and co-workers were able to trace the problem to the tall smokestacks of industries in the U.S. Midwest. The acidity was mostly in the form of sulfuric acid and nitric acid, formed when high-sulfur coal was burned in power plants.

Yet these power plants had been burning coal for a long time. Why did the damage not appear until the 1960s? It turned out to be an unintended consequence of well-meaning but damaging regulations. Originally, the air pollution from the smokestacks settled down and formed smog in the local region, choking people and causing illness and death. So the regulations were written to demand taller smokestacks that would disperse the pollutants higher in the atmosphere. However, this did not solve the problem; it only spread it to more people, since this practice dispersed pollutants over a wider area.

The Hubbard Brook research and studies by numerous other scientists were accumulating in the scientific literature, but there was little public awareness of the issue. Then in 1971 Swedish meteorologist Bert Bolin, a pioneer in acid rain research, led a panel sponsored by the Swedish government for the United Nations Conference on the Human Environment. The report was the first to present a comprehensive case of how acid rain formed, what effect it had on the environment, and how it could cross international borders and become a serious political issue. Soon there were hundreds of rigorous peer-reviewed studies published in top journals over the 1970s, as well as studies commissioned by leading governmental agencies, including the EPA in the United States. By the time Gene Likens wrote a summary paper in 1976 in *Chemical and Engineering News* (the magazine of the American Chemical Society), there was little doubt in the scientific community. Yet the journal editors wrote a caption above his article, "The acidity of rain and snow falling on parts of the U.S. and Europe has been rising—for reasons that are still not entirely clear and with consequences that have yet to be well evaluated."[4]

What the heck did that caption mean? Likens's article, and all the research it summarized, showed that acid rain was unquestionably real, as was the evidence of the damage it caused, and evidence it was caused by industrial processes, especially coal burning. Not every single dot had been connected, but there was more than enough good science published to show the seriousness of the problem, and the need for action. Yet the editors who wrote this wishy-washy caption were employing the strategy of the deniers discussed in chapter 3: "We're not 100% certain, so we cannot act," or "There might be other causes we haven't found yet," or other dodges that create doubt and uncertainty in the public mind and force delays on actions that should have been carried out immediately.

One criticism leveled by the deniers was that scientists could not be sure that the sulfur was due to power plants, since it could have been generated by volcanoes or sea spray. Then Bolin showed that the acid rain in Northern Europe must be due to coal burning, since there were no active volcanoes and sea spray did not travel far. The most conclusive evidence came when scientists analyzed the sulfur isotopes of the acid rain; sulfur mined from the ground has a very different isotopic ratio of sulfur-34 than does natural sulfur from volcanoes or other sources. In 1978 Canadian scientists showed that the sulfur isotopes in Canadian acid rain exactly matched those of the mining in the Sudbury district, and not those of natural sources of sulfur. By the time an article on acid rain appeared in the widely read popular journal *Scientific American* in 1979, or by the time an eight-year Norwegian study integrating all previous research had appeared in *Nature* in 1981, acid rain had reached the status of accepted scientific consensus. The jury was in, the public had been notified (via popular journals such as *Scientific American*), and there should have been no reasons (at least scientific reasons) for further delay.

Some governmental organizations (especially international organizations such as the UN) acted immediately. In 1979, The UN Economic Commission for Europe passed a Convention on Long-range Transboundary Pollution, an international treaty that made it illegal for one country to generate acid rain pollution that then crossed international boundaries. Meanwhile, the United States and Canada began negotiations in 1979 about their own boundary-crossing acid rain problems. The pollution was due mostly to smokestacks in the U.S. Midwest, whose pollution was blown north or northeast into eastern Canada. According to one study by Environment Canada, at least half of their pollution came from the push by Jimmy Carter's chairman of the Council on Environmental Quality to restrict U.S. emissions, noting that in previous years the polluters had claimed that the taller smokestacks "could remedy the problem by dispersing pollutants high in the atmosphere where they would finally come down as harmless traces."[5] They had claimed that the environmental standards were absurdly strict and were costing

the regions jobs and hurting the economy (the polluters' favorite man-
tra). In this case, the environmentalists had been right, and the taller
smokestacks had created the new problem of long-distance acid rain.
If the polluters had done the right thing in the first place and installed
equipment in their power plants to prevent the pollutants from reaching
the atmosphere, the problem would have been averted—but that was
expensive, and polluters are interested only in profit. They do not care
about the commons (our air and water supply) until the goverment or
others force them to.

In 1980, his final year in office, Carter signed into law the Acid Precipitation Act, which established the National Acid Precipitation Assessment Program (NAPAP) and created the Acid Rain Coordinating
Committee that continued negotiations with Canada to mitigate the
cross-border problem.

Then in November 1980, Ronald Reagan was elected, and polluting industries suddenly had a friend in the White House who had campaigned
against government regulation and interference with business. By 1983,
administration representatives backpedaled on carrying out the agreements that their own scientists had said were necessary. During their
meetings with Canadian representatives, they fell back on the old dodge
of emphasizing the uncertainty of the research, and claiming that if scientists were not 100% certain that a process was occurring, there would
be no grounds for action. Once again, they confused the way science
operates (nothing is 100% certain, but we have to operate within the
context of levels of probability) with the public misconception of science
as final truth: only when we reach 100% certainty can we act. The scientists, of course, were outraged, because their conclusions were very clear
and persuasive, even if they were couched in the language of scientific
uncertainty. The chair of the scientific committee wrote, "The facts about
acid deposition are actually much clearer than in other environmental
causes celebres."[6]

Once the report was released, the scientists found that their words had
been changed and toned down as they were redrafted by Reagan's people. Summaries were much more ambiguous about the acid rain problem
than the actual texts. The final report did not accept that cause and effect
had been established, since there were other minor possible causes out

there (the same strategy the tobacco companies had been using for years, as we saw in chapter 3). The Canadians were outraged, but stopped short of accusing the U.S. representatives of tampering with evidence. By the time the bill went before Congress, it was rejected, and we still had no treaty with Canada about acid rain. Gene Likens noted that the federal agencies were very reluctant "to do anything that would jeopardize their positions in the Reagan White House."[7] The chair of the National Clean Air Coalition, Richard Ayres, who had fought to get many acid rain laws passed, wrote, "This was during the Reagan years, when acid rain almost as verboten as global warming under George W. Bush."[8]

The same shenanigans occurred when the National Academy of Sciences (NAS) recommended in 1981 that there was "clear evidence of serious hazard to human health and the biosphere," and that a continuation of past policies was "extremely risky from a long-term economic standpoint as well as from the standpoint of biosphere protection."[9] The NAS is the most famous, exclusive, and respected body of scientists in the United States, and it does not take on controversial positions lightly unless the scientific evidence is overwhelming. Its mission is to advise the government about scientific reality and consensus so the best possible policy decisions can be made. Then in 1982 the EPA released a strongly worded, 1,200-page report, compiled over two years by forty-six industry, government, and university scientists, which was to give a "scientifically unimpeachable assessment" of the issue.

But what did the Reagan administration do in response to all this advice from the scientific community? It rejected all these reports and commissioned yet another study of the issue, headed by none other than William Nierenberg, one of the founders of the pro-business, antienvironmental Marshall Institute (see chapter 3). Like Seitz, Teller, and others who served as front men for right-wing causes, Nierenberg was a physicist (not an atmospheric chemist—nor was he trained any other field relevant to understanding acid rain), who had worked on the atomic bomb in 1942 and spent most of the 1950s working on Cold War defense projects. His hawkish anticommunist history made him very pro-business and antienvironmentalist. He put together a panel that included scientists who regarded acid rain as a threat (such as Gene Likens), as well as some who were sympathetic to his views.

The Reagan White House added one more: Fred Singer, yet another nuclear physicist who became a front man for right-wing causes. He had started as an environmentalist, but by the 1980s he was affiliated with the conservative Heritage Foundation, and became worried that the cost of environmental protection was too much compared to its benefits. As documented by Oreskes and Conway, the battles within the walls of the commission became complicated and political as Nierenberg and Singer tried to manipulate the results to suit their views.[10] When the scientists in the peer review panel wrote a report urging immediate action on the acid rain problem, Nierenberg and Singer altered and toned down the language until the final report emphasized the uncertainties of acid rain research, and that the technologies to fix the problem were costly and unreliable. Therefore, the issue needed even more study before any action was taken. Much of what was put in the final report was propaganda from lobbyists for the electrical power industry, who were doing everything they could to avoid their responsibility for acid rain. For the rest of Reagan's term (which finally ended in 1989), the official policy of the government was that we do not know for sure what is causing acid rain, it is too expensive to fix, and so it needs more study—the same delaying tactics pioneered by the tobacco industry as discussed in chapter 3.

Even though the Reagan administration stalled and dithered and prevented anything from becoming law, the scientific research into acid rain continued and became even more compelling. When Reagan left office, the pressure to do something increased. In 1990 Congress passed, and George H. W. Bush signed into law, amendments to the Clean Air Act that put in a cap-and-trade system for reducing emissions. Sure enough, it worked like a charm, and sulfur dioxide pollution levels dropped 54% between 1990 and 2007, and 65% since 1976.[11] However, the European Union regulations managed to reduce sulfur dioxide by 70% in Europe between 1990 and 2007.

Meanwhile, the NAPAP that was commissioned back in 1980 issued a 1991 report that documented serious problems due to acid rain and urged strong actions. As the cap-and-trade system has slowly worked on the problem, the amount of acid rain had decreased markedly. By 2010, the total emissions of sulfur dioxide were well below NAPAP's original target, years ahead of their legal deadline.[12] According to EPA estimates, the overall costs of the environmental modifications that the laws have required have cost businesses about $1–$2 billion per year—less than a quarter than what was originally predicted.[13]

What lessons might we draw from this battle? First of all, this was a clear-cut case of nearly unanimous scientific consensus that was (for at least eight years of Reagan's term) hijacked by a handful of activists who wanted to serve a right-wing administration and the business interests that supported them. It took an election that brought in a different administration and Congress to change U.S. policy. Second, the end result was that the scare tactics of big business (the power industries in this case) about the unacceptable costs and the supposedly inconclusive science were just smokescreens for delay. Some two decades later the acid rain issue is no longer a real problem. It was fixed more quickly and cheaply than anyone predicted, and we are all better off for it. Third, as we hear about the battles over cap and trade with regard to carbon dioxide and other greenhouse gases, we might look at how well the system worked in this case before we condemn it as impractical.

THE OZONE HOLE:
ANOTHER ENVIRONMENTAL CRISIS RESOLVED

You could say that the ozone situation is stabilizing at a low level. We are approaching the maximum of ozone depletion, it is kind of leveling off, but it is still too early to say that the situation is improving.

Geir Braathen

Those of us who are old enough to remember the political events of the 1970s and 1980s might recall the great public concern about the ozone layer. Like the debate on acid rain, this environmental crisis arose from earlier research that suggested a problem, then ran into huge opposition from conservative business and antienvironmental interests during the 1980s, when Reagan's cronies were in power. Like the debate on acid rain, the evidence for the hole in the ozone layer only increased until pressure from scientists and governments around the world overcame the resistance of the affected industries and resulted in an eventual global agreement to curb the causes of this pollution.

The story began in the late 1960s when the United States and Europe were both engaged in a race to develop a supersonic transport (SST) for civilian passengers, which would whisk people across huge distances in much shorter times (primarily between Europe and the United States). I vividly remember this series of events, because my father worked for Lockheed Aircraft at the time, and spent many months working very long hours to help develop a huge multivolume proposal for how the Lockheed SST would be built. Huge numbers of man-hours and millions of dollars building two working prototype aircraft were wasted by Lockheed; the contract went to Boeing Aircraft instead. Ironically, the Boeing SST was eventually canceled as well. The only SST that was built was the Anglo-French Concorde, which traveled between New York and Paris from 1976 until it was retired in 2003 due to low demand after the 2000 crash, low air traffic after 9/11, and high costs—since most of the electronics in the aircraft were over thirty years old and obsolete.

One of the concerns that caused the cancellation of the Boeing SST project (besides the technical problems, delays, huge cost overruns, and the concern that it might never be profitable enough to justify its existence) was the possibility that as the SST flew through the ozone layer in the stratosphere, it might cause environmental damage. Ozone (O_3) is a chemical made of three oxygen molecules bonded together; it has a distinctive smell most commonly noticed when a lightning strike or electrical discharge occurs. It is formed when O_2 in our stratosphere is bombarded by radiation and splits up, and some of those free oxygen radicals join other O_2 molecules to form O_3. At ground level, it is not good for us, but up in the stratosphere, it performs an important role in screening out excessive ultraviolet-B (UVB) radiation from the sun, which can cause skin cancers and blindness.

The issue of the possible damage of SST travel in the ozone layer was raised in the late 1960s as the Boeing and Concorde SST projects were well along (previously there had been worries that SSTs might contribute to greenhouse gas warming). A crucial 1970 paper by atmospheric chemist Harold Johnston raised the possibility that nitrous oxides from jet engines might break down the ozone layer, and the story soon reached national attention. However, by the time the paper was fully published, the House of Representatives had already canceled funding for the Boeing

SST program, so it was only applicable to the European Concorde. Nevertheless, Congress funded another program, Climate Impact Assessment Program (CIAP), which involved nearly one thousand scientists across many different agencies and universities over a three-year period; they assessed the possible impact of the SST. When the 7200-page report came out in 1975, it suggested that 500 Boeing-type SSTs could deplete the ozone layer by 20%, especially over the heavily traveled North Atlantic corridor.[14] Yet the executive summary, written by Department of Transportation bureaucrats who wanted the SST to go forward, claimed just the opposite, and suggested that a newly modified SST would not damage the ozone layer. The scientists who had worked so long and hard on the research were outraged, but their corrections appeared only in the scientific literature, whereas the mass media reported only the summary that falsely claimed that the SST was safe for the ozone layer.

Meanwhile, the focus on SSTs and nitrous oxide shifted to another culprit when scientists testing the engines for the upcoming space shuttle missions discovered that it emitted chlorine, which was much more reactive and capable of destroying ozone. Paul Crutzen, an important contributor to the nuclear winter debate, presented a paper on the chlorine problem at a 1974 NASA conference on the topic in Kyoto. Soon thereafter F. Sherwood Rowland and Mario Molina published a historic paper in *Nature* showing that the most abundant source of stratospheric chlorine was chlorofluorocarbons (CFCs), which were commonly used in refrigerators and air conditioners as a coolant, and in spray cans as a propellant.[15] (They received the 1995 Nobel Prize in Chemistry for their discovery.) CFCs were particularly nasty and rapid destroyers of ozone. In the stratosphere, they break up due to solar radiation to release a free radical chlorine atom, which then bonds to ozone and breaks it apart. Once this reaction occurs, however, the chlorine atom is freed up to break up more and more ozone.

With the news that everyday products like hairspray and refrigerators were dangerous to the environment, the National Academy of Sciences, the National Research Council, and Congress soon convened panels and commissioned research to look into the problem. As Oreskes and Conway describe in detail, the battle between scientists and the CFC industry soon became bitter and nasty.[16] The main industry trade associations and their lobbyists, the Chemical Specialities Manufacturers Association and the Manufacturing Chemists' Association, soon set up a PR campaign to try to discredit the research, while pouring over $5 million in research grants to scientists (in hopes they might find results that would dispute Molina and Rowland's conclusions). They hired a British professor of theoretical mechanics, Richard Scorer, to plug their viewpoint. He argued that humans could not cause a problem that might affect the entire atmosphere (even as he gave speeches in Los Angeles, which was experiencing dangerous smog alerts). Once a *Los Angeles Times* reporter exposed his links to the CFC industry and called him a hired gun, his propaganda was discredited, but others soon took his place.

Then the industry made a big fuss about the idea that volcanic eruptions might be a bigger source of CFCs than humans, and therefore hairspray was not to blame. They spent money on a so-called research program monitoring an Alaskan volcano, which erupted as expected in January 1976. But when the research did not show what they wanted it to show, they quickly toned down the PR machine and claimed the results were "inconclusive." Yet the lie that volcanoes were a bigger source of ozone damage continued in their PR campaign. As Harold Schiff put it, the CFC industry "challenged the theory every step of the way. They said there was no proof that fluorocarbons even got into the stratosphere, no proof that they split apart to produce chlorine, no proof that, even if they did, the chlorine was destroying ozone."[17] Then scientists went out in 1975 and 1976 and disproved every one of the industry's claims, and showed that CFCs were indeed a severe threat to the ozone layer and needed to be taken seriously.

When the National Academy of Sciences finally released its long-delayed report on September 15, 1976, it was devastating in its clear-cut conclusions: CFCs were indeed a serious threat to the ozone layer. Despite the efforts of the aerosol industry, the federal regulatory machinery jumped into action, and the FDA and EPA both began to work on regulation of CFCs. Ironically, by the time the FDA announced regulations in 1977, the bad publicity surrounding hairspray had already had an effect on consumer buying patterns. People had discovered that there were numerous CFC-free products that sprayed their contents without dangerous chemicals, such as roll-on deodorants and pump sprays for most kitchen cleaners. The sales of CFC propellants had dropped by 75%, and when the ban took effect in 1979, it was merely the final step in the process already underway.

Meanwhile, NASA began to devote more and more of its satellite time to look into the problem. By the 1980s, their satellites had documented an alarming hole in the ozone layer that arose over the Antarctic at the beginning of each austral spring in September and October, as the warming stirred up the stratospheric clouds and sped up the chemical reactions. The hole was huge and persisted for months, and the ozone levels were alarmingly low. There were even hints of an Arctic ozone hole as well, although it was not as predictable and well established. Both of these discoveries were alarming, since it meant that people living at high latitudes (southern South America, New Zealand, Australia), as well as the wildlife of these areas and the Antarctic, were being exposed to dangerous levels of UVB during the austral spring. Not only does high UVB cause skin cancer, but at high enough levels it can cause eye damage as well. This was a threat to be taken seriously.

As this research emerged through the 1980s, international conferences were held to reach agreements for a worldwide ban on CFCs. The final result of years of negotiation was the 1987 Montreal Protocol on Substances That Deplete the Ozone Layer, which was ratified by all the signatory nations by 1988. Not only was the research supporting the agreement scientifically impeccable, but even the manufacturers had a member on the panel. They could see the market trends away from CFC use, and the risks they took by fighting regulation and scientific consensus. Finally, on March 18, 1988, DuPont (the largest maker of CFCs) announced that it would cease production of CFCs within a few years.

The battle should have been over. The scientific research had reached a consensus, governments around the world had agreed on a solution to the problem, and since CFCs were easily replaced, the industry had complied and actually done better financially without CFCs. Case closed. But the right-wing organizations that challenge any science restricting business (even a business no longer fighting the science) were not ready to give up. The major conservative think tanks (the Heritage Foundation, American Enterprise Institute, the Competitive Enterprise Institute, and the Marshall Institute again) wanted to keep fighting regulations that violated their free market philosophy, even if the industry being regulated did not make the polluting product anymore, and had agreed to the regulation—and improved their bottom line thereby. Some of these people were prominent in the Reagan administration, such as Secretary of the Interior Donald Hodel. His "protection plan" called for people under the ozone hole to wear hats and long-sleeved shirts! The ridicule that he received soon led to his resignation, but there were many other conservatives both in and out of the Reagan White House who were working both publicly and quietly to deny the already settled problem of ozone depletion.

Among the most prominent of these critics was none other than Fred Singer of the Heritage Foundation, whom we already met in our review of his efforts to deny the science of acid rain. In 1987, he wrote an article for the *Wall Street Journal* claiming that the ozone scare was not credible. He claimed that CFCs were not responsible, and that the ozone simply moved somewhere else. Of course, if he had not been a long-retired physicist with no experience in satellites since the 1960s, he would have known that this claim was ridiculous. The newer generation of satellites have global coverage, so if the ozone had moved somewhere else, it would have been detected.

Singer's writings are full of the same dodge, deny, and divert tactics pioneered by the tobacco industry. For example, he pulled the old distraction of "other causes": there are other reasons for skin cancer, therefore we should not worry about the ozone hole! As we saw before, this is irrelevant: if the ozone hole causes skin cancers, we do not want to add it to the list of other known carcinogens, but try to eliminate it.

His main stratagem was a familiar one: scientists have changed their minds in the past, therefore we should not take them seriously. He brought up the ancient 1960s debate over whether the SST might deplete ozone, and laughed at scientists when this proved to be wrong. But he never mentioned that scientists themselves had corrected this error decades ago, and the current evidence of CFCs causing ozone was based on a huge amount research in the ensuing 20 years.

In 1988, Singer misinterpreted valid scientific research by V. Ramanathan about greenhouse gases to claim that the fluctuations of chlorine and stratospheric cooling were just "natural variations" and humans did not cause them. But in his original paper, Ramanathan argued just the opposite: humans were warming the troposphere that was causing cooling of the stratosphere. Singer did the same to James Hansen, director of the Goddard Institute of Space Studies, and an early prophet of the global warming problem. Singer pulled a graph out of context from one of Hansen's publications to argue that the warming trend was part of a natural cycle. Of course, any objective look at those papers would show that this is a deliberate distortion to pervert their meaning. Ramanathan's and Hansen's research was arguing in the clearest possible terms that the changes in the troposphere and stratosphere were not cyclic and were due to human-induced greenhouse gases.[18]

Soon Singer's view of the world—that CFC-ozone science was incomplete and uncertain, that scientists had made mistakes in the past, that it would be expensive to fix the problem, and that scientists were corrupt and money grubbing—was picked up by the right-wing media, including the ultraconservative *Washington Times* (founded and owned by Sun Myung Moon and his Unification Church), and business publications such as the *Wall Street Journal, Forbes,* and *Fortune.* William F. Buckley published an article by Singer in his conservative journal *National Review.* There, Singer blamed rejection of his writings by major scientific journals not on his own incompetence and bad science, but on a global conspiracy by scientists to shut out dissenting points of view. According to Singer, "It's not difficult to understand some of the motivations behind the drive to regulate CFCs out of existence. For scientists: prestige, more grants for research, press conferences, and newspaper stories. Also the feeling that maybe they are saving a world for future generations."[19]

To working scientists, this entire statement sounds bizarre and absurd. Yes, scientists are motivated to publish research that will be noticed and have some importance. They are human, after all. And why is it a bad thing to save the planet? Unlike biased think tanks that publish their own opinions over and over, and push a political-economic agenda, scientists cannot get away with claiming just anything. The peer review process is very strict, and if their data or conclusions do not pass muster, they will be quickly refuted by other scientists eager to shoot them down. Outsiders like Singer (long retired from doing any real science) or the creationists love to propagate this myth of scientific conspiracy, but as any working scientist knows, it is a lie. The scientific community are sharpening their knives to critique each other through peer review and checking published results with later follow-up research to prove someone wrong, and they are about as far from a unified conspiracy as one could imagine.

And the charge that scientists do their research just to get rich is equally absurd. Most of them are in relatively low paying teaching positions, where they rarely reach a six-figure salary even after twenty or more years of hard work. I've been teaching for over thirty years now, and I still have not made it to a six-figure salary, despite publishing numerous books and over 250 scientific papers. To reach our goals of working in science, we had to make many hard sacrifices of long hours and living in near poverty in five to seven years of grad school (for a total of ten to twelve years in college) to earn Ph.D.s. Then we go through the brutal process of teaching for for to six years or more on starvation wages as lowly assistant professors, all the while under the threat of not getting tenure and losing our jobs forever. All the scientists I know have made these sacrifices willingly, because they love what they do and want to discover new and important things about the natural world. As scientists, we were typically bright students near the top of our class, capable of going in a number of directions. Had we wanted to make real money, we would have gone into business or law, where the grad program is only a few years long, and huge salaries are often available at the other end.

For Singer or any of the other conservative antienvironmentalists to claim scientists are corrupt and trying to earn big money and better reputations is a clear case of projecting their own motivations, or the proverbial pot calling the kettle black. Singer and his cohort are certainly not hurting for monetary support. For example, Singer's foundation, the Science and Environmental Policy Project (SEPP), was originally affiliated with a Moonie organization, and now receives funding from Arco, Unocal, Shell, and ExxonMobil.[20] SEPP netted $226,443 in 2007 and had accumulated assets of $1.69 million.[21] We do not know how much Singer himself gets paid for shilling for all these different conservative foundations, but it is safe to say he earns a lot more (as do most of the rich businessmen who fund these foundations) than academics. And Singer, a former scientist with an agenda, gets a lot more publicity and coverage when his causes are trumpeted by right-wing media than do any scientists involved in the debate.

What motivates this bizarre perspective of Singer and his conservative cohorts? As Oreskes and Conway show, Singer sees the environmental science community as "technology-hating Luddites" with a goal to regulate and change our economic system; according to Singer, scientists have a "hidden political agenda" against "business, the free market, and the capitalistic system."[22] The real agenda of scientists is, apparently, to overthrow capitalism and replace it with communism.

I am not sure what scientists Singer hangs out with, but this sounds like the Red Scare paranoia of the 1950s, and it is about as far from reality as possible. I know hundreds of natural scientists (geologists, biologists, chemists, and physicists in many subspecialties), and if there is one thing almost all share, it is a lack of interest in politics and economics, let alone a unified socialist-communist agenda. Many got into science specifically because they were not interested in economics and politics, and had a gift or love for doing science instead. What they are committed to is a sincere love of the truth, and a willingness to make sacrifices of their time, money, and even comfort and personal safety to find out what is really true about nature, no matter whose agenda it might support. Only rarely do most of us think about the possible political or economic implications of our research. Typically scientists try to downplay those aspects because they do not want to attract attention or controversy! If you doubt this, just look at all the negative comments that scientists heaped on Carl Sagan or Stephen Jay Gould because they were willing to be public figures.

Nor are we all Commies. I know of large numbers of both conservative and liberal scientists (but no outright Communists or Socialists), despite the claim that we are all left-wingers. There are some scientists who do have strong political opinions, but as scientists we try our best to prevent our political biases from influencing our scientific results. We are human, of course, so occasionally research with a political agenda does get published—but then the rest of the scientific community will jump in and criticize it, so we do not get away with our biases for long.

But let us get back to the ozone depletion issue and what this example tells us. Scientists originally looking for something else accidentally discovered the problem. Then it was found to be serious and generated a huge volume of conclusive scientific evidence. From this, governmental agencies finally took action, but long after consumers had almost stopped using CFCs. Soon the industry stopped making them because demand had dropped, and they were not necessary, and other materials and methods for refrigeration and propellants worked cheaper and better. Since the Montreal Protocol, the ozone layer has been gradually recovering and CFCs have been gradually vanishing from the stratosphere, although they may not be gone until 2050 or later. Yet the antienvironmentalist movement kept beating a dead horse, filling the right-wing media with false or misleading stories, and claiming that even something like curbing CFCs (which was a good economic decision for both DuPont and the planet) was somehow leading us to communism.

RACHEL CARSON AND DDT: HOW FAR WILL THE ANTIENVIRONMENTALISTS GO?

We stand now where two roads diverge. But unlike the roads in Robert Frost's familiar poem, they are not equally fair. The road we have long been traveling is deceptively easy, a smooth superhighway on which we progress with great speed, but at its end lies disaster. The other fork of the road—the one less traveled by—offers our last, our only chance to reach a destination that assures the preservation of the earth.

Rachel Carson, Silent Spring

As a final coda to this story, let us consider one more example briefly. One of the most startling cases documented by Oreskes and Conway is the demonization of Rachel Carson.[23] Most people consider her to be one of the pioneers and a heroine of the environmental movement. Her 1962 *Silent Spring* was an environmental classic and helped galvanize the early environmental movement in the United States. It popularized the research that showed that DDT sprayed indiscriminately to kill mosquitoes and other insects was also killing a wide variety of other harmless wildlife. In particular, it was destroying bird populations because of the increasing concentrations of the poison as it went up the food chain, so that apex predators like hawks, falcons, and eagles were dying off at alarming rates. Historians regard the banning of DDT as an early environmental success story. At the time, the ban had widespread support from both Republicans and Democrats in Congress, and DDT was finally made illegal under Nixon's Republican administration. Carson herself died in 1964, too soon to see how her efforts led to the banning of DDT and the birth of the EPA and Environmental Defense Fund and many other environmental movement organizations.

Most of us consider Carson's legacy settled. But one can never underestimate the antienvironmentalists and their ability to write revisionist history and to create villains out of heroines. In a modern world with openly partisan networks like Fox, and industry-funded conservative think tanks generating their own propaganda, and with the crazy garbage that flies across the internet, even environmental saints like Rachel Carson cannot escape the abuse.

As Oreskes and Conway document in detail, since 2007 the right-wing and libertarian organizations have been calling Rachel Carson a mass-murderer.[24] What? Did they even read about her life? This shy, humble scientist a mass murderer? Their reasoning is this: because her work led to the banning of DDT, thousands of Africans died of malaria, which might not have happened if DDT were available to them. I will not rehash the entire ill-informed and crazy, convoluted thinking of these people, since Oreskes and Conway have done it already. The reality is that even if DDT had not been banned its use would have stopped anyway, because insects had evolved resistance to it. DDT was already being phased out at the time of the ban, and other pesticides that worked better and did not damage too many harmless animals were being used instead—because DDT did not work! If, as these people propose, DDT had been sprayed across the waterways of Africa, it would not have saved any lives whatsoever because of the evolution of resistance. In fact, many other pesticides that have since been used over the years are now useless because insect pests (especially mosquitoes) evolve resistance so quickly. Yet these people manage to distort history as badly as any Holocaust denier—except instead of trying to exonerate the Nazis of genocide, they turn Rachel Carson into a mass murderer.

Such strange revisionist thinking would not even be worth mentioning if it were not so common in the public discourse these days. Especially on the internet, where there is no peer review or fact checking, crazy is the norm. Just look at the huge number of sites that support demonstrably false ideas, from Holocaust denialism to creationism to the 9/11 Truther movement, or the idea that Obama is a Kenyan citizen and a Muslim. As many people have pointed out, the universality and democratization of the internet means that anyone can post anything and anyone can believe a crazy, false idea (such as the anti-vaxxers discussed in chapter 7). There is no longer any strong probability that they will be confronted with moderate, fact-checked media (like the old evening news with Walter Cronkite or the *New York Times*) whose judgment the entire American public used to accept. And so, we have a polarized America, with a significant number of people who get their information only from the conservative echo chamber and no longer have a reality check from more moderate, unbiased media sources. In a worldview like this, the polluters can be made into saints, and Rachel Carson into a demon. It is fortunate she died almost fifty years ago, and never had to suffer through the besmirching of her reputation that many living scientists have endured.

FOR FURTHER READING

Alm, L. R. 2000. *Crossing Borders, Crossing Boundaries: The Role of Scientists in the U.S. Acid Rain Debate.* New York: Praeger.

Dotto, L., and H. Schiff. 1978. *The Ozone War.* New York: Doubleday.

Gribbin, J. 1988. *A Hole in the Sky: Man's Threat to the Ozone Layer.* New York: Bantam.

Jacobson, M. Z. 2002. *Atmospheric Pollution: History, Science, and Regulation.* Cambridge: Cambridge University Press.

McCormick, J. 1989. *Acid Earth: The Global Threat of Acid Pollution.* London: Earthscan.

Likens, G. E., R. F. Wright, J. N. Galloway, and T. J. Butler. 1979. Acid Rain. *Scientific American* 241 (4): 43–51.

Roan, S. 1990. *The Ozone Crisis: The 15-Year Evolution of a Sudden Global Emergency.* New York: Wiley.

Vallero, D. 2007. *Fundamentals of Air Pollution.* 4th ed. New York: Academic.

5

Hot Enough for You?
The Heated Debate over
a Warming Planet

Reality must take precedence over public relations, for Nature cannot be fooled.

Richard Feynman

POLITICAL HOT AIR

On April 8, 2010, former vice-presidential candidate and half-term Alaska governor Sarah Palin spoke to the Southern Republican Leadership Conference. She mocked the scientific community, calling global warming research "this snake-oil science stuff that is based on the global warming, Gore-gate stuff."[1] As blogger Stephen Webster wrote on Raw Story the next day,

> Up yours, scientists. That's essentially the message sent by former politician Sarah Palin during a recent speech to the Southern Republican Leadership Conference, where she disparaged the work of thousands of the world's top minds to the delight of a large crowd that laughed, clapped and cheered her on the whole way....

To her credit, Palin has at least been remarkably consistent on this point, actually calling on President Obama to insult the international community and boycott the 2009 Copenhagen climate summit over emails stolen from the University of East Anglia. Even then, in December 2009, she was whipping up her fans with the term "snake oil" and claiming that because a small group of people had a dispute over data methodologies, the entire body of knowledge generated by tens of thousands from around the world was suddenly void.[2]

The evidence for global climate change has been accumulating since the 1950s, and was a minor political topic in the 1970s and 1980s—but there was no concerted effort to deny its reality. Even when James Hansen, the head of NASA Goddard Institute for Space Studies, and other prominent scientists brought it to the attention of Congress and the public starting in 1988, and continuing into the 1990s, very little denial or criticism occurred. There was no chance that Congress would act upon it, or that George H. W. Bush would sign a global warming bill if Congress did act. But ever since the late 1990s, the political debate has heated up. In 1997, the Clinton administration tried to take a leadership role in the Kyoto Protocol about climate change. During the early 2000s, the George W. Bush administration actively censored government scientists and allowed oil company lobbyists to tamper with and rewrite government scientific reports describing the evidence of global climate change, or the role of oil and coal companies in contributing to it. Al Gore's 2006 documentary *An Inconvenient Truth* brought the issue to the forefront, and convinced many more people than any scientific report could. The world community took notice and awarded Gore and the film's producers the 2007 Oscar for Best Documentary, and the 2007 Nobel Peace Prize for Gore and the IPCC (Intergovernmental Panel on Climate Change) scientists who discovered, compiled, and reported the evidence. Since 2009, when the Obama administration and the Democratic majority in Congress tried to act on some sort of bill, the debate has reached white-hot intensity, and the public is more confused than ever about what to believe. The political battle is largely polarized along party and cultural lines, with the right-wing media and their followers uniformly opposed and critical, and the rest of the developed world largely accepting the scientific evidence. The 2009 Copenhagen climate change conference may not have accomplished many of its lofty goals, but at least all the world's nations agreed that global climate change is real, and that something should be done about it (even if they fell short on acting upon it during the meeting).

In the midst of all this noise and confusion, how does the average person decide whom to believe? Is there really no consensus among climate scientists, so that any opinion is as good as the next? Let us look at the scientific side of the question first.

GLOBAL CLIMATE CHANGE: THE SCIENTIFIC EVIDENCE

[Carl] Sagan called [the earth] a pale blue dot and noted that everything
that has ever happened in all of human history has happened on that tiny
pixel. All the triumphs and tragedies. All the wars. All the famines. It is our
only home. And that is what is at stake—to have a future as a civilization. I
believe this is a moral issue. It is our time to rise again to secure our future.

Al Gore, An Inconvenient Truth

The story goes back over a century to John Tyndall's 1850 discovery
that greenhouse gases such as water vapor and carbon dioxide absorbed
solar radiation and could warm the planet. Swedish scientist Svante Arrhenius, who received the Nobel Prize in Chemistry for his work, made
the next major breakthrough in 1896. Arrhenius discovered that carbon
dioxide was an important greenhouse gas. When the earth gets energy
from the sun, the solar radiation arrives in shorter wavelengths (mostly
visible and ultraviolet light) that penetrate our atmosphere. After the
earth absorbs this energy, it radiates it back out as longer-wavelength
infrared radiation (which we call heat), which greenhouse gases prevent
from escaping. Since more heat comes in to the planet than can leave it,
the earth's atmosphere warms up. Gases like carbon dioxide, nitrous
oxide, and methane are similar to the glass ceilings of a greenhouse or
the glass windows in your car when it is shut; they hold in heat but let the
light through. Originally Arrhenius calculated that doubling the level of
atmospheric carbon dioxide would cause global temperatures to rise by
5–6°C. This is remarkably close to the current estimates of scientists in
the IPCC report in 2007.

The next major step occurred when Charles Keeling invented one of
the first devices for measuring atmospheric carbon dioxide. In 1958 he
began to take measurements in places isolated from major cities (thus
minimizing local effects), and ran experiments in Antarctica and Mauna
Loa on Hawaii. The Antarctic project ran out of grant funds after a few
years when the NSF decided that he had proved his point, but the Mauna
Loa Observatory has been running continuously for the over fifty-five
years, and has collected one of the longest sets of atmospheric data ever.
By the 1960s, Keeling and his colleague, the legendary Scripps oceanog-

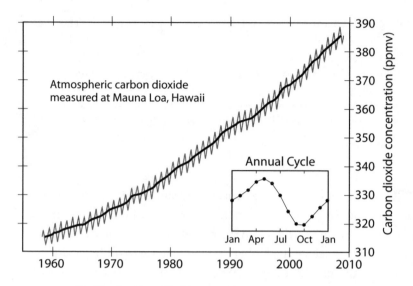

FIGURE 5.1. The "Keeling Curve" of the increase in atmospheric carbon dioxide since 1958. Superimposed on the steady upward increase each year is the annual fluctuation of the seasons. During fall and winter in the Northern Hemisphere, plants die back, decay, and release carbon dioxide; during the spring and summer, plants grow and pull carbon dioxide out of the atmosphere. *Modified from image at Globalwarmingart.com; redrawn by Pat Linse.*

rapher Roger Revelle, could see the dramatic increase in carbon dioxide (fig. 5.1). There are also the annual cycles of decreasing carbon dioxide, when the Northern Hemisphere spring plant growth takes in CO_2, and increasing carbon dioxide in the fall, when the trees lose their leaves in the north.

From Keeling's initial data to every data set that has been collected since then, the trend is clear (fig. 5.2). Carbon dioxide in our atmosphere has increased at a dramatic rate in the past two hundred years. Not one data set collected over a long enough span of time shows otherwise. Mann and Kump (2008) compiled the past nine hundred years' worth of temperature data from tree rings, ice cores, corals, and direct measurements of the past few centuries, and the sudden increase of temperature of the past century stands out like a sore thumb. This famous graph is now known as the "hockey stick" because it is long and straight through most of its length, then bends sharply upward at the end. Other graphs show that climate was very stable within a narrow range of variation through the past one, two, or even ten thousand years since the end of the last Ice Age. There were minor warming events during the Climatic Optimum about seven thousand years ago, the Medieval Warm Period, and the slight cooling of the Little Ice Age from the eighteenth and nineteenth centuries. But the magnitude and rapidity of the warming represented by the last two hundred years is simply unmatched in all of human history. More revealing, the timing of this warming coincides with the Industrial Revolution, when humans first began massive deforestation and released carbon dioxide by burning coal, gas, and oil.

If the data from atmospheric gases were not enough, we are now seeing unprecedented changes in our planet. The polar icecaps are thinning and breaking up at an alarming rate. In 2000, my former graduate advisor Malcolm McKenna was one of the first humans to fly over the

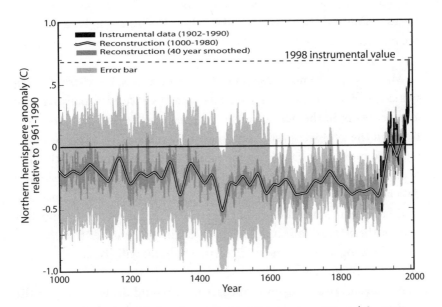

FIGURE 5.2. The record of the last thousand years of temperature change (after M. E. Mann and L. R. Kump, *Dire Predictions: Understanding Global Warming* [New York: DK, 1999]). Temperature was fairly stable until the late nineteenth century, when it suddenly shot upward in response to greenhouse gases released by the Industrial Revolution. Because of its long straight line with the sudden kick upward it is nicknamed the "hockey stick curve." *Modified from image at Globalwarmingart.com; redrawn by Pat Linse.*

North Pole in summertime and see no ice, just open water. So much for Santa's Workshop! The Arctic ice cap has been frozen solid for at least the past three million years and maybe longer,[3] but now the entire ice sheet is breaking up so fast that by 2030 (and possibly sooner) less than half of the Arctic will be ice covered in the summer.[4] In the fall of 2012, satellite data showed a record amount of melting in Greenland, with nearly all the surface ice melting for at least part of the summer (and much of it never freezing again, but pouring into the oceans).[5] As you can see from watching the news, this is an ecological disaster for everything that lives up there, from the polar bears to the seals and walruses to the animals they feed upon. The Antarctic is thawing even faster. In February–March 2002, the Larsen B ice shelf, over 3,000 square km (the size of Rhode Island) and 220 m (700 feet) thick, broke up in just a few months, a story typical of nearly all the ice in Antarctica. The Larsen B shelf had survived all the previous ice ages and interglacial warming episodes of the past three million years, and even the warmest periods of the ten thousand years—yet it and nearly all the other thick ice sheets in the Arctic, Greenland, and Antarctic are vanishing at a rate never before seen in geologic history.

Many people do not care about the polar ice caps, but there is a serious side effect worth considering: all that melted ice eventually ends up as more water in the ocean, causing the sea level to rise, as it has many times in the geologic past. At present sea level is rising about 3–4 mm per year, more than ten times the rate of 0.1–0.2 mm per year that has occurred over the past three thousand years (fig. 5.3). Our geological data show that the sea level was virtually unchanged over the past ten thousand years since the present interglacial began. A few millimeters here or there does not impress people, until you consider that the rate is accelerating and that most scientists predict it will rise 80–130 cm in just the next century.

A sea-level rise of 1.3 m (almost 4 feet) would drown many of the world's low-elevation cities, such as Venice and New Orleans, and low-lying countries such as the Netherlands or Bangladesh. A number of tiny island nations such as Vanuatu and the Maldives, which barely poke out above the ocean now, are already vanishing beneath the waves. Their entire population will have to move someplace else.[6] If the sea level rose

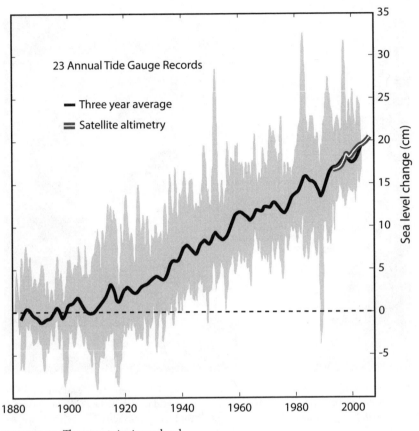

FIGURE 5.3. The recent rise in sea level.
Courtesy Globalwarmingart.com; redrawn by Pat Linse.

by just 6 m (20 feet), nearly all the world's coastal plains and low-lying areas (such as the Louisiana bayous, Florida, and most of the world's river deltas) would be drowned. Most of the world's population lives in coastal cities such as New York, Boston, Philadelphia, Baltimore, Washington, D.C., Miami, Shanghai, and London. All of those cities would be partially or completely underwater with such a sea-level rise. If all the glacial ice caps melted completely (as they have several times before during past greenhouse worlds in the geologic past), sea level would rise by 65 m (215 feet)! The entire Mississippi River valley would flood, so you could dock your boat in Cairo, Illinois (fig. 5.4). Such a sea-level rise

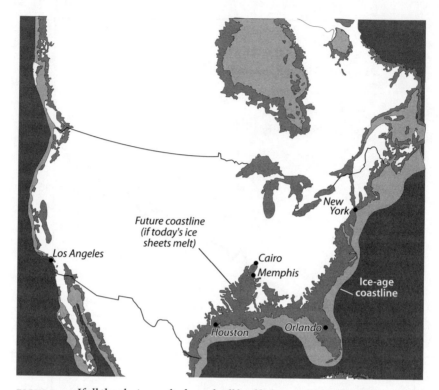

FIGURE 5.4. If all the glaciers melted, nearly all land below 215 feet in elevation would be drowned—and coastal plains, cities, and harbors would vanish. *From D. R. Prothero and R. H. Dott, Jr., Evolution of the Earth, 8th ed. (New York: McGraw-Hill, 2009); redrawn by Pat Linse.*

would drown nearly every coastal region under hundreds of feet of water, and inundate New York City, London, and Paris. All that would remain would be the tall landmarks, such as the Empire State Building, Big Ben, and the Eiffel Tower. You could tie your boats to these pinnacles, but the rest of these drowned cities would be deep under water.

One of the chief congressional critics of global warming research is Senator James Inhofe of Oklahoma. Ironically, nearly every rock in Oklahoma is a product of seas which drowned that state during past greenhouse worlds. If his activities against global warming legislation lead to the eventual drowning of his property, I hope he can swim! I wonder whether future residents of the drowned Oklahoma (much like Kevin

Costner in the science fiction film *Waterworld*) will curse their political representative who denied the evidence all around him, and did his best to bring on the disaster they were facing.

The changes occur not only in polar ice and in rising sea level. It has effects on all the climates around the world. Glaciers are all retreating at the highest rates ever documented. Many of those glaciers, especially in the Himalayas and Andes and Alps and Sierras, provide most of the fresh water that the populations below the mountains depend upon—yet this fresh water supply is vanishing. The permafrost that once remained solidly frozen even in the summer has now thawed, damaging the Inuit villages on the Arctic coast and threatening all our pipelines to the North Slope of Alaska. Not only is the ice vanishing, but we have seen record heat waves over and over again, killing thousands of people, as each year joins the list of the hottest years on record. 2010 topped that list as the hottest year, surpassing the previous record year of 2009. 2011 was the twelfth warmest year on record, even though it was supposed to be a cooler La Niña year, and 2012 looks to break the record based on the incredible summer heat waves in North America that broke all local records. Natural animal and plant populations are being decimated all over the globe as their environment changes.[7] Many animals respond by moving their ranges to formerly cold climates, so now places that once did not have to worry about disease-bearing mosquitoes are infested as the climate warms and allows them to breed further north.

Climate deniers try to distort or obfuscate the evidence about the changing atmosphere, and it is not always easy to give overwhelmingly conclusive data that would convince them. In some cases the data are tricky to analyze, or do not have the well-documented long-term histories necessary to answer every concern about whether recent weather events are truly unprecedented. The atmospheric system is very complicated, with many different processes operating on short-term, medium-term, and long-term time scales, and not all of it is as well understood as we would like. Thus, the arguments over changes in earth's atmosphere often reach an impasse.

Not so for the oceans. Although oceans are an even larger system than the atmosphere, we understand them much better. More importantly, we have an excellent long-term record of how the oceans have changed over millions of years based on thousands of deep-sea cores and the paleontological record of marine fossils that goes back over 700 million years. And unlike the atmospheres, oceans change very slowly over time, since the thermal inertia of water makes the seas very resistant to change, except on long-term time scales. In addition, most ocean currents move slowly compared to atmospheric currents. So no matter what you want to make of the data showing atmospheric change, the changes in the oceans are more alarming, since oceans require immense stimuli to cause such change.

A few years ago, marine biologist and filmmaker Randy Olson (famous for his film *Flock of Dodos*, which lampoons not only creationists but also arrogant scientists who refuse to communicate with the public) founded a web-based effort to publicize the destruction of the oceans. Named "Shifting Baselines," it refers to the fact that many ecological systems have shifted to a "new norm" or "new baseline," and conditions no longer return to those they exhibited only thirty years ago.[8] For example, longtime divers and marine biologists have all documented dramatic changes in the oceans, especially coral reefs. When Olson and most senior marine biologists began diving, coral reefs were thriving around the world, and these same people are now documenting the rapid deterioration of reefs around the world in a single lifetime. Thus, the "baseline" of what is considered normal marine diversity has changed in just a few decades, and biologists being trained today have a very different concept of "normal" marine diversity than those just thirty years ago. As my friend and colleague Jeremy Jackson of the Smithsonian put it, "Every ecosystem I studied is unrecognizably different from when I started. I have a son who is 30, and I used to take him snorkeling on the reefs in Jamaica to show him all the beautiful corals there. I have a daughter who is 17—I can't show her anything but heaps of seaweed."[9] Or as marine biologist Steve Miller of the University of North Carolina, Wilmington, wrote,

> Caribbean coral reefs of the 1970s changed my life. But the reefs I first knew and loved are gone, casualties of disease, coral bleaching, and overfishing. The reefs I study now in Florida are only a shadow of their former glory. My tourist friends go snorkeling and marvel at the colors and structure, but little do they know they're looking at the ghost of a coral reef. While I can tell my friends about all that we

have lost, I am saddened that my children can't have the same personal experience I had, just 25 years ago.[10]

Although overfishing and disease are certainly important problems in the oceans, the biggest problem seems to be that the oceans are becoming warmer and more acidic as they absorb the excess heat and carbon dioxide from the atmosphere and turn it into carbonic acid. For a long time, some people argued that we did not need to worry about carbon dioxide, because the oceans would serve as a big buffer and absorb it all. Well, if that were ever true, it is no longer. The evidence is overwhelming that the acidity of the ocean is changing faster than it has in 300 million years.[11] This, more than any other factor, is responsible for the worldwide dying of the tropical coral reefs. Known as "bleaching," it occurs when the individual coral polyps (which look like tiny sea anemones) cannot tolerate the environmental conditions, such as excess heat or acid ocean waters, any longer. They shed their symbiotic algae (zooxanthellae), which in normal times help them metabolize carbon dioxide and build their skeletons, and thus lose their color. Eventually, the coral polyps die off and leave behind their huge stony skeletons, which gradually turn white. Although some reefs, like the Great Barrier Reef of Australia, are also suffering from problems like out-of-control predation by the crown-of-thorns sea star, the worldwide bleaching and dying of coral reefs can only be attributed to a global oceanographic change—and only ocean warming and acidification fits that description. Certainly, there are marine organisms that thrive in warmer, more acidic oceans (such as the algae that cause the deadly red tide, or encrusting algae growing on rocks uncropped, plus sand fleas, some less calcified crustaceans, and sea urchins),[12] but the vast majority of marine species are negatively affected. Once the reef corals themselves die, nearly all the hugely diverse community of animals and plants vanishes soon thereafter, leaving a mass of dead stony coral rock covered by algae, where once a gloriously beautiful and diverse reef community lived.

If the loss of the coral reefs and their huge effect on diversity were not worrisome enough, there is even more direct evidence of what ocean acidification is doing to the marine realm.[13] Several studies have just reported new data that shows the shells of sea creatures are now dissolving faster than they can be grown. First spotted in the thin-shelled planktonic mollusks known as pteropods (or "sea butterflies") in the Antarctic waters (where colder water allows higher carbon dioxide concentrations), this is an alarming sign. Once the rest of the world's oceans become acidic enough, most calcareous shelled invertebrates (especially the world's population of clams and snails, plus echinoderms, some sponges, and corals) will literally dissolve away as larvae before their shells can grow. In addition, the loss of the planktonic pteropods (and most other calcareous plankton, such as foraminifera and cocco-lithophorid algae) will wipe out the marine plankton that are the base of the food chain throughout the world's oceans. Once the plankton vanish, so do their predators higher up, leading eventually to most of the world's fish and whales, all of which feed on smaller animals from lower in the food chain. This would cause a dramatic extinction in the world's oceans. It would have adverse effects not only on our need for seafood to help provide protein for some of the seven billion people on the planet, but dead oceans have a huge effect on the atmosphere as well. Once the calcareous planktonic algae vanish, they remove our largest absorber of carbon dioxide from the atmosphere, since the world's planktonic algae have a much bigger effect on atmospheric carbon dioxide than do the land plants in rainforests and elsewhere (which are also diminishing due to deforestation).

Even more alarming is how quickly this is all happening. In one life-time, marine biologists have witnessed widespread mass extinction in the coral reef community, and the first signs of oceans so acidic that the marine shelled organisms are dissolving before our eyes. As many stud-ies have shown, this is faster than at any time in geologic history—even the famous "methane burp" event 55 million years ago that caused a sudden spike in carbon dioxide and worldwide mass extinction in the ocean.[14]

As I mentioned above, we have 700 million years of ocean history recorded in the fossil record, especially in the deep-sea cores that record the past 100 million years in great detail. We can analyze the carbon isotopic composition of shells of planktonic microfossils and show how the ocean chemistry has changed. We can look at the patterns of diversity and extinction of acid-sensitive marine fossils, and find out when the ocean has experienced this kind of "acid bath" before. As a recent article by Hönisch and others pointed out, the current episode of mass extinction and rapid acidification of the ocean has no precedent.[15] The closest we can come to is the worst mass extinction in earth history, the "Great Dying" at the end of the Permian Period, about 250 million years ago. The extinction was so severe that about 95% of marine species vanished, and a similar number of land species as well. Although the complete causes are complex and still under discussion, there is a clear signal from the chemical isotopes that there was a global warming event, as well as too much carbon dioxide in the seawater (hypercapnia). It is thought to have been driven by the largest volcanic eruption in earth history, which occurred in northern Siberia. As these eruptions released greenhouse gases, they drove the delicate chemical balance in the oceans to supersaturation in carbon dioxide and highly acidic conditions. Between the toxicity of hypercapnia and the effects of dissolving shells, nearly every group of animals in the oceans vanished 250 million years ago. These included many groups, such as rugose and tabulate corals, trilobites, and blastoid echinoderms, that had survived many previous oceanic mass extinctions. Other groups, such as the brachiopods, the bryozoans, the crinoids, the bivalves and gastropods, and the ammonoid cephalopods nearly vanished, with only a few subgroups surviving to repopulate the world later.

The fossil record provides us with a sobering lesson: what we are doing to our atmosphere is bad enough, but what we do to the oceans is even deadlier, even if it is less visible to us landlubbers. Previously, all the focus has been on the mass extinction in land animals caused by humans and their associated animals, but the devastation of the oceans is far worse. The last time it was this bad, life nearly vanished from this planet.

If you have seen the documentary *An Inconvenient Truth,* or any of the other documentaries on the topic, the long litany of "things we have never seen before" and "things that have never occurred in the past 3 million years of glacial-interglacial cycles" is staggering. Still, there are many people who are not moved by the dramatic images of vanishing glaciers, or by the forlorn polar bears starving to death. Many of these people have been fed lies, distortions, and misstatements by the global warming deniers who want to cloud or confuse the issue. Let us examine some of these claims in detail:

"It's Just Natural Climatic Variability"

No, it is not. As I detailed in my 2009 *Greenhouse of the Dinosaurs,* geologists and paleoclimatologists know a lot about past greenhouse worlds, and the icehouse planet that has existed for the past 33 million years. We have a good understanding of how and why the Antarctic ice sheet first appeared at that time, and how the Arctic froze over about 3.5 million years ago, beginning the twenty-four glacial and interglacial episodes of the so-called Ice Ages that have occurred since then. We know how variations in the earth's orbit (the Milankovitch cycles) control the amount of solar radiation the earth receives, triggering the shifts between glacial and interglacial periods. Our current warm interglacial has already lasted ten thousand years, the duration of most previous interglacials, so if it were not for global warming, we would be headed into the next glacial any time now. Instead, our pumping greenhouse gases into our atmosphere after they were long trapped in the earth's crust has pushed the planet into a super-interglacial (fig. 5.5), already warmer than any previous warming period. (This is why some deniers try to discredit the evidence by saying scientists predicted global cooling in the 1970s. In

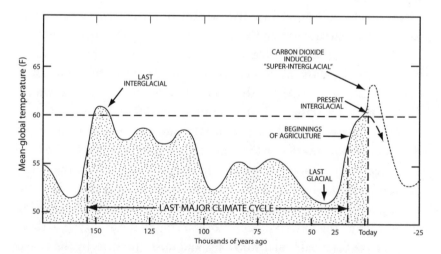

FIGURE 5.5. The last 130,000 years of glacial-interglacial cycles, showing the typical 10,000-year duration of the previous interglacial 125,000 years ago and the predicted end of our current interglacial after 10,000 years. *After Prothero and Dott 2009; redrawn by Pat Linse.*

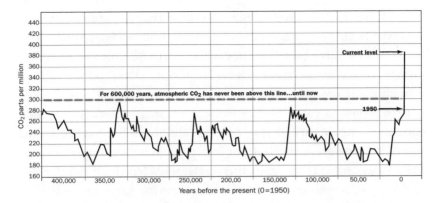

FIGURE 5.6. The climate record from EPICA core from Antartica. It shows the normal range of climate variability over the past 650,000 years and the last six glacial-interglacial cycles. At no point in any previous interglacial was the carbon dioxide level higher than 300 ppm, or the temperatures so high, yet we are almost to 400 ppm today. This is ironclad evidence that our present episode of warming is not "normal fluctuations." (Shown here are the last three glacial cycles.)

fact, it was mostly the media making this prediction. The peer-reviewed scientific literature consistently pointed to evidence of warming.)

We can see the "big picture" of climate variability most clearly in the EPICA cores from Antarctica (fig. 5.6), which show the details of the last 650,000 years of glacial-interglacial cycles. *At no time during any previous interglacial did the carbon dioxide levels exceed 300 ppm, even at their very warmest.* Our atmospheric carbon dioxide levels are already close to 400 ppm today. The atmosphere is headed to 600 ppm within a few decades, even if we stopped releasing greenhouse gases immediately. This is decidedly *not* within the normal range of climatic variability, but clearly unprecedented in human history. Anyone who says this is normal variability has never seen the huge amount of paleoclimatic data that show otherwise.

"It's Just Another Warming Episode, Like the 'Medieval Warm Period,' or the 'Holocene Climatic Optimum' or the End of the 'Little Ice Age'"

Untrue. There were numerous small fluctuations of warming and cooling over the last ten thousand years of the Holocene. But in the case of the Medieval Warm Period (about 950–1250 BCE), the temperatures were only 1°C warmer than today, much less than the temperature changes since the beginning of our current global warming (fig. 5.7). This episode was also only a local warming in the North Atlantic and Northern Europe. Global temperatures over this interval did not warm at all, and actually cooled by more than 1°C. Likewise, the warmest period of the last ten thousand years was the Holocene Climatic Optimum (5000–9000 BCE), when warmer and wetter conditions in Eurasia caused the rise of the first great civilizations in Egypt, Mesopotamia, the Indus Valley, and China. Once again, this was largely a Northern Hemisphere Eurasian phenomenon, with 2–3°C warming in the Arctic and Northern Europe. But there was almost no warming in the tropics, and cooling or no change in the Southern Hemisphere.[16]

To the Eurocentric world, these warming events seemed important, but on a global scale the effect is negligible. In addition, neither of these warming episodes is related to increasing greenhouse gases. The Holo-

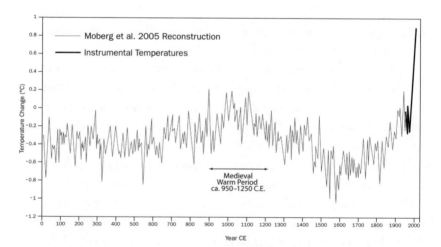

FIGURE 5.7. Plot of the details of the last thousand years of earth's average surface temperature, which shows over eight hundred years of relative stability followed by the rapid warming of the past two centuries, giving it the shape of a "hockey stick" (compare with fig. 5.3). The slight warming of the "Medieval Warm Period" is much smaller compared to the warming of the last hundred years, and this is a Northern Hemisphere graph; there is no Medieval Warm Period in the global data set. A. Moberg, D. M. Sonechkin, K. Holmgren, N. M. Datsenko, and W. Karlén, "Highly Variable Northern Hemisphere Temperatures Reconstructed from Low- and High-Resolution Proxy Data," Nature 433, no. 7026 (February 10, 2005): 613–617; updated from the graph in Mann and Kump 1999.

cene Climatic Optimum, in fact, is predicted by the Milankovitch cycles, since at that time the axial tilt of the earth was 24°, its steepest value, meaning the poles got more solar radiation than normal—leading to the warmest period of the interglacial. By contrast, not only is the warming observed in the last two hundred years much greater than during these previous episodes, but it is also *global and bipolar,* so it is not a purely local effect. The warming that ended the Little Ice Age (from the mid-eighteenth century to the late nineteenth century) was due to increased solar radiation prior to 1940. Since 1940, however, the amount of solar radiation has been dropping, so the only candidate for the post-1940 warming has to be carbon dioxide.[17]

"It's Just the Sun, or Cosmic Rays, or Volcanic Activity or Methane"

Nope. Sorry. The amount of heat that the sun provides has been decreasing since 1940,[18] just the opposite of the deniers' claims (fig. 5.8). Cosmic radiation causes an increase in cloud cover on the earth, so increased

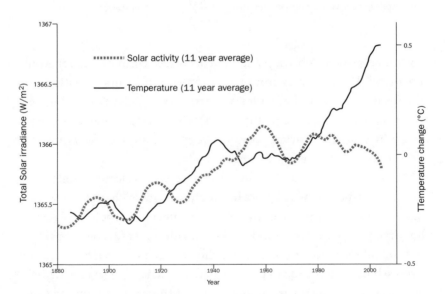

FIGURE 5.8. Plot of solar energy input to the earth versus temperature over the last century. The two tend to track each other until the last thirty years, at which time the earth warmed dramatically even as solar input went down.

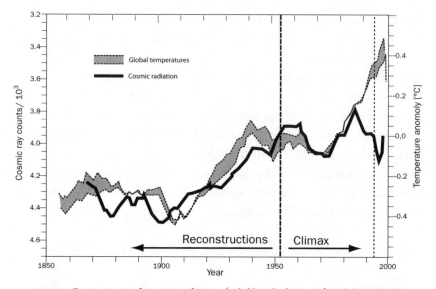

FIGURE 5.9. Reconstructed cosmic radiation (solid line before 1952) and directly observed cosmic radiation (solid line after 1952) compared to global temperature (dotted line). All curves have been smoothed by an eleven-year running mean. *Redrawn from N. A. Krivova and S. K. Solanki, "Solar Total and Spectral Irradiance: Modelling and a Possible Impact on Climate," Max-Planck-Institut für Sonnensystemforschung website, September 2003, www.mps.mpg.de/dokumente/publikationen/solanki/r47.pdf.*

cosmic rays would cool the planet, and decreased cosmic radiation would warm it.[19] There are numerous measurements of cosmic radiation, and the result is clear: in the last forty years, cosmic radiation has been increasing (which should cool the planet) while the temperature has been rising (fig. 5.9), the exact opposite of the effect expected if cosmic radiation contributed to recent warming.[20]

Nor is there any clear evidence that large-scale volcanic events (such as the 1815 eruption of Tambora in Indonesia, which changed global climate for about a year) have any long-term effect that would explain two hundred years of warming and carbon dioxide increase. Volcanoes erupt only 0.3 billion metric tons of carbon dioxide each year, but humans emit over 29 billion metric tons a year;[21] clearly, we have a bigger effect. Methane is a more powerful greenhouse gas, but there is two hundred times more carbon dioxide than methane, so carbon dioxide is still the most important agent.[22] Every other alternative has been looked at, but the only clear-cut relationship is between human-caused carbon dioxide increase and global warm. We just cannot squirm out of the blame on this one.

"The Climate Records since 1995 (or 1998) Show Cooling"

That's a deliberate deception. People who throw this argument out are cherry-picking the data.[23] Over the short term, there was a slight cooling trend from 1998–2000 (fig. 5.10A), because 1998 was a record-breaking El Niño year, so the next few years look cooler by comparison. But since 2002, the overall long-term trend of warming (fig. 5.10B) is unequivocal. This quotation is a clear-cut case of using data out of context in an attempt to deny reality. Likewise, you might hear people say that 1934 was the hottest year ever in the United States. That may be true for a local region, but globally it was nowhere near the warmest year on record.[24] All of the seventeen hottest years ever recorded on a global scale have occurred in the last twenty-one years. They are (in order of hottest first): 2010, 2009, 1998, 2005, 2003, 2002, 2004, 2006, 2007, 2011, 2001, 1997, 2008, 1995, 1999, 1990, and 2000.[25] In other words, every year since 2000 has been in the Top Ten hottest years list, and the rest of the list includes 1995, 1997, 1998, 1999, and 2000. Only 1996 failed to make the list (because of the short-term cooling mentioned already).

"We Had Record Snow in the Winter of 2009–2010"

So what? This is a classic case of how the scientifically illiterate public cannot tell the difference between *weather* (short-term seasonal changes) and *climate* (the long-term average of weather over decades and centuries and longer). Our local weather tells us nothing about the next continent, or the global average; it is only a local effect, determined by short-term atmospheric and oceanographic conditions.[26] In fact, warmer global temperatures mean *more moisture* in the atmosphere, which increases the intensity of normal winter snowstorms. In this particular case, the climate deniers forget that the early winter of November–December 2009 was actually very mild and warm, and then only later in January and February did it get cold and snow heavily. That warm spell in early winter helped bring more moisture into the system, so that when cold

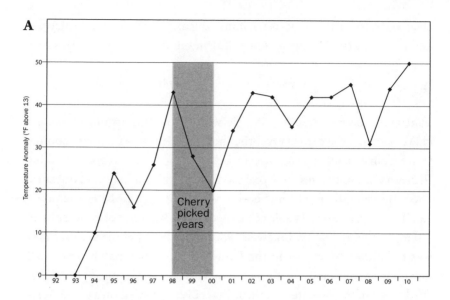

FIGURE 5.10. (A) Detailed plot of the past twenty years of global mean temperatures, showing how anomalous 1998 was. If you cherry-pick 1998 and the two years that followed, it appears that climate is cooling. However, if you pick any two points other than 1998–2000, or any rolling average, it is clear that climate is warming. Indeed, most of the years from 2002 onward are as warm or warmer than 1998, so any claim that "it has been cooling since 1998" is false. (B) The plot of global mean temperature over the past century, showing the yearly data (wiggly lines) and the smoothed curve using a five-year rolling average (gently curved line). Clearly the trend has been dramatically increasing, and individual data points from one year do not tell the whole story. The anomalous El Niño warm year of 1998 is one of those outliers. *Redrawn from GISS data by Pat Linse.*

weather occurred, the snows were worse. In addition, the snows were unusually heavy only in North America; the rest of the world had different weather, and the global climate was warmer than average. And the summer of 2010 was the hottest on record, breaking the previous record set in 2009. Anyone who mentions this silly argument is clearly ignorant of basic science.

"Carbon Dioxide Is Good for Plants, So the World Will Be Better Off"

Who do they think they are kidding? The people who promote this idea clearly do not know much global geochemistry, or are trying to play on

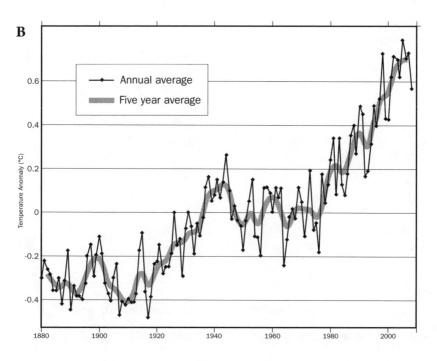

the fact that most people are ignorant of science. The Competitive Enterprise Institute (paid for mostly by money from oil and coal companies
and conservative foundations)[27] has run a series of shockingly ignorant
and misleading ads that insult the intelligence of any educated person,
concluding with the tag line "Carbon dioxide: they call it pollution, we
call it life." Anyone who knows the basic science of earth's atmosphere
can spot the deceptions in this ad.[28] Sure, plants take in carbon dioxide
that animals exhale, as they have for millions of years. But the whole
point of the global warming evidence (as shown from ice cores) is that
the delicate natural balance of carbon dioxide has been thrown out of
whack by our production of too much of it, way in excess of what plants
or the oceans can handle. As a consequence, the oceans are warming and
absorbing excess carbon dioxide, making them more acidic.[29] Already
we are seeing a shocking decline in coral reefs (due to bleaching) and extinctions in many marine ecosystems that cannot handle too much of a
good thing. There is strong scientific evidence that the so-called Mother
of all Mass Extinctions (which wiped out 95% of marine species about

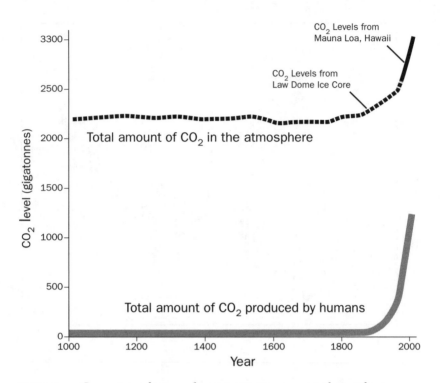

FIGURE 5.11. Comparison of measured temperature increases over the past few centuries and the amount of carbon released by humans into the atmosphere. As the plot shows, carbon dioxide tracks human emissions. *From "The Human Fingerprint in Global Warming [Intermediate]," Skeptical Science website, www.skepticalscience.com/its-not-us-intermediate.htm.*

250 million years ago) was due to excess carbon dioxide (*hypercapnia*) in the oceans, which not only dissolves shells and corals but also suffocates marine life.[30]

Meanwhile, humans are busy cutting down huge areas of rainforest every day, which not only means there are fewer plants to absorb the gas, but the slash-and-burn practices are releasing more carbon dioxide than plants can keep up with. There is much debate as to whether increased carbon dioxide might help agriculture in some parts of the world, but that has to be measured against the fact that other traditional breadbasket regions (like the North American Great Plains) are expected to get too hot to be as productive as they are today. The latest research actually

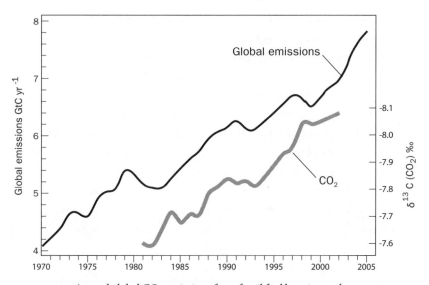

FIGURE 5.12. Annual global CO_2 emissions from fossil fuel burning and cement manufacture in GtC yr[1] (longer line to left), annual averages of the $^{13}C/^{12}C$ ratio measured in atmospheric CO_2 at Mauna Loa from 1981 to 2002 (shorter line to right). *Redrawn from "The Human Fingerprint in Global Warming [Intermediate]."*

shows that increased carbon dioxide inhibits the absorption of nitrogen into plants, so plants (at least those that we depend upon today) are NOT going to flourish in a greenhouse world.[31] Anyone who tells you otherwise is either ignorant of basic atmospheric science, or is trying to con a public that does not know science from bunk.

"I Agree that Climate Is Changing, but I'm Skeptical that Humans Are the Main Cause, So We Should Not Do Anything"

This is just fence sitting. A lot of reasonable skeptics deplore the climate denialism of the right wing, but still want to be skeptical about the cause. If they want proof, they can examine the huge array of data directly point to humans causing global warming.[32] We can directly measure the amount of carbon dioxide humans are producing, and it tracks exactly with the amount of increase in atmospheric carbon dioxide (fig. 5.11). Through carbon isotope analysis, we can show that this carbon dioxide in the atmosphere is coming directly from our burning of fossil fuels, not from natural sources (fig. 5.12).

We can also measure oxygen levels that drop as we produce more carbon that then combines with oxygen to produce carbon dioxide. We can also examine the spectrum of the gases in the atmosphere, and they exactly match the spectrum expected if human-caused gases were increasing.[33] We have satellites out in space that are measuring the heat released from the planet and can actually *see and measure* the atmosphere get warmer. The most crucial proof emerged only in the past few years: climate models of the greenhouse effect predict that there should be cooling in the stratosphere (the upper layer of the atmosphere above 10 km (6 miles) in elevation, but warming in the troposphere (the bottom

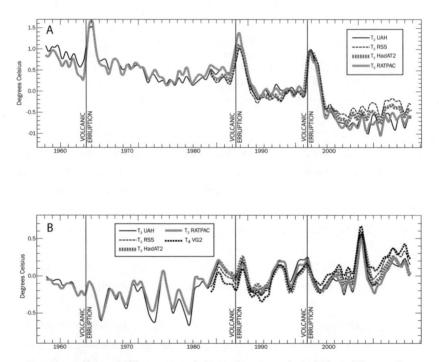

FIGURE 5.13. Change in lower stratospheric temperature, observed by satellites and weather balloons relative to period 1979 to 1997, smoothed with seven-month running mean. Major volcanic eruptions indicated by dashed lines. *Redrawn by Pat Linse from T. R. Karl et al., "Temperature Trends in the Lower Atmosphere: Steps for Understanding and Reconciling Differences," Climate Science Watch website, April 2006, www.climate science.gov/Library/sap/sap1–1/finalreport/sap1-1-final-all.pdf.*

layer of the atmosphere below 10 km [6 miles]). In contrast, an increase in solar radiation would warm the stratosphere and cool the troposphere. In fact, our space probes have measured stratospheric cooling and upper troposphere warming (fig. 5.13), just as climate scientists had predicted, and proving it is due to greenhouse gases, not the sun.[34] Finally, we can rule out any other culprits (see above): solar heat is decreasing since 1940, not increasing, and there are no measurable increases in cosmic radiation, methane, volcanic gases, or any other potential cause.

Face it—it is our problem.

THE GLOBAL DENIER CONSPIRACY

As I said on the Senate floor on July 28, 2003, "much of the debate over global warming is predicated on fear, rather than science." I called the threat of catastrophic global warming the "greatest hoax ever perpetrated on the American people."

James Inhofe, January 2005

Thanks to all the noise and confusion over the debate, the general public has only a vague idea of what the debate is really about, and only about half of Americans think global warming is real or that we are to blame.[35] As in the debates on evolution and creationism, the scientific community is virtually unanimous on what the data demonstrate about anthropogenic global warming. This has been true for over a decade. When historian of science Naomi Oreskes surveyed all peer-reviewed papers on climate change published between 1993 and 2003 in the world's leading scientific journal, *Science,* she found that there were 980 supporting the idea of human-induced global warming and *none* opposing it.[36] In 2009, Doran and Kendall Zimmerman surveyed all the climate scientists who were familiar with the data.[37] They found that 95–99% agreed that global warming is real and that humans are the reason. In 2010, the prestigious *Proceedings of the National Academy of Sciences* published a study that showed that 98% of the scientists who actually do research in climate change are in agreement about anthropogenic global warming.[38] Geology James Lawrence Powell searched the Web of Science for peer-reviewed articles mentioning global climate change between 1991 and 2012; only 24 out of 13,950 (less than 0.17%) reject global warming.[39] Every major scientific organization in the world has endorsed the idea of anthropogenic climate change as well. This is a rare degree of agreement within such an independent and cantankerous group as the world's top scientists. This is the same degree of scientific consensus that scientists have achieved over most major ideas, including gravity, evolution, and relativity. These and only a few other topics in science can claim this degree of agreement among nearly all the world's leading scientists, especially among everyone who is close to the scientific data and knows the problem intimately. If it were not such a controversial topic politically, there would be almost no interest in debating it, since the evidence is so clear-cut.

If the climate science community speaks with one voice (as in the 2007 IPCC report, and every report since then), why is there still any debate at all? The answer has been revealed by a number of investigations by diligent reporters who got past the PR machinery denying global warming, and uncovered the money trail. Originally, there were no real dissenters to the idea of global warming by scientists who are actually involved with climate research. Instead, the forces with vested interests in denying global climate change (the oil and coal companies, and the conservative free market advocates) followed the strategy of tobacco companies: create a smokescreen of confusion and prevent the American public from recognizing scientific consensus. As the famous memo from the tobacco lobbyists said, "Doubt is our product."[40]

The deniers generated an antiscience movement entirely out of thin air and PR. The evidence for this PR conspiracy has been well documented in numerous sources. For example, Oreskes and Conway (2010) revealed from memos leaked to the press that in April 1998 the right-wing Marshall Institute, SEPP (Fred Seitz's lobby that aids tobacco companies and polluters), and ExxonMobil met in secret at the American Petroleum Institute's headquarters in Washington, D.C. There they planned a $20 million campaign to get respected scientists to cast doubt on climate change, get major PR effort going, and lobby Congress that global warming was not real and was not a threat. In 2012, leaked documents showed that the Heartland Institute, a libertarian think tank and the major sponsor of denialist propaganda and phony "scientific meetings," planned to try to get schools to teach its propaganda instead of the science of climate change.[41]

They beat the bushes to find scientists—*any* scientists—who might disagree with the scientific consensus. As investigative journalists and scientists have documented over and over again, the denier conspiracy essentially offered bribes to anyone who could be useful to them.[42] The day that the 2007 IPCC report was released (February 2, 2007), the British *Guardian* reported that the conservative American Enterprise Institute (funded largely by oil companies and conservative think tanks) had offered $10,000 plus travel expenses to scientists who would write negatively about the IPCC report.[43] We are accustomed to the hired gun experts hired by lawyers to muddy up the evidence in the case they are fighting, but this is extraordinary—buying scientists with outright bribes to act as shills for organizations trying to deny scientific reality. With this kind of money, however, you can always find a fringe scientist or crank or someone with no relevant credentials who will do what they are paid to do.

Oklahoma senator James Inhofe (who gets nearly all his campaign money from oil and gas and other energy companies)[44] and other others bragged about having their own group of scientists who dispute global warming and publishing a list of their names. By doing a little digging, the Center for Inquiry discovered that fewer than 10% of the names on the list had any appropriate credentials or direct research experience in climate research. The rest were a mix of scientists with no relevant training or experience. Over 80% had no refereed publications in climate science at all. About 4% of the deniers on the list protested their inclusion because they *supported* the IPCC 2007 consensus that global warming is real and man-made. Dr. Stuart Jordan, formerly a climate scientist for NASA and now with the CFI, wrote, "As a result of our assessment, Inhofe and other lawmakers using this report to block proposed legislation to address the harmful effects of climate change must face an inconvenient truth: while there are indeed some well respected scientists on the list, the vast majority are neither climate scientists, nor have they published in fields that bear directly on climate science." Dr. Ronald Lindsay of CFI wrote, "Sen. Inhofe and others have had some success in conveying to the media the impression that the number of scientists skeptical about man-made global warming is swelling, yet this is demonstrably not true." Inhofe had falsely claimed that the number of dissenting scientists was thirteen times more than the number of UN scientists (52) who authored the 2007 IPCC. "But those 52 U.N. scientists were in fact summarizing for policymakers the work of over 2,000 active research scientists, all with substantially similar views on global warming and its causes. This is the kind of broadside against sound science and scientific integrity that we at CFI deplore."[45]

There are polls and petitions circulated by groups like Arthur Robinson's tiny home office in Cave Junction, Oregon, known by the grandiose name the "Oregon Institute of Science and Medicine," claiming he has a list of thousands of dissenting scientists. If you look closely at the list, however, you will find that there are virtually no climate scientists or people with direct research experience in climate science on it.[46] The list consists mostly of people without relevant scientific background, nonscientists, and lots of TV weathermen who do not have any research experience in climate science. (The Oregon Institute's list of names includes many duplications, cartoon characters, fake names, people with no qualifications, and people who did not consent to have their names used because they believe global warming is real.)

Fishing around to find anyone with some science background who will agree with you and dispute a scientific consensus is a tactic employed by the creationists (as we shall see in chapter 6). It may generate lots of PR and a smokescreen to confuse the public, but it does not change the fact that *scientists who actually do research in climate change are unanimous in their insistence that anthropogenic global warming is a real threat.* Most scientists (including my many friends in the climate science community) I know and respect work very hard for little pay, yet they still cannot be bribed to endorse some scientific idea they know to be false.

If this is not convincing enough, let us use the rules about evaluating expert opinions that we discussed in chapter 2: *relevant credentials,* and *conflict of interest.* If they do not have their Ph.D. *in climate science,* and are not actively doing research *in climate science,* and publishing *in respected journals of science,* they are just rank amateurs in that topic and do not deserve to be taken seriously. This applies to many books and other writings that claim to show that there is no problem with global warming.

For example, Bjorn Lømborg has made a big splash with two books[47] that argue that global warming is no big deal, and we should not take measures to stop it. First of all, Lømborg is an economist, not a scientist of any kind, and the reviews of his book have ripped his arguments to shreds, because he is abysmally ignorant about the climate data he attempts to interpret.[48] More recently, people have carefully fact checked his footnotes and sources, and found that he has been quoting out of context (like a creationist), and most of his sources do not in fact support the claims he makes in his book.[49] Either Lømborg cannot understand what he is reading, or he is dishonestly trying to distort the meaning of his sources for his own purposes. Now Lømborg has come out in favor of the seriousness of global warming, and says the world's governments need to spend $100 billion to stop it.[50]

Ian Plimer's *Heaven and Earth: Global Warming, the Missing Science* received more plaudits from deniers because it came from a geologist.[51] But Plimer is a *mining* geologist, *not* a climate scientist, and he (along with oil and coal geologists) would be expected to have a conflict of interest that would bias him toward *not* understanding climate data that threatens his livelihood. Numerous scathing reviews of the book by both climate scientists and other kinds of earth scientists revealed his complete incompetence in climate science,[52] yet it is hailed by the denier community as some sort of exposé from the geological community.

The climate deniers have many other things in common with creationists and other antiscience movements. They, too, like to quote people out of context (quote mining), finding short phrases in the works of legitimate scientists that seem to support their position. But when you read the full quotations in context, it is obvious that they have been used inappropriately. The original authors meant things that do not support the deniers' goals. The Climategate scandal is a classic case of this. It started with a few stolen e-mails from the Climate Research Unit of the University of East Anglia. If you read the complete text of the actual e-mails and comprehend the scientific shorthand of climate scientists who are talking casually to each other, it is clear that there was no great conspiracy or that they were faking data.[53] The phrase "neat trick," for example, talks about an inventive method to process and display the data, not a deliberate deception. The phrase "hide the decline" refers to the well-known problem with recent tree-ring data that are not showing the directly measured global increase in temperature, so the scientists have replaced a flawed tree-ring data set with the actual observed temperature records. Read in context, these and other quotations make perfect sense and show no evidence of deliberate attempts to deceive the public. Yet climate deniers and politicians never read these letters in context, but grab just the quotations and use them as political weapons. All six subsequent investigations have cleared Philip Jones and the other scientists of the University of East Anglia of any wrongdoing or conspiracy.[54]

Even *if* there had been some conspiracy on the part of these few scientists, there is no evidence that the entire climate science community is secretly working together to generate false information and mislead the public. If there is one thing that is clear about science, it is about competition and criticism, not conspiracy and collusion. Most labs are competing with each other, not conspiring together. If one lab publishes a result that is not clearly defensible, other labs will quickly correct it. Only when every scientist in a community comes to the same conclusion independently would you arrive at the type of consensus shown by the IPCC 2007 report, or every subsequent report. In other cases, the climate deniers have claimed that the conspiracy is motivated by money. This is so bizarre and contrary to reality that it is laughable. Most scientists are just hardworking people who are willing to survive on a measly researcher's or professor's salary because they love the thrill of discovery of the truth about the world, not because they have some economic or political agenda. If they had really wanted to become rich, they would have gone into law or business or oil jobs, where the big bucks are (as many of the climate deniers have done). Yes, scientists try to win grants to support their research, but that money is minuscule compared to the huge amounts made in the oil industry, for example. You could not find a better example of the pot calling the kettle black.

This attempt to smear the hardworking scientists is one of the slimiest and most dishonest tactics of all, because the quote-mining climate deniers are either deliberately trying to mislead their audience by distorting the evidence, or they are not intelligent enough to understand the quotations and their context in the first place.

Even more bizarre is that the alarms over global climate change is some sort of left-wing conspiracy to foist Big Government on us. In fact, scientists come in every political color and stripe, but most try to rigorously exclude politics from their science. For example, Kerry Emanuel of MIT, who showed the connection between climate change and more intense hurricanes, is a Republican, as are many less famous climate scientists. Yet he does not let his political views contaminate his science. As James Lawrence Powell wrote,

> Scientists . . . show no evidence of being more interested in politics or ideology than the average American. Does it make sense to believe that tens of thousands of scientists would be so deeply and secretly committed to bringing down capitalism and the American way of life that they would spend years beyond their undergraduate degrees working to receive master's and PhD degrees, then go to work in a government laboratory or university, plying the deep oceans, forbidding deserts, icy poles, and torrid jungles, all for far less money than they could have made in industry, all the while biding their time like a Russian sleeper agent in an old spy novel? Scientists tend to be independent and resist authority. That is why you are apt to find them in the laboratory or in the field, as far as possible from the prying eyes of a supervisor. Anyone who believes he could organize thousands of scientists into a conspiracy has never attended a single faculty meeting.[55]

The climate deniers have many other traits in common with the creationists, Holocaust deniers, and others who distort the truth. They pick on small disagreements between different labs as if scientists cannot get their story straight, when in reality there is always a fair amount of give and take between competing labs as they try to get the answer right before the other lab can do so. The key point here is that when *all* these competing labs around the world have reached a consensus and get the same answer, there is no longer any reason to doubt their common conclusion. The antiscientists of climate denialism will also point to small errors by individuals in an effort to argue that the entire enterprise cannot be trusted. It is true that scientists are human, and do make mistakes, but the great power of the scientific method is that *peer review weeds these out*, so that when scientists speak with consensus, there is no doubt that their data are carefully checked.

Finally, the most convincing evidence of the fact that this is a purely political controversy, rather than a scientific debate, is that the membership lists of the creationists and the climate deniers are highly simi-lar. Both antiscientific dogmas are fed to their overlapping audiences through right-wing media like Fox News, Glenn Beck, and Rush Lim-baugh. Just take a look at the intelligent design creationism website for the Discovery Institute. Most of the daily news items lately have nothing to do with creationism at all, but are focused on climate denialism and other right-wing causes.[56]

"IT'S ALL POLITICS"—AND OUR PLANET IS THE HOSTAGE

We're in a giant car headed toward a brick wall, and everyone's arguing over where they are going to sit.

David Suzuki

The conclusion is clear: there is science, and then there is the anti-science of the global warming deniers. As we have seen, there is a nearly unanimous consensus among climate scientists that anthropogenic global warming is real and that we must do something about it. Yet the smokescreen, bluster, and lies of the right-wing media have created enough doubt that less than half of the American public is convinced the problem requires action. Ironically, the United States is almost alone in their denial of this scientific reality. International polls of thirty-three thousand people in thirty-three nations in 2006 and 2007 show that 90% of their citizens regard climate change as a serious problem,[57] and 80% think that humans are the cause of it.[58] Just as in the case of creation-ism, the United States is out of step with much of the rest of the world in accepting scientific reality. In this case, however, the main driving force is not religion, but the fear of the consequences of cutting back on our wasteful use of oil and coal and other sources of carbon dioxide, plus the conservative and libertarian political attitudes that the government should not interfere with a corporation's right to foul our planet and destroy it for future generations. Some of these people sound like foam-ing-at-the-mouth loonies when you read what they say. For example, BBC News reported on a May 21, 2010, gathering of libertarian global warming deniers organized by the Heartland Institute (a right-wing think tank funded largely by oil companies to question the evidence for global climate change).[59] Their speakers repeated the message that global warming is a hoax to allow governments to control businesses and people. Most of the scientific speakers had no relevant credentials in climate science—or even worse for the deniers, told them that the data do indeed support climate change.[60]

In some cases, the right-wing fringe has gone to extreme lengths in their hostile attitude toward legitimate science. The FBI has reported a sharp increase in threats and hate mail and intimidation against prominent climate scientists Michael Mann, James Hansen, and others. The transition from conservative climate denier to a dangerous antisemitic hate group is not difficult; one white supremacist website posted Michael Mann's picture and those of other climate scientists and labeled it "Jew." (In fact, most climate scientists are not Jewish, but the facts do not matter to racists and antisemites.) Another climate scientist told ABC News that he found a dead animal placed on his doorstep, and now he must travel with a bodyguard. As Mann said, "Human-caused climate change is a reality. There are clearly some who find that message inconvenient, and unfortunately they appear willing to turn to just about any tactics to try to suppress that message."[61]

Even more despicable are the right-wing politicians and pundits who target prominent scientists. These demagogues use persecution of scientists to further their own political careers, all but inviting some of their crazy followers to gun them down. We have already heard the story of the crosshairs over the names of certain Democratic members of Congress on Sarah Palin's website and Palin's telling her followers to "Reload."[62] These targeted members received numerous death threats, and many now require bodyguards. James Inhofe of Oklahoma is equally brazen. He listed the names of seventeen prominent climate scientists and claimed that they engaged in "potentially criminal behavior" for violating the Federal False Statements Act.[63] This is the classic tactic of McCarthy-style witch-hunting, or analogous to how conservative authorities (the Inquisition) threatened Galileo with torture when he dared to speak scientific truth to power. It has a tremendously chilling effect on science, not to mention what it does to the personal lives of hard-working scientists and their families. Of course, it is an entirely baseless charge, since the truth lies with the scientists, and it is Inhofe who is distorting scientific reality. Nevertheless, an antiscientific denier such as Inhofe is capable of wasting a lot of scientists' time and money fighting and defending charges in court or in Congress, not to mention the fact that all these scientists are now targets of gun-toting crazy right-wingers.

But the most extreme of all is Virginia Attorney General Ken Cuccinelli. Even before his election in 2008, he was known to be an extreme right-winger, and now he is abusing the powers of his office to push his extremist agenda. He is suing to release all the raw data and e-mails collected by Michael Mann when he worked at the University of Virginia.[64] (Mann is now at Penn State, so Cuccinelli cannot touch him there.) Cuccinelli hopes to find some sort of smoking gun along the lines of the East Anglia Climategate scandal. This is despite the fact, as we showed above, there was nothing amiss in the e-mails, and no conspiracy was discovered—just careless language quoted out of context. Given the right wing's scientific incompetence and misinterpretation of the East Anglia data, there is no reason to think that they will have any better ability to interpret Mann's data, should they release it. Instead, we can expect that they will find stuff that fits their preconceptions and lack of scientific expertise to judge the data in the first place. Cuccinelli is trying to claim that Mann had committed fraud, and should return all the research money, along with legal fees and triple damages. This is really just a right-wing witch hunt by an extremist politician who is using his relatively obscure position as state attorney general to further his political career. It is consistent with all the other ways he is using his office for political gain and street cred in the right-wing fringe. His crusades have ranged from the silly (he tried to cover the naked breast of the crude sketch of the goddess of virtue on the Virginia state seal) to the serious. The latter include directing public universities to remove sexual orientation from their antidiscrimination policies, attacking the Environmental Protection Agency, filing a lawsuit challenging federal health care reform, and trying to reverse George Mason University's policy on concealed weapons on campus. Polls show that the voters of

Virginia are tired of his antics and want him to work on the job that most
state attorney generals are paid to do: prosecuting criminals and corporations on the behalf of the state and enforcing state laws, not tilting at
right-wing windmills.[65]

Even though the right-wing media and politicians and oil companies
have managed to bamboozle almost half of the American public, a very
strong climate bill was once approved by the House, and similar bills
were in discussion in the Senate.[66] Such bills may not pass for a while due
to the current political stalemate in Congress, but eventually they will.
After years of doubt in the American public thanks to the deniers' PR
campaigns, recent polls are beginning to show that the American public
is coming to accept the reality of climate change as well. The day before
the November 6, 2012, election, a poll revealed that 68% of Americans
now regard climate change as a "serious problem," up from only 48% in
2011, and 46% in 2009.[67] A few weeks later, another poll found that 80%
of Americans accept that climate is changing (compared to 73% in 2009),
and 57% say the U.S. government should do something about it.[68] Even
Republican politicians like New Jersey governor Chris Christie and New
York City mayor Michael Bloomberg were warning about the dangers of
climate change. What accounts for this change in attitude? Apparently,
the extreme climate events of 2012 (from the summer's record-breaking
heat waves to Hurricane Sandy) are much more persuasive than anything said by scientists or politicians.

As paleontologist (and now climate activist) Tim Flannery pointed
out in a talk at the Natural History Museum of Los Angeles County in
October 2009, the good news is that the entire debate over global warming in the United States is largely a rearguard action and irrelevant to
where the political winds are blowing now. As we showed above, most
of the rest of the world's population accepts the reality, and the fact that
even Kyoto holdouts such as China, India, and the United States agreed
to the basic science of global warming in the 2009 Copenhagen climate
summit is a big step forward (quite a bit was actually accomplished, even
though they could not get binding agreements on everything).[69] And
it is not just the liberals and environmentalists who are taking climate
change seriously. Historically conservative institutions (big corpora-

tions such as General Electric, the insurance companies, and the military) are already planning on how to deal with global warming. Many of my friends high up in the oil companies tell me of the efforts by those companies to get into other forms of energy, because they know that oil will be running out soon and that the effects of burning oil will make their business less popular. BP officially stands for British Petroleum, but one of its ad campaigns states that it stands for "Beyond Petroleum." (After its 2011 spill in the Gulf of Mexico, people were saying that BP stood for "Biggest Polluter.")[70] Although oil companies still spend relatively little of their total budgets on alternative forms of energy, they still can see the writing on the wall about the eventual exhaustion of oil (see chapter 11)—and they are acting like any company that wants to survive, by getting into a new business when the old one is dying.

The Pentagon (normally not a left-wing institution) is also making contingency plans for how to fight wars in an era of global climate change, and what kinds of strategic threats might occur when climate change alters the kinds of enemies we might be fighting, and water becomes a scarce commodity. The *New York Times* reported that in December 2008, the National Defense University outlined plans for military strategy in a greenhouse world.[71] The entire May 2004 issue of *Monthly Review* is full of articles about how the Pentagon is planning for climate change. This issue was a summary and analysis of an October 2003 Pentagon report commissioned by Peter Marshall, director of the Pentagon's Office of Net Assessment. The report laid out the grim scenarios that the military must consider in a greenhouse planet, and discusses the likelihood of agricultural decline and extreme weather conditions that would overtax energy demand throughout the globe. Rich countries with resources, like the United States and Australia, might build defensive fortresses around themselves to keep hordes of immigrants out, while the rest of the world fights over resources: "Violence and disruption stemming from the stresses created by abrupt changes in the climate pose a different type of threat to national security than we are accustomed to today. Military confrontation may be triggered by a desperate need for natural resources such as energy, food and water rather than by conflicts over ideology, religion, or national honor. The shifting motivation for confrontation would alter which countries are most vulnerable and the existing warning signs for security threats."[72] To the Pentagon, the big issue is global chaos and the potential of even nuclear conflict. The world must "prepare for the inevitable effects of abrupt climate change—which will likely come [the only question is when] regardless of human activity."[73]

Insurance companies have no political axe to grind. If anything, they tend to be on the conservative side. They are simply in the business of assessing risk in a realistic fashion so they can accurately gauge their future insurance policies and what to charge for them. Yet they are all investing heavily in research on the disasters and risks posed by climatic change. In 2005, a study commissioned by the reinsurer Swiss Re said, "Climate change will significantly affect the health of humans and ecosystems and these impacts will have economic consequences."[74]

Right-wingers may still try to deny scientific reality, but big businesses such as oil and insurance, and conservative institutions such as the military, cannot afford to be blinded or deluded by phony science. They must plan for the real world that we will be seeing in the next few decades. They do not want to be caught unprepared and harmed by global climatic change when it threatens their survival. Neither can we as a society.

FOR FURTHER READING

Alley, R. 2000. *The Two-Mile Time Machine: Ice Cores, Abrupt Climate Change, and Our Future.* Princeton, N.J.: Princeton University Press.

Archer, D. 2009. *The Long Thaw: How Humans Are Changing the Next 100,000 Years of Earth's Climate.* Princeton, N.J.: Princeton University Press.

Barnosky, A. D. 2009. *Heatstroke: Nature in an Age of Global Warming.* Washington, D.C.: Island Press.

Broecker, W. S., and R. Kunzing. 2008. *Fixing Climate: What Past Climate Changes Reveal about the Current Threat—and How to Counter It.* New York: Hill and Wang.

Flannery, T. 2006. *The Weather Makers: How Man Is Changing the Climate and What It Means for Life on Earth.* New York: Atlantic Monthly Press.

Gore, A. 2006. *An Inconvenient Truth.* Emmaus, Pa.: Rodale Press.

Hansen, J. 2009. *Storms of My Grandchildren: The Truth about the Coming Climate Catastrophe and Our Last Chance to Save Humanity.* New York: Bloomsbury.

Hoggan, J. 2009. *Climate Cover-Up: The Crusade to Deny Global Warming.* Vancouver: Greystone.

Linden, E. 2006. *The Winds of Change: Climate, Weather and the Destruction of Civilizations.* New York: Simon and Schuster.

Mann, M. E., and L. R. Kump. 2008. *Dire Predictions: Understanding Global Warming.* New York: DK.

Mooney, C. 2006. *The Republican War on Science.* New York: Basic.

———. 2007. *Storm World: Hurricanes, Politics, and the Battle over Global Warming.* New York: Harcourt.

Oreskes, N., and E. M. Conway. 2010. *Merchants of Doubt: How a Handful of Scientists Obscured the Truth on Issues from Tobacco Smoke to Global Warming.* New York: Bloomsbury.

Pearce, F. 2007. *With Speed and Violence: Why Scientists Fear Tipping Points in Climate Change.* Boston: Beacon.

Prothero, D. R. 2009. *Greenhouse of the Dinosaurs: Evolution, Extinction, and the Future of our Planet* New York: Columbia University Press.

Schneider, S. H. 2009. *Science as a Contact Sport: Inside the Battle to Save Earth's Climate.* Washington, D.C.: National Geographic Society.

6

Gimme That Old Time Religion: Creationism, Intelligent Design, and the Denial of Humanity's Place in Nature

Nothing in biology makes sense except in the light of evolution

Theodosius Dobzhanky

THE BATTLE THAT NEVER ENDS

Item: Ten Republican presidential candidates are debating on May 4, 2007. The moderator asks for a show of hands, and three of the ten candidates (former Arkansas governor Mike Huckabee, former Colorado representative Tom Tancredo, and Kansas senator Sam Brownback) admit that they do not believe in evolution.[1] Three men, one of whom who conceivably could become one of the most powerful leaders in the world, do not accept an idea that is at the foundation of science. Four years later, during the GOP presidential races in 2011–2012, eleven of the twelve candidates said they do not accept evolution.[2] Michele Bachmann got her start fighting evolution in Minnesota schools, and Rick Santorum tried to push intelligent design repeatedly during his political career. And in August 2011 Governor Rick Perry ignited a firestorm when he said: "It is a theory that is out there. It is got some gaps in it. In Texas we teach both creationism and evolution."[3] Eventual nominee Mitt Romney believes in both creation and evolution, as does Newt Gingrich. Only Jon Huntsman, a long shot who was eliminated after the first Iowa caucus, clearly supported the science when he tweeted, "To be clear. I believe in evolution and trust scientists on global warming. Call me crazy."[4]

Item: During the presidential campaign in 2000, candidate George W. Bush says, "I'd make it a goal to make sure that local folks got to make the decision as to whether or not they said creationism has been a part of our history and whether or not people ought to be exposed to different theories as to how the world was formed." Later, during the battle over intelligent design creationism in August 2005, he comments that creationism should be taught in public schools alongside evolution (in violation of the constitutional separation of church and state, and going against every court decision on the topic since the Scopes trial). His remarks trigger worldwide derision, shock, and disbelief by political leaders and citizens of every developed country and by the entire scientific community. The topic even makes the cover of *Time* magazine as a result of his comments.[5]

Item: In his campaign ads for gubernatorial primary in Alabama in March 2010, candidate Judge Roy Moore attacks his opponent Bradley Byrne for believing in evolution. Byrne replies with ads boasting that he too is a creationist. As Jason Linkins wrote,

> In most . . . races, when a Republican group wants to assail its Democratic opponents, it is a good bet that it will lead off by mentioning support for tax increases. Not so in Alabama! In that state, taxes take a back seat to your opponent's claim that science is real. . . .
>
> Of course, Byrne has degrees from Duke University and the University of Alabama, and he most recently served as the Chancellor of Alabama's Community College System, so I'm sure that he is going to respond to this nonsense in a manner befitting a modern educator who matriculated at some serious academic institutions, right?
>
>> As a Christian and as a public servant, I have never wavered in my belief that this world and everything in it is a masterpiece created by the hands of God. As a member of the Alabama Board of Education, *the record clearly shows that I fought to ensure the teaching of creationism in our school textbooks.*
>
> Well, fantastic![6]

Item: At any given time in the United States, at least a dozen states have creationist legislation or creationist educational plans pending before state legislatures or school boards (respectively). Each time one of these efforts gets past the initial stages, it is stopped cold in its tracks by the legal process. As soon as the creationists get to court, they are slapped down by the fact that creationism is a sectarian religious viewpoint and has nothing to do with science, and therefore it is unconstitutional to teach it in public institutions. Nevertheless, these creationist efforts have occurred every year since the 1960s.

Item: The news recently has been full of shocking and disconcerting remarks by members of Congress. Senator Marco Rubio of Florida was quoted as saying that "the age of the earth is a great mystery,"[7] then had to backtrack when the storm of controversy descended upon him. The most outrageous is by Representative Paul Broun of Georgia (an M.D., even!),[8] who said (in a recent speech at the Liberty Baptist Church Sportsman's Banquet):

> God's word is true. I've come to understand that. All that stuff I was taught about evolution, embryology, Big Bang theory, all that is lies straight from the pit of hell. It's lies to try to keep me and all the folks who are taught that from understanding that they need a savior. There's a lot of scientific data that I found out as a scientist [note: Broun is NOT a real scientist] that actually show that this is really a young Earth. I believe that the Earth is about 9,000 years old. I believe that it was created in six days as we know them. That's what the Bible says. And what I've come to learn is that it's the manufacturer's handbook, is what I call it. It teaches us how to run our lives individually. How to run our families, how to run our churches. But it teaches us how to run all our public policy and everything in society. And that's the reason, as your congressman, I hold the Holy Bible as being the major directions to me of how I vote in Washington, D.C., and I'll continue to do that."[9]

Or take the now-infamous Todd Akin, congressman from Missouri who lost his bid for a Senate seat from that state. It was bad enough that he believed and spouted some myth from a discredited anti-abortion doctor that a women's body can "shut down" and prevent impregnation from rape. But at a recent Tea Party meeting, he said,

> I don't see it [evolution] as even a matter of science because I don't know that you can prove one or the other. That's one of those things. We can talk about theology and all of those other things but I'm basically concerned about, you've got a choice between Claire McCaskill and myself. My job is to make the thing there. If we want to do theoretical stuff, we can do that, but I think I better stay on topic.[10]

These statements are shocking enough by themselves, but they are even more alarming because Broun, Akin, and many others who deny evolution are on the House Science and Technology Committee, charged with overseeing science funding and priorities in this country! More than any example of reality denial discussed in this book, evolution denialism or creationism is the most widespread in the United States, and the most difficult to overcome. All other examples of denial in this book deal with scientific concepts that are opposed by the political or economic agendas of certain groups, mostly right-wing political organizations and powerful business interests. Many of them (acid rain, global ozone hole, AIDS denialism, and others) were resolved over time, and scientific reality overcame ignorance and political obstructionism. But creationism is a horse of a different color: its support comes entirely from religious dogma and the misunderstandings fostered by religious zealots who view it as a threat to their worldview. It is also virtually restricted to the United States. Although there are small creationist movements in Canada, the United Kingdom, Australia, and a few other countries, at least 75–95% of the population of most developed countries accept the reality of evolution (fig. 6.1). This is a result of their educational systems that value well-documented science over religious dogma, and the fact that these countries have no strong fundamentalist institutions to interfere with the educational system. The United States is down near fortieth on the list, along with Turkey, Cyprus, and other nations that have strong dogmatic national religions (Islam or Greek Orthodox). None of these other countries have a fraction of the economic might of the United States, nor anywhere close to the amount of money spent on education in the United States—yet that is where the United States places in science literacy, as well as acceptance of evolution.

It is one thing to poll the population of underdeveloped countries, many of whom are poorly educated or illiterate. But in this book, we are talking about ideas that have reached consensus of greater than 95% acceptance among the scientific community. If ever there was an idea established within the realm of science, it is the fact that life has evolved. The evidence is so overwhelming that it is accepted by virtually all scientists (especially those who have any training or exposure in the topic). Every legitimate scientific organization in the United States has published strong statements affirming the reality of evolution, including the prestigious National Academy of Sciences, the American Association for the Advancement of Science, the American Geological Institute, the Geological Society of America, the Paleontological Society, and the Society of Vertebrate Paleontology. In 1987, *Newsweek* found that less that 0.15% of 480,000 biologists and earth scientists polled doubted evolution.[11] Another study found that 99.9% of biologists agreed that evolution has occurred.[12]

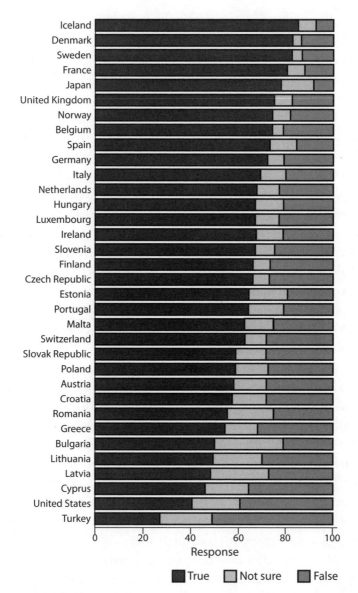

FIGURE 6.1. Percentage of population in each country that accepts evolution as true (dark bar to left) and regards it as false (solid bar to right); the undecideds are indicated by the light bar in the middle. Note that nearly every developed country in Europe and Asia has at least 75% or more acceptance of evolution. The United States is down at the bottom with Cyprus and Turkey, countries with heavy Islamic fundamentalist or Greek Orthodox influence, yet the United States has a much higher GNP and spends much more per child than do the countries at the bottom of the list (see fig. 13.5). *Modified from "Public Acceptance of Evolution by Country," Blame It on the Voices website, March 13, 2009, www.blameitonthevoices.com/2009/03/public-acceptance-of-evolution-by.html.*

Naturally, creationists dispute this consensus by claiming they have lists of scientists who disagree with evolution. If you scan their lists carefully, however, you will find that it is mostly people who do not have Ph.D.s (or any degree at all) in any field relevant to evolution; they are largely in physics or engineering or mathematics, where they have never had the chance to study the evidence of evolution firsthand. Since they have never actually worked with fossils or living organisms, they have zero credibility when they argue against evolution. If your car needed repair, you would not trust it to anyone but a trained auto mechanic—nor should you trust the opinions of an engineer or a physicist on the topic of evolution, in which they have no formal training or experience. (Some of those on the list actually agree with evolution, but were duped into signing and have never gotten the creationists to remove their name from the list.)[13] The other peculiar thing about the list is that when you do research on the individuals, you will find that every one of them came to their creationist positions through religious motivation.[14] *Not one* came to doubt evolution purely based on scientific evidence and without the influence of religious dogma.

The National Center for Science Education lampooned the ludicrous nature of these lists of scientists who doubt evolution by organizing Project Steve, a list of scientists who accept evolution with the names Steve, Stephen, Stephanie, or a related name.[15] *There are almost 1,200 members of Project Steve as of this writing—more than twice the number of scientists that the creationists claim on their side.* Based on the number of people named Steve or Stephanie in the population (less than 0.1%), that means that if every scientist were allowed to sign, there would be more than 114,000 on the list who accept evolution, which is nearly all of the active biological and geological researchers in the United States.

It is not just scientists and scholars, and nearly everyone else in the developed world, who accept the reality of evolution. The Clergy Letter Project currently has over 11,111 ministers, priests, and rabbis who have signed a statement indicating that they accept evolution and that it is not in conflict with their faith.[16] Of the twelve largest Protestant denominations, almost 90% of their members accept evolution. (The Southern Baptists are the exception.)[17] The Catholic Church has no problem with evolution. Pope John Paul II wrote, "The Bible teaches you how to go to heaven, not how the heavens go" (a statement that originally come from Galileo), and "Today, almost half a century after publication of the encyclical, new knowledge has led to the recognition of the theory of evolution as more than a hypothesis."[18] Contrary to the efforts of creationists and fundamentalists to portray evolution as atheism, most religious people except the fundamentalists in the United States have no problem with evolution.

One of the strongest and clearest statements was made by a list of 177 top biologists, including many with Nobel Prizes:

> There are no hypotheses, alternative to the principle of evolution with its "tree of life," that any competent biologist of today takes seriously. Moreover, the principle is so important for an understanding of the world we live in and of ourselves that the public in general, including students taking biology in high school, should be made aware of it, and of the fact that it is firmly established, even as the rotundity of the earth is firmly established.[19]

WHY DO WE SAY THAT EVOLUTION IS REAL?

Speak to the earth, and it shall teach thee.

Job 12:8

Life has evolved. It is still evolving and we can watch it happen. Scientists, scholars, and virtually all educated people in the developed world (except for fundamentalist Protestants, Jews, and Muslims) recognize this fact. On what evidence is this acceptance based? I do not have room in this short chapter to cover every line of evidence in detail, since I have done so at length in my 2007 book. For the purposes of this chapter, I will summarize briefly some of the major lines of evidence that make the reality of evolution undeniable.

Let us first examine of major lines of evidence that Darwin and his contemporaries knew about. Many of these were discovered before Darwin wrote about them, but it was *On the Origin of Species* (1859) that pulled all these lines of evidence together into a coherent argument and made his case so convincing:

The Tree of Life

Since the days of Linnaeus's pioneering classification of animals in 1758
(a full century before Darwin's book), it was apparent that all organisms
formed a branching pattern of smaller groups (species grouped into genera, genera grouped into families), which then clustered into larger and
larger groups (orders, classes, phyla, kingdoms). If you lay out the classification of life in tabular form, it clearly shows the hierarchical bushy
branching pattern of the tree of life. Back in 1758, Linnaeus did not make
the connection between this branching pattern and the interrelatedness
of life, but by Darwin's generation, it was already considered very powerful evidence not only that similar organisms should cluster together but
that they might also be closely related. Darwin's arguments of the tree
of life were made strictly on external anatomical features that scientists

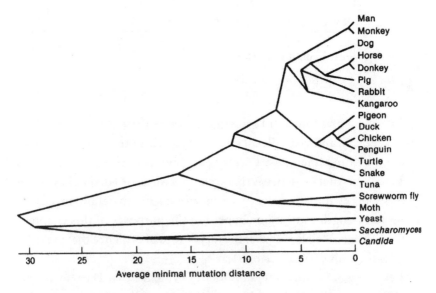

FIGURE 6.2. Molecular similarities of cytochrome c from different organisms. The
branching pattern of this and all other molecules matches the pattern seen by looking at
the anatomy of these organisms, something that Darwin could never have known, but
is a powerful fulfillment of his predictions. *Modified from W. M. Fitch and E. Margoliash,
"Construction of Phylogenetic Trees," Science 155 (1967): 279–284, available at Eugene
Garfield homepage, garfield.library.upenn.edu/classics1988/A1988N888200001.pdf.*

were documenting 150 years ago. In the late twentieth century, the most amazing confirmation of the tree of life has occurred. When the molecular sequence of biochemicals in cells was determined (from RNA to DNA to any protein), they also show the same pattern of a branching pattern of similarity that the external anatomy suggests (fig. 6.2). In other words, every cell in your body (and in the bodies of all organisms) proclaims the handiwork of evolution! Such discoveries make absolutely no sense unless we were all interrelated and descended from common ancestors, a reality that creationists continually deny.

Homology

If you look at the basic building blocks of the vertebrate skeleton, you find that we all have roughly the same number of bones attached to each other in the same arrangement, no matter whether you look at a human or a mole or a dolphin. For example, the forelimb of all tetrapods (amphibians, reptiles, birds, and mammals) is composed of a single upper element (humerus, or "funny bone"), two more bones side by side (radius and ulna), and then a very predictable pattern of bones in the wrist and fingers (fig. 6.3). Yet all these different organisms use this basic homologous pattern of forelimb bones, muscles, nerves, and blood vessels in extremely different ways, from the paddle of the dolphin to the digging hand of the mole to the wings of a bird or a bat (built in two completely different ways) to the single-toed long limbs of a horse. These are by no means the best or most efficient ways to build these structures (the wings of insects are built in an entirely different pattern, which works better than any bat or bird wing). Clearly, they are evidence against the idea of perfect design for the functions these forelimbs now perform, but they show their common ancestry through the common building blocks (homologous bones) that are now modified for entirely different purposes.

Vestigial Structures and Other Imperfections

If nature were perfectly engineered by an all-powerful supernatural creator, life should be perfectly engineered and constucted. There should be no signs of shoddy workmanship or less-than-optimal engineering.

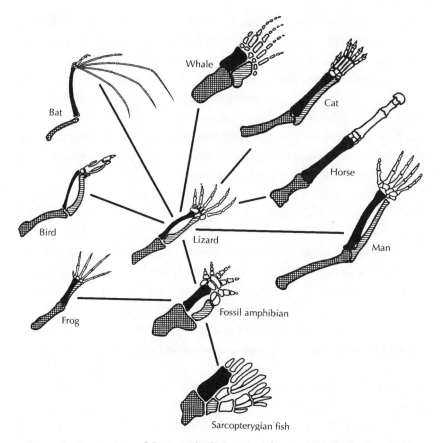

FIGURE 6.3. Comparison of the upper limb elements of various mammals, showing how homologous bones are used for entirely different functions. *Drawing by Carl Buell.*

However, natural selection will not get rid of every imperfection or functionless left over from ancestral forms unless there is strong adaptive pressure to do so. The list of imperfect or vestigial structures goes on and on. They include many features of our own bodies that no longer have a function (tail bones, male nipples), or are suboptimally designed (like our problems with our poorly designed bipedalism and upright posture, or the retinas in our eyes that have the sensing cells beneath all the nerves and blood vessels, obscuring vision), or are nonessential but detrimental to us when they get infected (appendix, tonsils). Then there are the examples of relict hind limbs in both whales and snakes, relict side toes in horses (the so-called splint bones that will cripple a horse if they break), and many, many others. There are numerous examples of jury-rigged or suboptimal structures that do not work very well, but they function just well enough to allow the organism to survive, from the panda's thumb to the crude fishing lures on anglerfish and the fresh-water clam *Lampsilis*.[20] These make absolutely no sense in the context of an intelligent designer (unless that designer was incompetent, cruel, or careless), but instead provide strong evidence that nature uses what-ever the organism inherited from its ancestor. Perhaps the most striking example of all is the course of the recurrent laryngeal nerve. Instead of taking the direct course from the spine through the neck to the larynx, it loops all the way down to the aorta in the chest cavity and then back up to the larynx, making it many times longer than it needs to be to carry nerve impulses from the spine to the voice box. In giraffes, it is over 15 feet long—more than ten times longer than needed—and travels down the long neck from the throat to the aorta and back again! This bizarre "design" is easily explained by evolution: the nerve is attached to the embryological gill arch that becomes the aorta, so when the heart and aorta migrate backward in development, so does the nerve.

Embryology

More than twenty years before Darwin's book was published, the great German embryologist Karl Ernst von Baer discovered a striking thing: the embryos of all vertebrates are extremely similar in their early stages. When our embryology begins, we all have a fishlike body including a long tail, gill arches, and many other features that do not occur in adults. This makes absolutely no sense whatsoever unless we all had a common ancestor that was fishlike as well. Knowing how striking this evidence is, creationists like the Moonie Jonathan Wells (2000) have tried to confuse the issue by pointing to the excesses and mistakes of later embryologists, such as Ernst Haeckel. But this does not refute the original evidence discovered by the devoutly religious creationist Baer, and Wells's entire argument has been completely dismantled by Alan Gishlick.[21] In more recent years, we have learned a tremendous amount about the details of embryology (what is known as evolutionary developmental biology, or

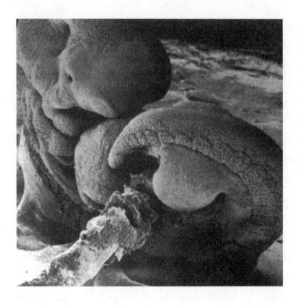

FIGURE 6.4. Five weeks after fertilization, you looked like this, with a long segmented fishlike tail and the precursors of gill slits. These are all modified or lost when you developed, but are clear evidence of your "inner fish." *Image from* IMSI *Photo Library.*

"evo-devo"). Now we can see exactly how certain parts of the embryo control the developmental process and produce such different adult body plans from similar embryos. If you have any doubts that you were descended from a fishlike ancestor, just look at any human embryo at five weeks after conception. You had pharyngeal pouches (which become gill slits in fish) and a long fishlike tail (fig. 6.4).

Biogeography

When the great scientific explorations of the eighteenth and early nineteenth century voyaged around the world, they documented the fact that nearly every part of the world has its own unique and distinctive set of animals and plants. The South American jungles have many kinds of creatures not found in Eurasia. The mammals of Australia are mostly marsupials or "pouched mammals," and are unrelated to those found anywhere else (fig. 6.5). Even more impressive, many of these marsupials had evolved into forms that converged on those found on the rest of the continents: marsupial "moles," marsupial "wolves" (Tasmanian wolf), marsupial "cats" (dasyurids), marsupial "flying squirrels" (phalangers), "groundhogs" (wombats), "anteaters" (mymecobiids), "mice" (dasycercids), and so on. These animals are not closely related to their

Placentals

Marsupials

Wolf
(Canis)

Tasmanian wolf
(Thylacinus)

Ocelot
(Felis)

Native cat
(Dasyurus)

Flying squirrel
(Glaucomys)

Flying phalanger
(Petaurus)

Ground hog
(Marmota)

Wombat
(Phascolomys)

Anteater
(Myrmecophaga)

Anteater
(Myrmecobius)

Mole
(Talpa)

Mole
(Notoryctes)

Mouse
(Mus)

Mouse
(Dasycercus)

FIGURE 6.5. Convergent evolution in Australian marsupials (right) with mammals from elsewhere in the world (left). *Drawing by Pat Linse.*

placental counterparts elsewhere in the world, so they could only be explained by convergent evolution on a continent isolated from Eurasia. Even Darwin's trip to the Galápagos Islands revealed that it had its own unique species not found anywhere else, although closely related to and probably descended from creatures of the nearby South American coast. None of these observations make sense in the Noah's ark model of biogeography. Instead of a pattern of creatures distributed away from Mount Ararat in Turkey, we find something that clearly makes sense only in the light of evolution. Otherwise, why did all those marsupials (but no other groups of mammals) migrate from Turkey to Australia?

These observations have only been amplified by a huge amount of new evidence that accumulated since Darwin died in 1882:

- We have sequenced the entire genomes of many animals, including ourselves and our close relatives, and see the remarkable evolutionary similarities. The most striking of these is the fact that roughly 98% of our genome is identical to that of the chimpanzees and gorillas, which is strong evidence that we are closely related. Jared Diamond (1992) points out that if extraterrestrial geneticists had only our genes and those of chimps, apes, and several other primate species (but no bodies), they would classify us as a third species of chimpanzee.
- Through evo-devo we now understand the mechanism by which genetic instructions are translated into bodies, and how just a small change in the regulatory genome can make huge changes in the phenotype. This allows for much more rapid evolutionary change than was once supposed and allows whole new segments and appendages to be added or modified with minimal genetic change.
- Genetics has also shown that a very high percentage (80–90% in many organisms) of the DNA we carry is functionally neutral and cannot be read during translation, or it is junk DNA inherited from distant ancestors, which no longer has a function and is no longer used. This would make no sense if a divine designer were composing a genome.

- The evidence from the fossil record has grown tremendously since Darwin's time (when it was relatively incomplete and not much help to him). As detailed in the second half of my 2007 book on evolution, we now have not only detailed evolutionary sequences in the fossil record for many different types of animals, but we also have the transitional fossils that link major groups together. Creationists have loudly tried to deny their existence again and again, but my book shows original photos and diagrams of hundreds and hundreds of examples.
- Finally, evolution is occurring right now, in real time. We see evolution every time a virus or bacterium modifies its external protein coating and eludes our immune system to emerge as a new infection each winter cold and flu season. We see evolution in the insect pests we battle so relentlessly, as they evolve resistance to each new poison we throw at them, one after another. As entomologist Martin Taylor put it, "It always seems amazing to me that evolutionists pay so little attention to this kind of thing, and that cotton growers are having to deal with these pests in the very states whose legislatures are so hostile to the theory of evolution. Because it is the evolution itself they are struggling against in their fields every season. These people are trying to ban the teaching of evolution while their own cotton crops are failing because of evolution. How can you be a creationist farmer any more?"[22] Field biologists, working long and hard in very difficult conditions for many years, have documented hundreds of examples of evolution occurring in real time in many different kids of organisms, from fish to lizards to birds to mammals.[23] It is one thing for creationists to try to deny something that occurred in the past and must be reconstructed through inference. *But you can watch evolution occur in real time*—only the most stubborn antirealist cannot accept that evidence. As Jonathan Weiner wrote, "Biologists finally began to realize that Darwin had been too modest. Evolution by natural selection can happen rapidly enough to watch. Now the field is exploding. More than 250 people around the world are observing and documenting evolution, not only in finches and guppies, but also in aphids, flies,

grayling, monkeyflowers, salmon and sticklebacks. Some workers are even documenting pairs of species—symbiotic insects and plants—that have recently found each other, and observing the pairs as they drift off into their own world together like lovers in a novel by D. H. Lawrence."[24]

In short, the evidence that life has evolved is overwhelming. We can watch evolution happening before our very eyes. In layman's terms, it is a fact—it is a reality. As the great geneticist Theodosius Dobzhansky said so eloquently in 1973, "Nothing in biology makes sense except in the light of evolution."[25] More than 99% of all scientists around the world agree with this consensus, and as we discussed above, those who do not have no firsthand training or experience with the evidence. In addition, we have already seen that virtually all educated peoples in developed countries accept the reality of evolution, except for fundamentalists. So the battle is not over scientific issues at all. It is about dogmatic religious ideas and how they clash with reality.

WHAT IS CREATIONISM?

Creation is, of course, unproven and unprovable by the methods of experimental science. Neither can it qualify as a scientific theory.

Duane Gish, 1973

Creation isn't a theory. The fact that God created the universe is not a theory—it's true. However, some of the details are not specifically nailed down in Scripture. Some issues—such as creation, a global Flood, and a young age for the earth—are determined by Scripture, so they are not theories. My understanding from Scripture is that the universe is in the order of 6,000 years old. Once that has been determined by Scripture, it is a starting point that we build theories upon.

Kurt Wise, 1995

Ours is a time of space telescopes, electron microscopes, supercomputers, and the worldwide web. This is not a time for parsing the lessons given to a few goatherds, tentmakers, and camel drivers.

Rev. Michael Dowd, Thank God for Evolution!

Creationism, as practiced in the United States and a few other countries, is a peculiar offshoot of fundamentalist religion. Its roots go back to the 1890s, when the fundamentalist movement started in the United States, primarily as a reaction to the rapid changes of the modern world. Although theologians for centuries (going back to Saint Augustine and earlier) had warned against taking the Bible too literally, one of the tenets of fundamentalist thinking is the literal interpretation of the Bible. Of course, most Bible scholars (and most non-fundamentalist people of faith) realize that the Bible contains parables and allegory. Many of its descriptions and images were stories passed around for generations in Middle Eastern cultures and were not intended to be modern works of science.

As I summarized in my recent book, every culture has its own creation myths, and the ancient Hebrews clearly cribbed theirs (with only minor changes) from Sumerian, Akkadian, and Babylonian myths, such as *Enuma Elish* and *The Epic of Gilgamesh,* that were handed down for centuries in the cultures of the Tigris and Euphrates.[26] The Hebrew version gave the ancient Israelites comfort and assurance about their place in the world, and their god's power over other gods, but was never intended as a scientific account of nature for the twenty-first century. Imagine trying to explain modern astrophysics or molecular biology to a team of Hebrew shepherds from 2000 BCE!

Scholars and clergymen who can actually read the Old Testament in its original Hebrew (as I learned to do) or the New Testament in its original Greek (as I can) have also found that it is impossible to take the Bible literally. It is immediately apparent if you read the original texts that it is full of mistakes, errors in copying and translation, and ambiguous passages whose meaning is still obscure. In addition, the Old Testament is clearly a composite of text written by a least four different groups of authors working from different traditions and in different times, all patched together. As any Hebrew scholar can tell you, the signature of these different sources is very clear in their choice of phrases and even spellings of words. This explains why the Bible often contradicts itself, since it is a composite of different accounts patched together without worrying about word-for-word consistency (which was not important to the ancient Hebrews, only to modern literalists). For example, Genesis 1 (largely from what is known as the "Priestly" source) gives this order of creation: plants, animals, man, and woman; but Genesis 2 (from the "J" source, whose signature is their use of "Jahweh" as the name of God) gives it as man, plants, animals, and woman. In Genesis 1:3–5, on the first day, God created light, then separated light and darkness, but according to Genesis 1:14–19, the sun (which separates night and day) was not created until the fourth day.

Genesis 6 gives the story of Noah twice, once from the J source and once from the P source, with verses from the two sources intermingled so they sometimes contradict each other. Genesis 6:5–8 are from the J source, but Genesis 6:9–22 are from the P source. Then Genesis 7:1–5 is from the J source, but Genesis 7:6–24 alternate between the J and P sources.[27] This leads to many contradictions, such as Genesis 7:2 (from the J source) saying that Noah took *seven* pairs of each beast in the ark, but Genesis 7:8–15 (from the P source) saying he took only *one* pair of each beast in the ark. In Genesis 7:7, Noah and his family finally enter the ark, and then in Genesis 7:13 they enter it all over again (the first verse from the J source, the second from the P source). According to Genesis 6:4, there were Nephilim (giants) on the earth before the Flood, then Genesis 7:21 says that all creatures other than Noah's family and those on the ark were annihilated—but Numbers 13:33 says there were Nephilim after the Flood.

As we have seen in chapter 1, there are obvious absurdities in literal belief in the Bible, such as the numerous mentions of the earth being flat, the sun moving around the earth, and even one place where the Bible gives the value of pi (the ratio of a circle's circumference to its diameter) at 3.0, not 3.14159. We have already mentioned the Flat Earth Society, which takes biblical literalism to this absurd extreme, but it is a good question why modern creationists selectively reject literalism about the biblical claim of a flat earth at the center of the solar system, but embrace literalism in every other part of the Bible.

This is the dilemma of creationists. It has been since the late nineteenth century, when fundamentalism arose as a reaction to the discoveries of biblical scholars (known as Higher Criticism) that discredited any attempt at literalistic interpretation. In fact, the early fundamentalists were not entirely opposed to the idea of evolution,[28] and even in the Bible Belt of the U.S. South, evolution was taught in most textbooks in the early twentieth century without much resistance. It was not until the conservative backlash of the 1920s that so-called monkey laws banning the teaching of evolution spread through the South, and fundamentalists began to routinely reject evolution.

As I described in my previous book, the Scopes Monkey Trial made creationism look ridiculous in the public eye, but it was legally inconclusive; anti-evolution laws remained on the books until the Supreme Court struck them down in 1967.[29] It was actually the successful Soviet launch of the first satellite, Sputnik, in 1957, that made the United States realize how bad its science education had become. This shock and urgency reinstated evolution and other modern scientific concepts to the science education around the nation. These modern biology texts, in turn, stimulated the modern creationist movement in the 1960s and ever since. And although creationists have become more prominent, and still have enormous influence in the conservative Bible Belt parts of the United States, they have lost every legal battle. This is because their dogma is clearly the narrow view of a specific religious sect, which is forbidden in public school science classes by the First Amendment to the Constitution and the separation of church and state.

By the 1970s, creationists were trying to circumvent the Constitution by claiming that creationism was science, and therefore deserved equal time in public school science classes. Of course, creationism bears no resemblance to science as we defined it in chapters 1 and 2. Creationism takes its conclusions as final, and then tries to bend the evidence to fit them. The Institute of Creation Research demands that its members sign an oath agreeing to their ironclad dogmas, which is completely in violation of the idea that science comes only to tentative conclusions that must be falsifiable. There is no chance that creationists would reject any part of their belief system if they were contradictory to or falsified by the evidence. Just look at the epigraphs at the beginning of this section: creationists themselves admit that they are not doing science, and they would not change their worldview, no matter how much evidence accumulates against it. Nevertheless, in the 1970s they published so-called public school editions of their textbooks that deleted the direct references to God and Jesus, but otherwise were identical in philosophy and content.

This smokescreen of calling themselves scientific creationists was dealt a deathblow in the 1982 case *McLean v. Arkansas,* where Federal Judge William Overton ruled that the Arkansas law "was simply and purely an effort to introduce the biblical version of creation into the public school curricula." According to Overton, the law "left no doubt that the major effect of the Act is the advancement of particular religious beliefs." The law requiring balanced treatment "lacks legitimate educational value because 'creation science' as defined in that section is simply not science."[30] The same conclusion was reached in a Federal court in Louisiana in 1985. When the Supreme Court struck down all scientific creationism/equal time laws in 1987, the scientific creationism version of trying to intrude into public schools with creationism was legally dead.

WHAT IS INTELLIGENT DESIGN CREATIONISM?

The evidence, so far at least and laws of Nature aside, does not require a Designer. Maybe there is one hiding, maddeningly unwilling to be revealed. But amid much elegance and precision, the details of life and the Universe also exhibit haphazard, jury-rigged arrangements and much poor planning. What shall we make of this: an edifice abandoned early in construction by the architect?

Carl Sagan, Pale Blue Dot

Almost as soon as the scientific creationist ruse was declared unconstitutional, the creationists tried to become even more subtle in avoiding the First Amendment. They began to talk about an intelligent designer who may or may not be the Judeo-Christian God. Thus was born the "intelligent design creationism" movement of the 1990s and early part of this century. Its main spokespeople kept saying (in public) that they were not pushing a specific religious viewpoint, and that the Designer could be any number of gods—or aliens, for that matter. As I described in my earlier book, they built whole institutes dedicated to pushing their propaganda and books for the class, primarily revolving around the Discovery Institute (a right-wing foundation based in Seattle).[31] They promoted a few legitimate scientists (such as Michael Behe) who posed problems they claimed could not be solved by evolution, but only by a supernatural designer. These so-called problems were easily and repeatedly refuted by other scientists, but the intelligent design (ID) creationists persisted in promoting these discredited arguments in the intelligent design literature and websites nonetheless. But it took very little digging to show that this was all a scam, a front for fundamentalists to get around the First Amendment.[32] All of the funding for the Discovery Institute comes from right-wing institutes and rich conservative donors who want to promote a religious viewpoint and push the ascendancy of a Christian theocracy. The Discovery Institute prepared a document about a "wedge strategy" that clearly admitted that its entire movement was purely a PR campaign to push its point of view, and really did not have any science to back it up. (The institute has since tried to deny the existence of the wedge strategy, but it is archived on the internet and can be easily found.)[33]

When they are speaking to religious audiences, the founders of intelligent design creationism openly admit that their whole movement is a smokescreen for their real agenda: to establish their religious beliefs in the schools and discredit evolution. In an article in the Christian magazine *Touchstone*, leading ID creationist William Dembski wrote: "Intelligent design is just the Logos theology of John's Gospel restated in the idiom of information theory."[34] In 1999, Dembski wrote, "Any view of the sciences that leaves Christ out of the picture must be seen as fundamentally deficient. . . . The conceptual soundness of a scientific theory cannot be maintained apart from Christ."[35] On February 6, 2000, Dembski told the National Religious Broadcasters, "Intelligent Design opens the whole possibility of us being created in the image of a benevolent God. . . . The job of apologetics is to clear the ground, to clear obstacles that prevent people from coming to the knowledge of Christ. . . . And if there's anything that I think has blocked the growth of Christ as the free reign of the Spirit and people accepting the Scripture and Jesus Christ, it is the Darwinian naturalistic view."[36] At the same conference, the founder of the intelligent design movement, Berkeley lawyer Phillip Johnson, said, "Christians in the twentieth century have been playing defense. They've been fighting a defensive war to defend what they have, to defend as much of it as they can. It never turns the tide. What we're trying to do is something entirely different. We're trying to go into enemy territory, their very center, and blow up the ammunition dump. What is their ammunition dump in this metaphor? It is their version of creation."[37] In 1996, Johnson said, "This isn't really, and never has been, a debate about science. . . . It's about religion and philosophy." One of the ID creationist authors, Jonathan Wells, is a follower of Sun Myung Moon and his Unification Church (which is vehemently antievolution). As Wells wrote, "When Father chose me (along with about a dozen other seminary graduates) to enter a PhD program in 1978, I welcomed the opportunity to prepare myself for battle."[38]

Through the 1990s and until 2005, the intelligent design smokescreen seemed to be working. Local creationists would often use it to force their books on school boards and teachers. The crucial test came when the school district of Dover, Pennsylvania, tried to force ID creationism into their public schools in 2004. The case went to trial, and the openly religious statements of several Dover school board members revealed their true intent. Even more revealing was the discovery of an early draft of a creationist textbook that had been suddenly modified after the 1987 Supreme Court decision outlawing scientific creationism. The editors had clumsily pasted in the phrase "design proponents" over "creationists" in such a way that it made the curious string "cdesign proponentsists," a transitional form linking the old creationist manuscript with the revised ID version.[39] The "ID creationists" also hurt their cause tremendously when their best scientific witnesses were forced under oath to admit that there was no scientific evidence for "ID creationism," that ID creationism was not really science, and that they had deliberately ignored mountains of scientific literature that proved them wrong.

Finally, in December 2005, Judge John E. Jones III, a conservative Christian appointed by George W. Bush in 2002 (not a so-called liberal activist judge), rendered his verdict. In his words, "The breathtaking inanity of the board's decision is evident when considered against the factual backdrop which has now been fully revealed through the trial. The students, parents, and teachers of the Dover Area School District deserved better than to be dragged into this legal maelstrom, with its resulting utter waste of monetary and personal resources." Judge Jones was particularly irritated by the hypocrisy of the ID creationists, who attempt to sound secular where the Constitution was concerned, but crowed about their religious motives when not in court: "The citizens of the Dover area were poorly served by the members of the board who voted for the intelligent design policy. It is ironic that several of these individuals who so staunchly and proudly touted their religious convictions in public would time and again lie to cover their tracks and disguise the real purpose behind the intelligent design policy." Another passage reads, "We find that the secular purposes claimed by the board amount to a pretext for the board's real purpose, which was to promote religion in the public school classroom." Still later he wrote, "Any asserted secular purposes by the board are a sham and are merely secondary to a religious objective."[40]

Since the Dover decision, intelligent design has definitely faded in importance. The Discovery Institute and its minions have been as noisy as ever, but no school district has dared introduce an intelligent design strategy in their attempts to slide creationism beneath the barrier of the First Amendment. In the past eight years, creationists have tried a "teach the controversy" strategy, whereby they try to push their viewpoint by claiming that students should hear both sides of the argument and make up their own minds. Of course, we do not waste students' time by teaching other outmoded, discredited ideas about the universe (such as biblical ideas that the earth is flat, and that it is the center of the universe), so there is no reason to allow creationism through that loophole either. It has been conclusively debunked by overwhelming scientific evidence for over 150 years now and has no place in a science classroom. If that strategy does not work, religious fanatics will pressure school boards to insert language about the "problems with Darwinism" into the curriculum; this allows them to insert the standard litany (discussed below) of creationist arguments that have long been debunked. Most recently, several states had tried to pass (and Louisiana and Kentucky have passed) laws to promote "critical thinking." This sounds innocuous enough. After all, who does not want our children to learn to think critically? However, the religious intent is clear when you read (in the Louisiana law) that teachers should be encouraged to bring in their own materials "including, but not limited to, evolution, the origins of life, global warming and human cloning."[41] This is not a list of subjects on which there is legitimate scientific controversy, or which is still is openly debated in political circles, but only topics which are the bugbears of the right-wing fundamentalists.

What is clear through this entire 154-year process is that creationists are fanatical about their belief system, and unwilling to listen to any scientific evidence that contradicts their worldview. They also have lots of time and money to pressure school boards and push candidates who support their viewpoint, and thus the attack is never ending. By contrast, most paleontologists and evolutionary biologists have busy careers doing research, teaching, publishing real science, and taking care of their families. Most are unwilling to waste their precious time on a fruitless uphill struggle against extremist fundamentalists who will never give up and never go away, no matter how many times the courts and the Constitution stop them. Fortunately, there are groups such as the National Center for Science Education (NCSE).[42] It has a tiny office in Oakland, California, with just a few paid staff and a minuscule budget compared to such huge wealthy creationist organizations as Answers in Genesis, the Institute for Creation Research, and the Discovery Institute. Nevertheless, the NCSE has been very effective at doing the job most scientists do not have the time or training or budget for: speaking against creationism, mobilizing resources and people to help local parents fight the efforts of creationists in school districts, and assisting legal organizations like the ACLU when creationist lawsuits go to trial.

THE CREATIONISTS' STANDARD
(DISCREDITED) ARGUMENTS

If you tell a lie big enough, and keep repeating it,
people will eventually come to believe it.

Josef Goebbels

What harm would it do, if a man told a good strong lie for the sake of the good and for the Christian church . . . a lie out of necessity, a useful lie, a helpful lie, such lies would not be against God, he would accept them.

Martin Luther

Lying lips are an abomination to the Lord.

Proverbs 12:22

So far we have reviewed the overwhelming evidence that life has evolved and is still evolving, and we have delved into the history and religious motivations of creationists to deny evolution. But do the creationists have any valid arguments of their own? Are there any reasons to take their arguments seriously?

If you look closely, none of the Young Earth creationists' arguments are scientific arguments in support of their theory. They are all critiques of evolutionary biology, geology, astronomy, and the world that science has revealed to us over the past four centuries. Creationists lose in court precisely because they have no scientific arguments in their support, only negative arguments against their opponents. They are working on the classic either- or fallacy—if you aren't with us, you're against us—on the false premise that anyone who disagrees with them is an atheist, or at least not a Christian. Of course, we have already documented the fact that nearly all Christians and Jews (except the extreme fundamentalists) accept evolution, so the false premises of their arguments are clear. Even if all their criticisms of evolutionary biology, geology, astronomy, and science were true, it would not lead to creationism. People believe in creationism based on religious dogmatism, not scientific evidence, and then must distort or discredit science which gets in their way.

In fact, what is truly bizarre is that the creationists trot out the same lines of argument that they have been using since the 1960s and even earlier. No matter how many times they are debunked, they keep repeating these discredited ideas over and over. In some cases, creationist debaters (like Duane Gish) have been caught using a fallacious argument or an outright lie and were forced to retract it during a debate—but Gish went on to repeat the same falsified argument and deliberate deceptions to a different audience the following night, expecting that no one would have heard him retract it.[43] Such dishonesty and deception is not allowed in real science, since if you make a mistake or your hypothesis is wrong, the scientific community expects you to correct it and start over; you cannot repeat arguments or data that are clearly discredited. Likewise, the fact that creationists never change their arguments or ideas, no matter how many times they are proven wrong, also speaks to their closed-mindedness and lack of scientific integrity. Scientific arguments are always changing because we learn new things and abandon those ideas that have been proved wrong.

The list of standard creationist arguments is not long, and every one of them has been debunked. If you encounter them, see the talkorigins. org website for instant clarifications on any creationist lie you read about. There is also the web resource "Answers to the Standard Creationist Arguments" by Ken Miller on the NCSE website,[44] and Michael Shermer's list of answers to twenty-five standard creationist arguments.[45] Some of the most common are these:

1. *The Second Law of Thermodynamics disproves evolution, because it says that order and complexity cannot arise from disorder.*

No, it does not. The Second Law says that *in closed systems (such as a sealed container of warm gases that gradually cools down) you cannot get an increase in order or complexity; this is known to scientists as entropy,* the tendency of systems to become disordered over time. But the earth and life are *not* closed systems—they have the energy of the sun pouring in, which helps drive the process against entropy and increases complexity. The universe as a whole is indeed slowly running down and increasing in entropy, but this does not prevent local systems like life from becoming more orderly. The creationists have been repeatedly caught in this lie about the Second Law, but they use it over and over again, because it impresses audiences who have never heard of thermodynamics (or know much about science at all).

2. *Life is too complex a system to have occurred by random chance. The odds of life assembling itself from non-life are astronomically impossible. It's like saying that a tornado in a junkyard could assemble a 707.*

There are numerous fallacies and flaws in this argument. The first is the creationist lie is about random chance; evolution is *not* just random chance. Chance in the form of mutation and recombination provides the raw materials on which a very non-random process, natural selection, can act. Creationists are fond of pointing to the analogy of a monkey with a typewriter, hunting and pecking away. They say, "What are the odds it could type a simple coherent sentence by chance?" But that analogy is false. A better one would be a monkey with a word processor, whose software is selecting for combinations that work and automatically eliminating or fixing misspelled words and bad grammar with its spell-checker even as it types. So, too, does natural selection eliminate molecular combinations that are not working and select for ones that work better, survive, and reproduce. Finally, this argument is entirely based on the fallacy of calculating a probability *after the fact* (in Latin, *a posteriori*). As any statistician would tell you, it proves nothing. I demonstrated this when I beat Duane Gish in a debate at Purdue University in 1983. Knowing he would raise this phony argument, I asked the audience to calculate the odds that they would be born on a certain day in a certain place, that they would go to a certain school, that they would have a certain number of siblings, and so on. If you stack up all those a posteriori probabilities one on top of the other, the odds that you would live the life you lived are so improbable that it is statistically impossible you are alive at all! Then combine that with all the probabilities of all the other people in that huge auditorium, and by arguing probabilities after the fact, you can point out that the audience is so statistically improbable that it could not exist!

3. *The earth cannot be millions or billions of years old, since radiometric methods are unreliable.*

The traditional Young Earth creationists (but not ID creationists, who typically accept the geological evidence of an old earth) are wedded to idea that the earth is only 6,000 years old—and therefore they must attack all of geology and physics, which provide abundant evidence that the earth and solar system are 4.5 billion years old. This claim is completely false.[46] Creationists will pick on a few isolated instances of individual dates that did not work, and claim that the whole system is flawed and cannot be trusted. I know many of the people who do radiometric dating, and they are among the most skeptical and demanding scientists alive. They are constantly on the lookout for mistakes (by their labs and others) to make sure that the published dates meet the highest standards. Nobody can publish a bad date and expect to get away with it for long. The most convincing line of evidence is that when geochronologists date a rock by several completely different methods (such as potassium-argon, rubidium-strontium, and both isotopes of uranium-lead), *all of these completely unrelated and independent methods give the same age.* This would be impossible if there were something that was biasing the results in a particular system. Yet creationists throw out the entire field of radiometric dating if they can find any single result that they think is wrong. As Dalrymple pointed out, that is like throwing out all clocks just because one clock does not keep time well. Instead, the clockmaker will learn to trust only the clocks that keep good time, and not rely on the bad clocks.[47] So, too, with geologists and geochronologists. Besides, the claim that the earth is only 6,000 years old is patently absurd and falsifiable on its own merits. There are dead bristlecone pines in the White Mountains of California that record over 7,000 years of tree rings, and some alive today over 4,700 years old, and there is a living spruce in Sweden almost 10,000 years old.[48] In fact, the archeological evidence from Mesopotamia shows that the Sumerian civilization was already well established and advanced enough to brew beer by 6,000 years ago. As the *Onion,* a satirical news website put it, the Sumerians must have been in shock and confusion, and downed a few more beers, when Yahweh suddenly showed up in their midst and decided to create the universe![49]

4. *There are no transitional fossils that show how life evolved from one "kind" to another.*

The second half of my 2007 book gives dozens and dozens of examples of one transitional fossil after another—from the many different "fishibian" fossils, such as *Ichthyostega, Acanthostega,* and *Tiktaalik,* that show how fish evolved into amphibians; to hundreds of different transitional bird fossils between *Archaeopteryx* and dinosaurs and modern birds; to the "frogamander," a transition between salamanders and frogs; to the turtle, with only half a hard shell and teeth; to the beautiful sequence of fossils that showed how mammals arose from reptilian ancestors, or whales and manatees from four-legged land animals. The evidence is overwhelming. How do creationists respond? Mostly they do not. They ignored my book for the first three years after it appeared. When you read creationist attempts to deny the fossil record, they usually resort to quotations taken out of context that seem to present paleontologists claiming that there are no transitional fossils. Once you read the quotation in context, however, you will see that the paleontologists said nothing of the sort. The creationist is deliberately leaving out the part of the quote that does not fit their biases. Almost no creationist has the training and ability to actually study the fossils themselves, and most cannot even tell one bone from another. When you read the efforts of creationists like Gish (1972, 1995) and Sarfati (2002) to discredit the incredible array of transitional fossils, they usually just take quotes out of context from kiddie books on fossils, or give a garbled interpretation of the anatomy of the fossils. As I carefully documented, they have no idea what they are talking about, let alone the skills to look at the actual fossils themselves and interpret them.

5. *The Cambrian explosion showed that life arose suddenly, as described in the Bible.*

No, it does not. The Cambrian explosion is an unfortunate misnomer from the early days of geology when there was very little known of fossils below the rocks of the Cambrian, which were the first to produce large hard-shelled animals such as trilobites. It only looks like an explosion in the context of millions of years of geologic time, since it took place over at least an 80-million-year period, but over 520 million years ago.[50] As usual, the creationists have not done their homework. Over the past 60 years, paleontologists have documented more and more fossils that predate the Cambrian explosion (fig. 6.6). Thus, we have single-celled bacteria reliably dated at 3.5 billion years old; at about 600 million years, we see the first multicellular but soft-bodied animals, the Ediacara fauna. At the beginning of the Cambrian, 550 million years ago (but before trilobites, which do not appear until the third stage of the Cambrian, the Atdabanian, 520 million years ago), we find fossils known as the "little shellies"—small (about 1 mm in size) fossils that show the first stages of hard skeletons on tiny creatures. Finally, at 520 million years ago, we get the trilobites and other hard-shelled fossils, the logical end of a long gradual process exactly as evolutionists would predict: from single cells to soft-bodied multicellular animals to tiny skeletonized fossils to large skeletonized fossils. This last part occurs over a time interval that hardly qualifies as an explosion, even to geologists who think millions of years are no big deal. Nevertheless, the public confusion on this issue (and the unfortunate habit of geologists to continue to use Cambrian explosion) gives the Young Earth and ID creationists opportunity to distort and lie about the fossil record, and confuse people who have not heard what paleontologists have discovered.

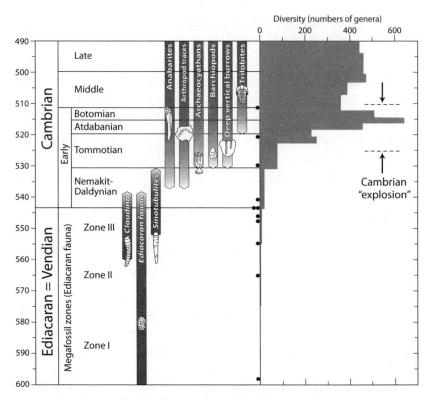

FIGURE 6.6. The so-called Cambrian explosion actually spanned the period 600–520 million years ago—hardly an explosion! *Redrawn from image by Carl Buell.*

6. Humans did not evolve from apes. If so, why are apes still here? And what about the Piltdown fraud, the "Nebraska Man" that was just a pig's tooth, all the "Peking Man" fossils were lost, the Neanderthal was based on a diseased skeleton—there is no fossil record of human transitional forms.

This classic creationist pack of lies and misunderstandings gets at the heart of the issue: creationists hate the idea that humans are another species of ape, and that we are all part of the animal kingdom. The first statement shows their fundamental misunderstanding of evolution. It is not a ladder of progress from amoebas to snails to rats to apes to humans to God, as was thought in the Middle Ages. Instead, life is bushy and branching, and descendants can branch off from their ancestors and life side by side for millions of years. Humans branched off from the ape lineage about 7 million years ago, but both groups survive today. It is analogous to saying that your father had to die the moment you were born, and that you could never overlap in time with your father or grandfather, just because they are your immediate ancestors. The second statement demonstrates the classic creationist tactic of cherry-picking a few anomalous ideas out of context and ignoring everything else the creationists do not want to admit. There are literally thousands of fossils now that show that there were once dozens of different hominid species, from the very ape-like *Sahelanthropus* at about 6–7 million years ago, through *Ororrorin*, *Ardipithecus*, the australopithecines, the paranthropines, and the early members of our genus *Homo*.[51] Yet all the creationists can do is nitpick a few examples pulled out of context. Yes, Piltdown Man was a fraud, but it was discovered to be fraudulent by scientists, not by creationists. "Nebraska Man" was based on a badly worn tooth fossil overinterpreted by a not-too-competent paleontologist, Henry Fairfield Osborn, but the mistake was quickly corrected by his colleagues. Besides, it is not the tooth of a pig, but of a fossil peccary, a New World family Tayassuidae only distantly related to the Old World pig family Suidae. Creationists do not even know enough biology to tell the difference. The original Peking Man specimens of *Homo erectus* from the Zhoukoudian caves near Beijing were lost in the confusion of World War II (American soldiers and Chinese scientists tried to smuggle them out ahead of the Japanese invaders). But there are hundreds of casts of the original specimens in museums all over the world. Since that time, the Chinese have gone back to the locality and found many more good specimens. The first nearly complete Neanderthal skeleton that was found happened to be suffering from diseases like rickets, which influenced early reconstructions. Since then many healthy Neanderthal skeletons have been found that show that it was clearly a different, much more robust, but even larger-brained species than ourselves. In short, this a classic propaganda tactic: cherry-pick a handful of cases that seem to embarrass your opponent while ignoring the vast body of evidence against you. This may work in propaganda, but it is unacceptable behavior among real scientists, who must address all the evidence and not cherry-pick their favorite data.

There is no need to go through all the rest of the arguments, since they are all refuted in my 2007 book and in the resources (especially the TalkOrigins website) listed at the beginning of this section. The overall picture, however, should be clear: creationist arguments consist largely of distortions, deceptions, and quotations out of context that reflect their fundamental lack of any training or understanding of science. At best, their arguments show that they read only kiddie-level science books, cannot or do not want to comprehend what the argument is really about, and deliberately pull out of context misleadingly edited quotations or examples that they think support their cause. Given their habit of repeating debunked and discredited arguments no matter how many times they are proven wrong, one has to conclude that they do not understand science and do not want to—or that they are deliberately deceiving their less scientifically sophisticated followers by lying about what science really says. Either way, their dishonesty and complete disregard for the rules of science and fair argument practice, let alone their inability to learn from debunked arguments and change what they present, speaks volumes about their motives and biases.

WHY SHOULD WE CARE?

We've arranged a global civilization in which the most critical elements profoundly depend on science and technology. We have also arranged things so that almost no one understands science and technology. This is a prescription for disaster.

Carl Sagan

In a democracy, it is very important that the public have a basic understanding of science so that they can control the way that science and technology increasingly affect our lives.

Stephen Hawking

So what is the big deal? What is the harm in allowing a few religious fanatics to believe whatever they want? Why does it matter?

As I discussed at length in the final chapter of my 2007 book, there are many reasons why creationism is a serious threat not only to the United

States, but also to civilization as a whole. Some of these arguments apply to the entire gamut of anti-scientific ideas covered in this book, so we will discuss many of these points in chapter 13. But some of these points are very specific to creationism as well:

· *Creationism is a narrow sectarian religious belief, and the Constitution forbids it to be taught in public schools as science.* This has been shown again and again, every time creationist ideas come before the legal system. Using ruses such as intelligent design and now "teach the controversy" or critical thinking or "teach the problems with Darwinism," they have tried to disguise their religious motivations for their opposition to evolutionary science. But the paper trail of their past actions, the clear link between the old strategies and the new, plus the publicly confessed religious motivations of each creationist group, belies their attempts to sound nonreligious. And there is a good reason why creationism should not be introduced to public school curricula. Not only is it antiscience, but it forces the views of a sectarian minority on people of other religious faiths and those of no faith. The Founding Fathers all remembered the problems with religious intolerance, theocratic states, and state-sponsored religion in Europe, and that is why they enshrined the separation of church and state in the First Amendment.

· *The attack on evolution is the leading edge of an attack on all of science.* The creationists have made no secret of the fact that although evolution is the easiest target, they intend to change scientific instruction in any other field that does not conform to a literal interpretation of Genesis. Goodbye, modern astronomy— the Bible says that we live on a flat earth at the center of the universe. Goodbye, anthropology—the entire human fossil record is ignored or distorted, as is any other evidence of our kinship to the apes and other animals. Goodbye, geology—we are going to have to interpret the entire rock record in the Noah's ark model (and never have a chance of finding oil, gas, coal that only real geology can find). Indeed, their religious dogma-driven approach to science undermines all scientific endeavors, even

those that do not directly threaten their literal view of the Bible.
We have seen this with the conservative/fundamentalist attacks
on many other areas of scientific research, such as stem cells and
global warming.

· *Creationism is a direct threat to our health and well-being.* If so-
called creationist science had its way, we would never be able
to fight the bacteria and viruses that evolve rapidly and infect
over and over again, or battle insects that evolve resistance
to pesticides and destroy our crops. Nearly all that molecular
biologists do to understand their molecules and how they change
requires an understanding of evolutionary principles. This is
especially true of combating human pathogens, such as the HIV
and H1N1 virus; we would be far sicker and more likely to die of
many more diseases if molecular evolutionary studies did not
help us with shortcuts to combating them.[52] In at least one case,
a creationist doctor is directly responsible for the death of an
innocent baby. This happened in 1984, when Dr. Leonard Bailey of
Loma Linda University transplanted a baboon heart into a "Baby
Fae," who had a defective heart at birth. A few days later, to no
one's surprise, the baby died of immune rejection of the baboon
heart. When reporters asked him why he did not use a chimp
heart, which is much closer to us in evolutionary terms, and thus
less likely to be rejected, Bailey said, "Er, I find that difficult to
answer. You see, I don't believe in evolution."[53] Most of us would
be shocked that such obvious medical malpractice did not result
in many lawsuits and Bailey's medical license being revoked—
except that Loma Linda University is a Seventh-Day Adventist
institution that preaches creationism.

· *Creationists actively harass and intimidate our public schools,
universities and museums.* As I documented in my 2007 book, the
creationists do not stop at trying to change school board policies
or curricula.[54] They are actively urging their young followers to
disrespect and harass their teachers, and then resort to shouting
and disobedience if the teacher does not cower before the student.
In most cases, K–12 teachers are so afraid of student and parent

interference that they do not even mention the topic of evolution at all, thus depriving all the rest of the students in their class with one of the fundamental ideas in science. Across the country, biology classes typically save evolution for the final unit at the end of the course, meaning that it often gets dropped when the course is behind schedule. Even sadder, polls have shown that a surprising number of biology teachers do not even understand or accept evolution, and as many as 20% of high school biology teachers are creationists. Creationists have repeatedly harassed and targeted universities and museums as well, especially when any exhibit dares mention evolution. In Kenya, evangelicals have demanded that the National Museums of Kenya (holders of most the world's best hominid fossils) remove displays that mention human evolution. It would be truly ironic if the country where our ancestors arose were to deny the existence of these glorious fossils. It would be even sadder if an unstable Kenyan political system led to chaos, and these same evangelicals tried to break into the museum and destroy these amazing fossils.

We will look at the implications of all the forms of science denialism (including evolution) when we summarize the dangers of anti-science in chapter 13. But creationism is one of the most pernicious and far reaching because it affects a higher percentage of Americans than global warming denialism or anti-vaxxers, and years of polls show that it never seems to diminish in influence and importance.

As the legendary paleontology George Gaylord Simpson on the Darwin Centennial in 1959, "One hundred years without Darwin are enough!" And since the Darwin celebrations of 2009, we can amend that to "One hundred fifty-four years without Darwin are enough!"

FOR FURTHER READING

Alters, B., and S. Alters. 2001. *Defending Evolution.* Sudbury, Mass.: Jones and Bartlett.

Brown, B., and J. P. Alson. 2007. *Flock of Dodos: Behind Modern Creationism, Intelligent Design, and the Easter Bunny.* Cambridge: Cambridge House.

Coyne, J. 2010. *Why Evolution is True.* New York: Penguin.

Dawkins, R. 2010. *The Greatest Show on Earth*. New York: Free Press.

Eldredge, Niles. 2000. *The Triumph of Evolution and the Failure of Creationism*. New York: W. H. Freeman.

Forrest, B., and P. R. Gross. 2005. *Creationism's Trojan Horse: The Wedge of Intelligent Design*. Oxford: Oxford University Press.

Friedman, R. 1987. *Who Wrote the Bible?* New York: Harper and Row.

Humes, E. 2007. *Monkey Girl: Evolution, Education, Religion, and the Battle for America's Soul*. New York: Ecco.

Isaak, M. 2006. *The Counter-Creationism Handbook*. Berkeley: University of California Press.

Loxton, D, and J. W. W. Smith. 2010. *Evolution: How We and All Living Things Came to Be*. Toronto: Kids Can Press.

McGowan, C. 1984. *In the Beginning: A Scientist Shows Why the Creationists Are Wrong*. Buffalo: Prometheus.

Miller, K. 1999. *Finding Darwin's God: A Scientist's Search for Common Ground between God and Evolution*. New York: Harper Collins.

Numbers, R. 1992. *The Creationists: The Evolution of Scientific Creationism*. New York: Knopf.

Pigliucci, M. 2002. *Denying Evolution: Creationism, Scientism, and the Nature of Science*. Sunderland, Mass.: Sinauer Associates.

Prothero, D. R. 2007. *Evolution: What the Fossils Say and Why It Matters*. New York: Columbia University Press.

Scott, E. C. 2005. *Evolution vs. Creationism: An Introduction*. Berkeley: University of California Press.

Shermer, M. 2006. *Why Darwin Matters: Evolution and the Case against Intelligent Design*. New York: Henry Holt/Times.

Shubin, N. 2009. *Your Inner Fish*. New York: Vintage.

Young, M., and T. Edis, eds. 2005. *Why Intelligent Design Fails: A Scientific Critique of the New Creationism*. New Brunswick, N.J.: Rutgers University Press.

7

Jenny's Body Count: Playing Russian Roulette with Our Children

If you want to save your child from polio, you can pray or you can inoculate.

Carl Sagan

THE "GOOD OLD DAYS"

Charles Darwin was distraught. His eldest daughter and second child, ten-year-old Annie, was slowly dying of a now-curable disease, scarlet fever. As historian and Darwin biographer Janet Browne wrote,

> Anne was . . . the apple of her proud father's eye, his favourite child, he confessed to [his friend and cousin William Darwin] Fox. More than any of the other children she treated him with a spontaneous affection that touched him deeply; she liked to smooth his hair and pat his clothes into shape, and was by nature self-absorbedly neat and tidy, cutting out delicate bits of paper to put away in her workbox, threading ribbons, and sewing small things for her dolls and make-believe world.[1]

When Annie finally died on April 23, 1851, Darwin wrote in his private memoir, "We have lost the joy of the household, and the solace of our old age. . . . Oh that she could now know how deeply, how tenderly we do still & shall ever love her dear joyous face."[2] Darwin was so utterly traumatized by the death of his favorite child just as she had reached the promising age of ten that it destroyed what little faith he had, and he never thought seriously about a wise, caring God again. Most of us are lucky enough never to experience the death of our young children, but some people have known the agonizing pain of a young life dying in innocence. The pain of that experience was capture in the last stanza of Frances Harper's poem "Thank God for Little Children":[3]

> Dear mothers, guard these jewels.
> As sacred offerings meet
> A wealth of household treasures
> To lay at Jesus' feet.

Charles and Emma Darwin had 10 children altogether, and 3 of them died in childhood (2 others died in early infancy). This was fairly typical only a century ago, before science and modern medicine changed the world. Today, when even one family loses a child to a disease or a gang shooting, it is so unusual that it often makes the local news. But back in 1851, it was so common even in wealthy, advanced nations such as England that many parents kept having more and more children just to offset the likely loss of a few of them. For most of human history, the death of a young child due to various diseases (and sometimes the death of the mother due to the dangers of childbirth) was a common occurrence. Scholars estimate that through ancient times and the Middle Ages, there were 200–300 child deaths per 1,000 live births (thus not counting children who are stillborn), so a family of 10 children could expect 2–3 deaths in childhood (exactly the number that Darwin's family suffered).[4] In the less developed world, the rate is still high. About 20 million children die each year, and half of these deaths occur in Africa, with its poverty and diseases (especially AIDS). The countries that have the highest rates (mostly poor African countries plus Afghanistan) still experience 200–300 deaths per 1,000 children each year.

Yet the rate has been declining steadily since the mid-twentieth century, when the worldwide rate for the 1950s was 152 per 1,000 births, to fewer than 50 worldwide (developed plus underdeveloped nations) in the past decade. In the developed world, the rates are dramatically lower— typically only 5–6 children per 1,000 births. The dramatic decline in this one statistic is clearly attributable to a single cause: the miracle of modern science and medicine. Whereas a child was once vulnerable to a whole host of diseases, today in the developed world most of the diseases have been conquered through a range of vaccinations, and other developments in sanitation and clean drinking water. Our improved health care has virtually eliminated most of the common reasons that a child died young.

The decline in the childhood mortality rate is just one of many miracles that we take for granted every day, yet they have radically changed the way we live our lives. No longer do parents try to have as many children as possible to offset the 2–3 per 10 births that might be expected to die young. In most of the developed world our health, survival rate, and life expectancy have increased dramatically since the middle of the twentieth century. And the most important cause for this improved survival and health of children is the miracle of vaccination.

My own family experienced this. My mother was stricken with polio when she was just a young girl in the early 1930s, and she was in an iron lung for months, and nearly died or became paralyzed (as did many other polio victims, such as Franklin D. Roosevelt). She fared better than most, surviving with just some loss of muscular strength in her torso, and since those years, her health has been fine. Since the polio vaccine was discovered by Drs. Salk and Sabin in the 1950s, the developed world has been spared the horrors of polio, because the vaccine is now routine and the disease is almost unknown in regions with good medical care. Nevertheless, there are almost a thousand cases a year in the developing world, where the vaccine is not widely available.[5]

The horrors of deadly diseases are well documented in history and epidemiology, as is the effectiveness of the vaccination programs that have nearly extinguished some of these diseases. Vaccines have eliminated smallpox, which once killed one in every seven children.[6] Vaccination has almost eradicated polio as well,[7] and the infection rate of bacterial meningitis has been reduced by 99%.[8] Cholera has long been one of the deadliest diseases in human history, killing everyone from kings and princes to presidents (Zachary Taylor) and musicians (Tchaikovsky). It is now virtually unknown in the developed world, even though in 2000 there were at least 140,000 cases, which killed at least 5,000 people in regions with poor sanitation and health care (mostly in tropical Africa and Asia).[9] Diphtheria used to kill thousands of children a year until vaccines were found.[10] Measles used to be common when I was a child, and infected nearly 4 million American kids and killed hundreds until

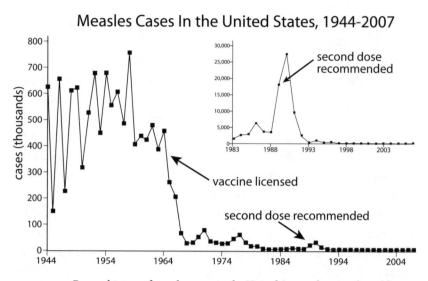

FIGURE 7.1. Recent history of measles cases in the United States, showing the wild swings in virulence and then the immediate suppression once the vaccine appeared. *Redrawn from "Creationism vs. Christianity (A Reprise)," Millard Fillmore's Bathtub website, October 30, 2012, timpanogos.wordpress.com/category/voodoo-science.*

the vaccine was introduced in 1963 (fig. 7.1). It is still deadly where vaccines are unavailable; in 2007 measles killed over 200,000 people in the underdeveloped world.[11] I remember being infected with mumps, a disease that infected a million children a year in the 1960s, during my own childhood. Normally it just meant a swelling of the salivary glands and lymph nodes in your throat and made you look like a chipmunk, but it killed hundreds of children a year until vaccines were found.[12]

Full vaccination of all children in the United States is estimated to save about 33,000 lives and prevent about 14 million infections.[13] In any sane society, there would be no reason to discontinue vaccination as long as the system is carefully controlled and monitored for safety (as the Center for Disease Control, Food and Drug Administration, and other federal agencies constantly do). Yet strangely enough there is a bizarre antiscientific movement that has arisen in some of the most affluent and best educated communities in the developed world: people who resist vaccinating their kids due to some bad science (now repudiated) and spurious word-of-mouth myth that vaccines might cause autism.

THE ANTI-VAXXERS

I do believe sadly it's going to take some diseases coming back to realize
that we need to change and develop vaccines that are safe. If the vaccine
companies are not listening to us, it's their fucking fault that the diseases
are coming back. They're making a product that's shit. If you give us
a safe vaccine, we'll use it. It shouldn't be polio versus autism.

Jenny McCarthy, 2009

The entire anti-vaccination scare began with a controversial study
conducted by Dr. Andrew Wakefield and twelve co-authors that was
published in the leading British medical journal *Lancet* in 1998. The study
reported on a tiny sample of only twelve children with developmental
disorders who were treated at the pediatric bowel unit at Royal Free
Hospital at Hampstead, North London, between July 1996 and February 1997. The short paper concluded some speculation that these cases
might be associated with the MMR (measles, mumps, rubella) vaccination that all the patients had received (as did all children in Britain then).
The paper itself made very modest claims, saying only this: "We did not
prove an association between measles, mumps, and rubella vaccine and
the syndrome described," and suggesting that much more study was
needed.[14] At a press conference during the release of the paper, Wakefield was less circumspect, recommending that the MMR shots not all
be given at one time, since the parents of some his twelve patients had
reported the symptoms about the time of the shots. Other medical experts immediately discounted Wakefield's suggestion. Nonetheless, just
the hint that some sort of connection between vaccines and autism led
to an overreaction by worried parents, who then began refusing to vaccinate their children.

Autism is a very scary and threatening developmental disorder, af-
fecting about one in eighty-eight children in the United States. The
condition delays speech and cognitive development in children ages
one through three, and leading to difficulty with social interactions and
other tics (such as repetitive and obsessive-compulsive behavior) when
the autistic child reaches adulthood (vividly portrayed by the Dustin
Hoffman character in the movie *Rain Man*). Autism is just the most

extreme part of what are known as autism spectrum disorders (ASD), including milder forms known as Asperger's syndrome. The causes are still unknown, although there is strong evidence of a genetic component, a strong possibility that it is caused by defective sperm from older fathers,[15] and possible connection to agents that cause birth defects.[16]

I know from the experiences of my own family how challenging life with ASD such as Asperger's syndrome can be, so I can sympathize with these parents to some extent. Because autism can be so difficult on a family, worried parents are always looking for answers and cures, even though autism is a complex range of problems with no single simple cause or cure, just as the causes of cancer are complex and there is no single cure. These parents can be fooled by con artists and snake-oil salesmen hawking their purported miracle cures. They are suckered by any report in the media or on the internet that promises them answers that science cannot give them.

At first, the media only mentioned Wakefield's study briefly and it had limited impact. But in 2001 and 2002, Wakefield published articles that were more strongly worded (even though no more real research had been conducted). At this point, the media and internet grabbed the 1998 Wakefield study and publicized it, even though it was inconclusive and very preliminary, as the authors themselves admitted. By 2002 there were over 1,500 stories on the topic in the British media, almost all written by commentators and journalists who had no background in medicine. They often mixed up the facts or printed outright false information. The panic was fueled in part by the refusal of British prime minister Tony Blair to confirm whether he had his son Leo vaccinated. Vaccination rates in the British Isles dropped sharply, followed by almost immediate outbreaks of the measles and mumps, and a number of deaths of children who had not been vaccinated.[17]

In the United States as well, rates of measles, mumps, and rubella infections increased as more and more parents read the news from Britain and refused to vaccinate their children. In this country, there are many internet sites that exchange information (and misinformation) between parents, so that the virus of fear spreads quickly through the population of parents who seldom spend much time talking to their overstretched doctors (required by the HMO system to hurry and see lots of patients), or reading reputable medical research. Much of this is propelled by the culture of celebrity. Former *Playboy* Playmate and T V and movie actress Jenny McCarthy (whose own son, Evan, developed autism), became the leading spokesperson for the anti-vaxxers. Due to her beauty, celebrity, and passion for her cause, she got a hearing in many American media outlets, from *Oprah* to *Larry King Live*—and these shows do not always bring in anyone with medical expertise to contradict her. On one Larry King broadcast, she stooped so low as to assert she was right because "there is [*sic*] angry mob on my side" (as if mob rule by ignorant people was the ultimate guide to truth). The three doctors on the show that disputed her claims tried to reasonably explain the scientific evidence, but she simply shouted, "Bullshit!" in response. When that tactic failed to silence the doctors, she shouted even louder. So much for reasonable, rational discussion, or acting on real scientific data rather than emotion! McCarthy is such a symbol of the dogmatic antiscience movement represented by anti-vaxxers that there is a website, www.jennymccarthy-bodycount.com, which keeps track of the number of new infections and deaths since June 3, 2007 (when she began her campaign).[18] At the time this book went to press, there were 110,000 new cases of preventable illness and over 1,000 deaths that could be attributed to failure to vaccinate.

Shamefully, it is not just airhead celebrities and talk show hosts who fall for the anti-vaxx arguments. Something with this much public interest is sure to attract the attention of politicians trying to curry favor with voters. Indiana congressman Dan Burton has been one of the most vocal politicians on their side, frequently attacking the FDA and many respected scientific organizations who have no axe to grind, no agenda, no conspiracy to push—they are just doctors and scientists trying to do their jobs. This much you would expect from an extreme right-winger like Burton, who has a long history of tilting at conservative windmills and advocating alternative medicine. Sadly, even though he retired in January 2013, he wasted taxpayer time and money again in December 2012 holding new hearings and flogging myths about autism and vaccination.[19] But moderate Republican California governor Arnold Schwarzenegger also fell for the pressure and ordered the withdrawal of certain vaccines from the state, even though there was no scientific evidence to back him up. Even more shocking was the activity of Robert F. Kennedy, Jr., an otherwise well-educated and progressive politician, who got sucked into the frenzy and misunderstandings over the vaccine and began harshly criticizing the top scientists and doctors who were doing their best to uncover the real truth behind the hysteria.

What the anti-vaxxers lack in data, evidence, and rationality, they make up for in passion. The sites that promote anti-vaxxer views are full of testimonials by anguished parents describing the regression of their children into autistic symptoms, and immediately jumping to the conclusion that their MMR shots were the bogeyman. All sorts of alternative medicine and fringe medical groups have jumped on this bandwagon, peddling their snake-oil cures for autism to parents who are most vulnerable to quackery. Any time the facts of the story are presented in the media or internet, there is a huge flood of angry e-mails and internet posts from parents disputing the science and claiming that they know for a fact that the MMR shot gave their child autism. Rallies have been held all around the country, often whipped to a frenzy by McCarthy, her former boyfriend comedian Jim Carrey, and other nonmedical speakers, all spouting the same discredited evidence and trying to push their viewpoint on federal agencies and medical institutions. When the scientific and medical community tries to be reassuring and clear in their explanations, the anti-vaxxers fall back into the same thought patterns of conspiracy nuts all over the world: the government and Big Pharma are part of a great conspiracy to protect their profits, and they are covering up; the doctors are all bribed by Big Pharma to lie to their patients; the medical establishment is out to persecute the alternative medicine movement; and on and on. Some of the unfortunate blunders by the pharmaceutical companies and their researchers on drugs such as Vioxx might have helped create this suspicion, although this remains an isolated case of a drug that was not pulled off the market as soon as the test data became available. Nonetheless, in this paranoid, conspiracy-tainted worldview, one mistake by one drug company is sufficient to show that they are all corrupt and out to get us.

This is classic example of health care by anecdote: my kid got this problem after this event. As discussed in chapters 1 and 2, this is classic *post hoc* thinking: one event followed by another does not prove a connection. *Correlation is not causation.* In our scientifically and medically illiterate society, however, such reasoning is alarmingly prevalent. A typical quotation from the website www.MomLogic.com reveals the depth of misinformation and confusion:

> Neither my husband nor anyone in his family had ever been vaccinated . . . and there isn't a single person in his family who has ever had anything worse than a cold. Myself and my family, on the other hand, were all vaccinated against every possible thing you could imagine. . . . Somehow we got the flu every single year. Somehow everyone in my family is chronically ill. And amazingly, when the people in my family reach 50 they are all old and deteriorated. In my husband's family they are all vibrant into their late 90s. My children will not be vaccinated.[20]

Anyone who knows a little about medicine will be appalled (but not surprised) by this amazingly ignorant statement. Let us dissect it piece by piece:

1. The husband's family (never vaccinated) was probably protected from getting sick (despite their lack of vaccinations) by *herd immunity* (also called *group immunity*). When a high enough percentage of a population is immunized, the disease cannot spread, and a handful of unvaccinated people will be protected as well. But when too many people fail to vaccinate, the entire group is threatened because the disease can spread.

2. The woman's family may have gotten sick more often, but almost certainly not from the deadly diseases that they were vaccinated against. Flu is *not* one of the diseases that we can become completely immune to after vaccination. We get shots every year during winter flu season in order to protect us from some of the more common flu viruses, but they do not provide 100% protection, since there are many different strains of flu that are constantly evolving to circumvent our latest vaccines. By contrast, MMR vaccinations have an almost 100% success rate in providing lifetime immunity against much deadlier diseases that do not evolve as fast as the flu or common cold viruses.

3. The issues of aging, living to a ripe old age, and resisting diseases has very little to do with childhood vaccinations. Instead, they are a product of many other things that influence our immune systems and

health. The only thing that the childhood vaccines do is protect us from specific diseases. They do not promise to make us healthy the rest of our lives.

This is the depth of public ignorance that faces doctors every day while they try to do their jobs. Let us take a step back and look at the actual scientific data about the alleged connection between vaccines and autism.

VACCINES AND AUTISM: IS THERE A LINK?

I have mentioned to you before that the prime objective is to produce unassailable evidence in court so as to convince a court that these vaccines are dangerous.

Lawyer Richard Barr to Dr. Andrew Wakefield, six months before the Lancet *paper*

Let us start by looking closely at Andrew Wakefield and his 1998 paper, the *only* study that has ever suggested a link. In normal medical science, it is customary *not* to publish such preliminary results (especially on such a tiny sample of twelve patients, only nine with the alleged link) until other labs have replicated them. Wakefield made no effort to do so, and the media jumped on the story and spread it so widely that it took a while for other studies to be conducted to check his claims and receive similar publicity. Even more suspicions were raised about the story when Brian Deer and other British investigative journalists began to do some digging into Wakefield and his background. It turns out that two years before he conducted the study, Wakefield had signed a contract with lawyer Richard Barr, who was digging around for evidence to file a class-action lawsuit against the major manufacturers of the MMR vaccine.[21] Barr paid Wakefield over £435,643 (about US$750,000) plus expenses at the extraordinary billing rate of £150 (US$218) per hour, funneled through accounts in his wife's name to hide his connection. This money was all paid from British taxpayers' money routed through the UK Legal Services Commission, meant to help poor people afford a lawyer. Ultimately, it came to £18 million (roughly US$26 million) paid to all the people in Wakefield's lab group. Even more astonishing is that

Wakefield predicted he would find a new syndrome, later dubbed "autistic enterocolitis," even before he began the research.[22]

All these revelations should set our baloney detector alarms ringing. First of all, it was an obvious conflict of interest for a scientific researcher to investigate a disease when he stood to get rich if he found what the lawyer wanted. Even a completely honest scientist could bias their results with that incentive. More damning is the correspondence that shows that Wakefield knew exactly how unethical this form of legal bribery was, because he tried to hide the money under his wife's name—classic proof of guilt and premeditation in a court of law. And finally, the fact that he had already decided what syndrome he would find in order to meet these legal obligations smacks of a scientist trying to find any data that would justify his predetermined conclusions—another big scientific faux pas.

The suspicious circumstances continued to accumulate. All twelve of the parents in this study were already clients of Barr in his class-action suit, and were already active in the anti-vaccine movement and on the record for their opposition to MMR. They were not just any twelve random patients drawn from the normal hospital population. This, too, is a violation of medical ethics, and it would invalidate any claims made because the sample size is preselected to get a specific result. The parents reported general problems with their child's health (later investigators discovered it was mostly simple constipation, not an exotic syndrome), and many of the children had received the MMR shots at widely different times before they reported symptoms. The children were put through some excruciating examinations, such as ileocolonoscopies and lumbar punctures (called "spinal taps" in the United States), over the course of five days, all of which turned out to be unnecessary since they were motivated by fraudulent goals. In addition, these dangerous procedures need to be approved by an institutional review board, and they were not. Just nine months before he announced his results in a press conference, Wakefield had filed a patent for a measles vaccine that would replace the MMR shot he was blaming for autism. This, too, was unethical, in that it was an attempt to create a medical scare to drive his commercial competitor out of business. None of these possible conflicts of interest or problems with the patient selection or the treatment was mentioned in the *Lancet* article. Such reporting of conflict of interest is required by most journals and usually testified to by a sworn statement that each author must sign (so Wakefield also lied in his sworn affidavit).

Journalist Brian Deer, who helped break the scandal, wrote,

> The apparent parental allegations, moreover, were written up in terms that misled. Where Wakefield spoke in the paper of behavioural symptoms within 14 days of MMR, the true position was that the parents [usually after advice from lawyers or activists, and always after being advised on the phone by Wakefield] had generally reported [if anything] common, benign, consequences of vaccination, such as crying, fever, rash, irritability, and even sometimes [also benign] febrile convulsions. No competent doctor, acting professionally, could describe these as "behavioural symptoms," in the context of autism, much less use them as evidence of regression.
>
> In fact, Wakefield's tabulated finding—linking MMR with the sudden onset of regressive autism in two thirds of a consecutive series of 12, seen routinely at a children's bowel unit within the space of a few months—was both biologically implausible and statistically impossible. It simply could not happen.[23]

The autism-MMR link soon began to unravel as other doctors looked into Wakefield's original patients and the data in the study, and found inconsistencies and serious lapses. By 2003, many doctors had come to the conclusion that the Wakefield study was flawed, and eventually the investigative reporters dug up the sleazy deal behind the hoax. Today, the investigation has finally run its course. Most of Wakefield's co-authors (who were just asked to consult on various parts of the study, but were not in on the planning or illegal back-door deals) have repudiated the original 1998 paper. On February 17, 2010, the UK General Medical Council (the highest review board in UK medicine) found Wakefield guilty of medical and ethical misconduct.[24] Five days later, *Lancet* took the extraordinary step of retracting the paper and declaring its conclusions to be false.[25] This is a very rare event in the annals of science, usually caused only by well-documented cases of research fraud. A paper that Wakefield published in the *American Journal of Gastroenterology* was retracted as well.[26] On May 24, 2010, Wakefield's medical license was revoked, so he can no longer practice medicine.[27] Of course, the anti-vaxxers regard him as a hero and martyr and claim he was unfairly persecuted for challenging the medical establishment. Never mind the preponderance of evidence that clearly shows he was a crook and a hoaxer.

So from a single fraudulent study that was reported too quickly and uncritically by the media and the internet, a gigantic medical scare arose. No matter how many studies and commissions that have been conducted since, the anti-vaxxer hysteria was entirely based on a hoax generated by a greedy doctor, lawyer, and their clients and patients, and no further evidence has emerged to support their position. Instead, huge amounts of money and person-months have been essentially wasted in study after study conducted to quiet public anxiety over a medical scare for which there was no actual valid scientific support. These included a huge 2008 study by the Centers for Disease Control (CDC), a 2004 report by the National Academy of Sciences, and a 2004 report by the UK National Health Services. A review of 120 other previous studies dealing with vaccines and autism concluded that there was no evidence of a link in any of the studies.[28] None of this impresses the anti-vaxxers, who believe their emotions and gut reactions, not careful scientific studies.

Many of the anti-vaxxers have claimed that the culprit is thimerosal (also spelled thiomersal), an organic mercury compound long used in vaccines as a preservative, with no documented bad side effects. Intuitively, having heard that mercury in its raw elemental form is toxic, some people naturally jump to the conclusion that any mercury compound is also dangerous (even though the amount in each vaccine is minuscule, only enough to keep bacteria and fungi from growing in the vaccine). The Institutes of Medicine conducted a study of tens of thousands of children[29] and found that there was no evidence that thimerosal was associated with autism, since autism developed both in children who had been given the MMR vaccine and those who had not. Furthermore, thimerosal was removed from vaccines by 2000, yet the number of autism cases did not drop in children inoculated with mercury-free MMR vaccine. In Canada, thimerosal was slowly phased out of vaccines between 1987 and 1998. Additional research found once again that autism was just in common in children who had MMR vaccines with no mercury in it as in those who had received MMR with thimerosal. Studies by the CDC and the FDA also found no evidence that thimerosal had anything to do with autism.[30] These are all classic examples of well-designed experiments with a *control group,* in which one examines the behavior with the active agent and without, to determine if the substance in question has any real effect.

If that were not conclusive enough, epidemiologists in Finland looked at the medical records of *two million children* in many different countries, and found that autism rates did not change when thimerosal was removed from the vaccine, and there was no difference in autism rates between children who did or did not receive the MMR vaccine. Similar studies have been conducted in Denmark, Sweden, the United Kingdom, Canada, and a number of other countries—all had the same conclusion: neither mercury in the vaccine, nor the vaccine itself, causes autism.[31]

Hundreds of studies. Millions of children. Thousands of person-hours and millions of dollars wasted on an effort to chase a bogeyman caused by a single fraudulent study by a corrupt doctor and lawyer, all to convince the public that there was nothing behind this hysteria. After the original Wakefield hoax, *not one case* shows a statistically significant link between vaccines and autism. Yet Congressman Dan Burton raged on the floor of the House about mercury in the vaccines, used his powerful committee chairmanship to hound the FDA and CDC demanding recall of the vaccines, and was quoted as saying, "My only grandson became autistic right before my eyes—shortly after receiving his federally recommended and state-mandated vaccines."[32]

Clearly, scientific evidence does not matter for the anti-vaxxers—only emotional responses to the anguish of autism and their gut feelings and hunches about seeing one event appear to follow another. Over and over again, we have to overcome our natural inclination to believe this false sense of connection between events, and remind ourselves that *correlation is not causation.* The sad cases reported by parents on the internet may grab our emotions, but we must put them aside and look at the clear-cut evidence and solid statistical analysis to help logic and reason overcome emotional knee-jerk reactions and jumping to false conclusions.

And when all these studies finally put the MMR-autism link to rest, what did the anti-vaxxers do? Did they admit they were wrong? No. They resorted to ad hoc hypotheses, claiming it must be some other vaccine besides MMR that caused the autism. They cannot come to terms with the idea that *the link is not real.* This is a classic case of special pleading—"moving the goalposts" to change the terms of the debate when your original hypothesis has been falsified. It may be common behavior among people who want to save their cherished ideas and respond to their gut feelings and emotional reactions, but it is not allowed in science.

Another common problem with the anti-vaxxers is that they then resort to all sorts of quack medical practices, often advocated by the same disreputable doctors who have been pushing the false connection between vaccines and autism to boost their own practice and income. A whole group of these quacks have been advocating chelation therapy, a process in which deadly chemicals are put in a patient to pull out the metals in their body (such as mercury) based on the false notion that mercury causes autism. Not only is this highly risky and might cause some harmful side effects, but the FDA has put out warnings about the dangers of this quack medicine.[33] Given that anti-vaxxers reject modern medicine, however, it is unlikely they will listen to the FDA, which they are convinced is in cahoots with drug companies. (Never mind that the quacks who advocate chelation therapy are themselves in a conspiracy to boost their own incomes by blaming autism on mercury.)

But some might say, "Where there's smoke, there's fire" (another false maxim that assumes guilt when nothing is proven). We have shown that there is no scientific or medical evidence connecting autism and vaccines, just gut reactions, anecdotes, and emotions of the anti-vaxxers. But what might explain the possible association in the first place? Why does it appear than autism increased at about the time vaccines became widespread?

As is the case of many examples of false correlation, there is often a third unrelated cause that might explain it. It turns out that children at about eighteen months old reach a critical stage in their development. This is when they form simple sentences and move from chewing or pawing toys to interactive forms of play. This is also the age at which a child who might already have inherited autism genetically will begin to show developmental delays, and thus parents would notice their child's autism for the first time. And it so happens that this is the age when the MMR vaccines are traditionally given. Once again, two unrelated events that happen to coincide in time might explain this false attempt at a causal connection.

Another explanation is that some other disease or condition within the environment (which we may not even know about) triggered the change in the children. Until every one of these supposed linked vaccines-autism cases is examined thoroughly by qualified experts in medicine, there is no justification for blaming the onset of autistic behavior on just the MMR vaccine. A number of other possible causes have to be ruled out before you can assert that MMR is the *only* possible agent.

A good example of this phenomenon was published by Thompson et al.[34] They looked at 1,047 children ages seven through ten, when their neuropsychology is well developed, and those with autism spectrum disorders that are fully diagnosed (there were 42 out of the 1,047, a typical percentage). They then tracked down the immunization records of each of these children to see if there was any association with thimerosal. As was seen in every other large study, there was no statistically significant linkage between autism and exposure to thimerosal. Most of the neurotypical children who had been exposed to it did not develop autism, whereas a few of the autistic children had never been exposed to thimerosal. Of the handful of cases, there appeared to be an association between unusual developmental outcomes, but there were just as many kids exposed to mercury who did *better* than average on measures of language, speech articulation, and motor coordination as there were students who did poorly. So are we to conclude that exposure to *thimerosal makes kids smarter*? No, the number of kids who did better and did worse make up a tiny percentage of the entire population, and their outcomes are more likely the results of wide variety of random environmental and genetic factors that control whether a child is developmentally advanced or delayed. We do not know what all those factors are, but we can rule out thimerosal for sure. Thus, the handful of parents who blame MMR or thimerosal on the web are basing their hysteria on the statistical anomaly. Even though their cases represent a tiny minority of children who are vaccinated, they are the proverbial squeaky wheels suffering through a tragedy and they want to find something to blame. By far, the overwhelming majority of kids who were exposed to thimerosal experienced only normal development, or even gifted development, but there is no justification for jumping to false conclusions or attributing their child's development (gifted, average, or autistic) to exposure to thimerosal.

One of the most important causes is clearly the genetic makeup of the parents, who pass the ASD genes to their children. Scientists have identified some of the genes that appear to cause ASD, particularly in boys (who exhibit ASD symptoms far more often than do girls). In fact, studies looking at identical twins (who were born from the same split fertilized egg and have identical genomes) show that as much as 90% of the symptoms of ASD may be genetic.[35] ASD runs in the males of my own family, and I myself have mild symptoms of Asperger's syndrome (but not diagnosable when I grew up in the 1950s and early 1960s). Recent studies have also that older fathers are more likely to have ASD children, possibly because they have more defective sperm as they get older.[36] We know that ASD are a complex of developmental disorders with multiple causes, but if it is 90% caused by genetics, then not only are environmental things like mercury or MMR shots unrelated to ASD, but nearly *all* environmental factors are irrelevant to the onset of the problem. So these anti-vaxxer parents like Jenny McCarthy have been wasting their time blaming mercury and vaccinations, when the fault lies in their own genes, *not* the stuff that happens in the environment after the child is born. Even if these anti-vaxxer parents had *never* vaccinated their ASD children, chances are that the ASD symptoms were preprogrammed in their own genes, and would have appeared anyway.

What about the increasing rates of autism in children over the past few years? Once again, when you look closely at the data, this assertion breaks down.[37] Contrary to the anti-vaxxers' claims, the increase in documented cases does *not* precisely track the timing of the introduction of vaccines, but lags some years behind it.[38] More importantly, the increase in reported cases is widely acknowledged by the medical community to be an artifact of the historical reality that autism was not even a recognized disorder in psychiatry until the late 1960s.[39] Naturally, once a disorder was formally diagnosed and named, it was gradually recognized and diagnosed by more and more doctors as it became more familiar. In recent years, child psychologists have developed even more sophisticated tests to spot it early, leading to even more diagnosed cases. Prior to the formal diagnosis of autism, there were many cases of people with this disorder, but they were just lumped in with the mentally retarded and their diagnoses were incorrect: they were called "adult psychopaths" or "schizophrenics." Once again, correlation does not prove causation. The increase in both diagnoses and vaccinations is an artifact of two unrelated events that occurred as medicine improved in the 1960s and the

1970s: the increase in childhood vaccinations, and the improved ability to diagnose different psychological disorders.

As the surgical oncologist and blogger David Gorski, aka Orac,[40] argued about the false correlation between autism and the rise of vaccinations in 1983,

> A lot of other things have happened since 1983 as well. For example, in the early 1990s, the diagnostic criteria for autism were broadened, and campaigns for greater awareness were begun. Diagnoses of autism in 1983 were made using the DSM-III, where the criteria for an autism diagnosis were much more restrictive than those in the DSM-IV, released in the early 1990s. Moreover, in 1983, categories of Asperger's and pervasive developmental disorder—not otherwise specified, both of which are lumped into the 1 in 150 figure for 2008, weren't recognized in the DSM-III. Of course, if I wanted to be snarky (and perish the thought that I would ever be snarky), I could point out that 1981 was the year that the IBM PC was released, followed by the Apple Macintosh in 1984, both of which led to the exponential growth of households owning and using personal computers. That's it! It must be computer use that led to the increase in autism in the 25 years since 1983! Wait, what about the compact disc? It just so happens that 1983 is the year that the CD was first released in the American market. Ergo, it must be CDs that cause autism.[41]

PLAYING RUSSIAN ROULETTE—WITH OTHER PEOPLES' CHILDREN

The decision to vaccinate is a decision for your child, but also a decision for society.

Jane Seward, deputy director, Centers for Disease Control

Some of the anti-vaxxers or parents scared by them might say, "Fine, there may be no link between vaccines and autism. But I want to be safe and avoid endangering my child, so I'm not going to let them be vaccinated." This may sound like a simple parental choice, but the issue is not strictly up to individual parents. The failure of enough parents in a population to get their children vaccinated against deadly diseases results in the return of the diseases. Once enough children have gone unvaccinated, the entire group loses its herd immunity, and those vulnerable kids can quickly spread the diseases all over again.

This is exactly what has happened since the vaccine-autism scare began. In the United Kingdom and the United States, vaccination rates have dropped from about 92% to as low as 63%, far too low to prevent the diseases from returning.[42] In the United Kingdom, the number of cases of measles went from 56 in 1998 to 449 in 2006—the first deaths from measles reported since 1992.[43] Mumps, which had been extremely rare in the United Kingdom before 1999, began spreading rapidly; there were over 5,000 cases in 2005.[44] In the ensuing years, both measles and mumps increased in the United Kingdom at levels thirteen to thirty-seven times what they had been in 1998. By 2008, both measles and mumps were declared endemic (in other words, they were fully established in the population).[45] There were also massive outbreaks around Europe in 2008, especially in Austria, Italy, and Switzerland. Hundreds of cases of measles were reported in the United States in 2008 as well, after measles had been considered eradicated throughout the entire Western Hemisphere in the 1990s. In 2010, California had an outbreak of pertussis (whooping cough) with over 910 cases—four times the number from the same time the previous year, and five deaths of infants under three months of age (the shot cannot be taken until three to four months after birth).[46]

In some cases, parents have been cherry-picking which vaccines they will consent to, and which ones they refuse. As reported in *Time*,

> that can be a high-stakes game, as Kelly Lacek, a Pennsylvania mother of three, learned. She stopped vaccinating her 2-month-old son Matthew when her chiropractor raised questions about mercury in the shots. Three years later, she came home to find the little boy feverish and gasping for breath. Emergency-room doctors couldn't find the cause—until one experienced physician finally asked the right question. "He took one look at Matthew and asked me if he was fully vaccinated," says Lacek. "I said no." It turned out Matthew had been infected with Hib bacteria that causes meningitis, swelling of the airway and, in severe cases, swelling of the brain tissue. After relying on a breathing tube for several days, Matthew recovered without any neurological effects, and a grateful Lacek immediately got him and his siblings up to date on their immunizations. "I am angry that people are promoting not getting vaccinated and messing with people's lives like that," she now says.[47]

If a population had been vaccinated before the autism hysteria, how does the virus return? We tend to forget that we now live in a highly mobile society, where people can fly around the world in a matter of hours, and be exposed to diseases in the less developed world, then bring them back to their vulnerable hometowns and infect the unprotected kids in

their schools or playgrounds. We also have high rates of travel and im-
migration by people from other parts of the world where these diseases
are still common, and they can be carriers, too. The only effective shield
that we have against another round of worldwide epidemics of formerly
vanquished diseases is vaccination.

Ironically, most of the parents who resist vaccination were raised in
the era when all these deadly diseases had been nearly vanquished. They
have no memory of measles, mumps, rubella, diphtheria, yellow fever,
and polio, so they would rather gamble on the false stories of the links
of autism than worry about diseases that had been virtually eradicated
since before they were children. Even then, they declare that it is their
choice, and they should not have to be forced to do something when it
concerns only their own family.

But it is *not* a purely private matter—it is a matter of public health as
well. Parents who refuse vaccination expose not only their own children
to the diseases, but they also allow the disease to infect newborns and
children too young to be vaccinated. At those early ages, infants are
particularly vulnerable to infections, yet they are not old enough yet to
receive the protection of vaccination. Their only safety against the hor-
rible days when infants and babies died at alarming rates is that the older
kids in their community are all vaccinated, so the disease cannot reach
the babies. When a parent refuses to protect their own kids, not only do
they endanger them but they also threaten even more vulnerable babies
and toddlers. In this context, refusal to vaccinate is highly selfish and
dangerous to society as whole.

In most societies, if you act in ways that harm others, you are liable
to be sued. The law has the authority to prevent you from doing dangerous things that affect other people, whether it be drunk driving or
pointing a gun at a crowd or exposing babies to deadly diseases. In 1905,
the Supreme Court recognized the importance of herd immunity and
ruled that states have the right to mandate vaccinations, even against the
parents' wills, because it is critical to the health of the community and
individuals cannot endanger other people with irresponsible behavior.

On an episode of the PBS series *Frontline* broadcast on April 27, 2010,
the camera crews went to the relatively affluent, well-educated com-

munity of Ashland, Oregon.[48] Unlike the poor immigrant communities in the big cities, which might be expected to have low vaccination rates due to ignorance and poor health care, Ashland has all the advantages— yet it has one of the lowest MMR vaccination rates in the country. The filmmakers interviewed numerous smug, self-absorbed yuppie parents who believed that they knew all that they needed to know from anti-vaxx websites, and refused to listen to the advice of doctor and nurses. The documentary also interviewed numerous medical professionals in the community who express their exasperation and frustration at these know-it-all parents who believe the garbage they have read online over doctors and nurses with actual professional training and up-to-date knowledge of the reality of vaccinations and infection. At the end, the consequences of this deliberate misconduct by the parents is apparent, as the film shows numerous babies stricken by diseases that had not been encountered in the United States in generations, and that few active doctors had ever seen or treated—all because these misguided parents had taken unnecessary risks—not only with their own children, but also with the vulnerable babies of their community who suffer from their ignorance and obstinance.

As reported by the *Los Angeles Times* on June 1, 2010, Dr. Pamela Nguyen described the latest outbreak:

> Measles is a serious public health threat. According to the Centers for Disease Control and Prevention, the disease remains the leading cause of vaccine-preventable deaths in children. In 2007, there were 197,000 measles deaths worldwide, 90% of them in children younger than 5. That is nearly 450 deaths every day.
>
> A study published in the April issue of *Pediatrics* examined a 2008 measles outbreak in San Diego. The index case was a 7-year-old unvaccinated child who was exposed to the virus while abroad. This case resulted in 839 exposed persons, 11 actual cases (all in unvaccinated children), and the hospitalization of an infant too young to be vaccinated. In total, the outbreak cost the public more than $175,000, which would have covered the costs of measles vaccinations for almost 180,000 children.
>
> And yet, many parents continue not to vaccinate their children. I see such children frequently. Last fall, when I entered an examination room, a 5-year-old patient loudly yelled, "Get out!" Her mother apologized, then explained. "Sorry, she's never gotten S-H-O-T-S before."
>
> Confused, I looked down at the chart to confirm that the patient was in for H1N1 and seasonal flu vaccines. Seeing that she was, I seized the opportunity

to offer her catch-up vaccines as well, but her mother declined. She explained matter-of-factly that it was because the flu was "going around" whereas the other vaccine-preventable diseases, she said, were no longer a threat.

She went on to tell me that she was a lawyer who had grown up in a country where measles is still endemic. Since moving to the United States, she had never known anyone to suffer from measles, but she did know several children who had autism. So, while she understood that vaccinations had not been definitively shown to cause autism, she felt that, here in America, the risk of autism was a bigger threat than that of vaccine-preventable diseases.

The parents in the San Diego outbreak also didn't vaccinate their children because they were afraid of autism. But exhaustive study has found no link between autism and vaccines. It's puzzling why well-educated, upper- and middle-income parents worry so much about a connection that doesn't exist while they ignore the very real risks of not vaccinating. Vaccine refusal is creating large reservoirs of susceptibility, primarily in private and charter schools that are generally free from state restrictions. I worry that we will soon see just how real that risk is. In addition to the case of measles, there also have been nine cases of mumps reported in L.A. County this year.

By choosing not to vaccinate, parents put not only their children but other peoples' children in harm's way. Immuno-compromised children, infants and pregnant women cannot be vaccinated, so they are put at increased risk when those who can be vaccinated are not.

As we saw in the recent H1N1 outbreak, it is often panic rather than education that moves a community to action. Not until parents saw children dying from swine flu did they move to vaccinate their children. Let's hope measles doesn't have to reach pandemic proportions again before we take notice. It is time to change our perspective and make the safety of all children our priority.

The first step is to demand stricter guidelines for personal-belief exemptions. Vaccinations should be mandatory for public school entry in all but the rarest of cases. The next step is to put pressure on private and charter schools to follow these same guidelines. It is selfish for parents who intentionally don't vaccinate to make other children vulnerable. We cannot afford to continue leaving the public's health in the hands of irresponsible parents.[49]

If none of this disturbs you, just go to the website vaccinecentral. wordpress.com. You can click through pictures of hundreds of babies who died from easily preventable diseases—whooping cough, measles, mumps, chicken pox—all because the anti-vaxxers have allowed the spread of these diseases to older children, and babies have no protection since they are too young for vaccination. For every heartrending story about the anguish of parents of a child who allegedly became autistic because of MMR vaccines (which we know is not true), think about parents of babies who *actually died* because of the dangerous situation that anti-vaxxers have created.

FOR FURTHER READING

Allen, A. 2008. *Vaccine: The Controversial Story of America's Greatest Lifesaver.* New York: W. W. Norton.

National Academy of Sciences. 2004. *Immunization Safety Review: Vaccines and Autism.* Washington, D.C.: National Academies Press. www.iom.edu/Reports/2004/Immunization-Safety-Review-Vaccines-and-Autism.aspx.

Offitt, P. A. 2010. *Autism's False Prophets: Bad Science, Risky Medicine, and the Search for a Cure.* New York: Columbia University Press.

Specter, M. 2009. *Denialism: How Irrational Thinking Hinders Scientific Progress, Harms the Planet, and Threatens our Lives.* New York: Penguin.

8

Victims of Modern Witch Doctors:
AIDS Denialism

The whole dissident idea attracted a lot of crazies. And then all of
a sudden, without realizing it, you've become one of them.

Peter Duesberg, 2009

THE STRANGEST DENIALISM OF ALL

For anyone who was an adult in the 1980s, the specter of AIDS was a
horrendous story that transformed the decade and brought an end to
the sexual freedom of the Swinging Sixties and Seventies. The issues of
the gay community went from being underground to the top of the head-
lines as gays died in disproportionate numbers. Within a few years after
the first AIDS diagnosis, the cause was identified by several labs in the
United States and France. Soon HIV (human immunodeficiency virus)
was one of the best known of all deadly pathogens, and the fact that AIDS
was caused by the HIV virus was as well established and uncontroversial
as gravity or the idea that the earth is round and goes around the sun.
It is stunning for most of us to hear that there are people who deny this
widely accepted reality, and believe in all sorts of nonsense cures rather
than the handful of drugs that truly help slow the progression of AIDS.
Every time I mention this story to people, they are astonished that there
is *anyone* who has not gotten the message that HIV causes AIDS. Yet
for many years political leaders in certain African countries were AIDS
deniers, feeding their people lies and preventing good information from
reaching them; they are directly responsible for the infection and deaths
of thousands of their citizens.

Thanks to its deadly nature, its effects on a wide spectrum of society, and a lot of funding, a huge amount has been learned about the nature of AIDS. By looking at the molecular sequences of HIV and similar viruses found in other animals, biologists have determined that HIV is a modified version of the SIV (simian immunodeficiency virus) found in chimpanzees and other apes and monkeys.[1] Apparently, this is one of the rare examples of virus hosted in one animal that has made a few mutations that enable it to take up residence in a new host (known as a *zoonosis*). The exact locus of this transfer from chimps to humans is not known, although the genetic evidence from chimp populations points to either Cameroon or (what is now the Democratic Republic of the) Congo as the original source.[2] By looking at the genetic code of old biopsy tissue samples of possible victims from the 1960s and earlier, and calculating backward using molecular clock rates of evolution, biologists have shown that it mutated from SIV to HIV and made this jump to humans sometime in the late nineteenth or early twentieth century—probably between 1884 and 1924, a time of rapid colonization and urbanization of tropical western Africa.

Most scientists think that the transfer occurred when West African hunters killed chimpanzees and ate them for bush meat (a practice still common today), which exposed their bodies to the infected tissues of a chimp.[3] The virus then spread among human populations many different ways including not only eating infected animals, but also unprotected sex and the reuse of infected needles from injecting antibiotics and antimalarial drugs.[4] Making the situation worse was the close contact of humans as colonialism and forced labor confined diseased individuals among healthy individuals. It is interesting that SIV in apes and monkeys is nonfatal, so the disease could have been endemic in tropical primate populations for a very long time and not caused deaths. Its virulence occurs only because of the mutations that made SIV into HIV. Many researchers are closely studying the differences between the two viruses to find out why HIV always kills and SIV does not. From this, they hope to find a cure.[5]

The specifics of how HIV spread through the human population are complicated. The earliest cases were not originally reported as AIDS, because the disease was not yet known or diagnosed, and only tracked down by looking at biopsied tissues long after the patient died. The two earliest conclusively documented cases are of a man in 1959 and a woman in 1960, both from what was then the Belgian Congo (later Zaire, and now Democratic Republic of the Congo).[6] The first documented U.S. case was in 1969, when a fifteen-year-old African American man who developed aggressive Kaposi's sarcoma died in a St. Louis hospital.[7] Kaposi's is a normally rare skin cancer whose red lesions soon became a signature of AIDS destroying the victim's immune system that suppresses the Kaposi's virus. A few more cases that occurred in the 1970s have been documented, but the first common infections in North America occurred in Haiti through their contacts with people in Zaire in the 1970s.[8] Because HIV takes a long time to incubate and show symptoms, the starting point of these infections is hard to track down. Although books like Randy Shilts's *And the Band Played On* and others blamed the infection in the United States on "Patient Zero" (Canadian flight attendant Gaëtan Dugas), it appears that the infection had already spread to almost 5% in some communities.[9] Haitian immigrants brought the first infections here long before 1983, when Patient Zero was connected with at least 40 of the 248 cases then known.[10]

The understanding of the AIDS epidemic emerged slowly in the United States. On June 5, 1981, the CDC noticed some unusual clusters of pneumocystis pneumonia (PCP) in Los Angeles, soon followed by clusters of more pneumonia, Kaposi's sarcoma, and other opportunistic infections in many major cities in the United States. These diseases are primarily known from people with compromised immune systems. In June 1982, there were so many cases in the Los Angeles gay community that it was called Gay-Related Immune Deficiency syndrome (GRID), but health authorities soon realized that it occurred in heterosexual men and women from Haiti and also in hemophiliacs, and so in August 1982 the CDC renamed it Acquired Immune-Deficiency Syndrome, or AIDS. During the 1980s it became epidemic in the gay community and soon began to kill off many famous people, some of whom (like Rock Hudson) were not yet openly gay. Compounding the problem was not only the difficulty of discovering the virus and determining its method of infection, but also trying to change the habits of sexually promiscuous people and encouraging them to use condoms. As Shilts's book documented, there is ample evidence that the Reagan administration and the conservatives in power at the time neglected and underfunded research into AIDS. Some of their religious-right supporters (especially fundamentalist ministers) called it the "gay plague" and thought it was God's wrath visited upon people they viewed as condemned sinners.

Eventually, the cases of famous, talented, and beloved people dying of AIDS began to change attitudes. It is one thing when mostly anonymous people die of everyday diseases and never make the news. But the list of famous people with HIV/AIDS was long and distinguished, including actors (Rock Hudson, Anthony Perkins, Denholm Elliott, Brad Davis); musicians (Liberace, Freddie Mercury of Queen, Tom Fogerty of Creedence Clearwater Revival, rapper Eazy-E of N.W.A, Jerry Herman of *Hello, Dolly!*, Howard Ashman of Disney's *Little Mermaid* and *Beauty and the Beast*); athletes (tennis star Arthur Ashe, Olympic gold medal diver Greg Louganis, figure skater Rudy Galindo, and professionals in every major sport); dancers and choreographers (dance company founders Alvin Ailey and Robert Joffrey, Michael Bennett of *A Chorus Line*, Rudolf Nureyev); and artists, photographers, writers, and fashion designers (Halston, Perry Ellis, Herb Ritts, Robert Mapplethorpe, Randy Shilts, Michel Foucault). The change in attitude really occurred when it turned out that AIDS was in the blood supply and was not screened for the virus until 1985, infecting many hemophiliacs and other heterosexuals who got blood transfusions (such as the AIDS poster child Ryan White, Tom Fogerty, and Isaac Asimov). Intravenous drug users were also transmitting it to one another by sharing dirty needles.

Nevertheless, the predominance of AIDS among the gay community and among drug users long made it a disease of shame in the United States, and most heterosexuals did not regard it as a serious threat to their own health. This was despite the fact that AIDS was easily transmitted among heterosexuals in Africa and Haiti. But when NBA star Earvin "Magic" Johnson announced that he had HIV in November 1991 from his long history of heterosexual encounters with basketball groupies, it shattered the general public's scorn and disdain for AIDS victims and shook people into realizing that it was a problem for heterosexuals, too.[11] From that point forward, there was much more acceptance and research into AIDS as it became a threat to all people, and a lot more emphasis

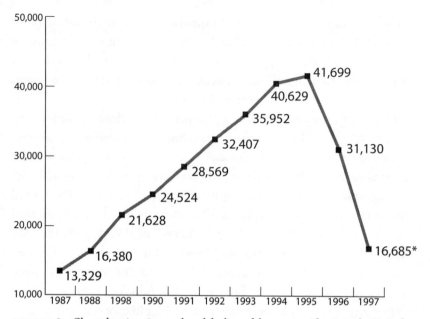

FIGURE 8.1. Chart showing rise, peak and decline of the AIDS epidemic in the United States in the period 1987–1997. *Redrawn from "Howard Zinn, A People's Champion? Part 29, Zinn and the Apocalyptic 90's," And Now for Something Serious blog, December 7, 2010, andnowforsomethingserious.wordpress.com.*

was placed on safer-sex practices. After a rapid increase in AIDS-related deaths in the 1980s and early 1990s, the annual number of deaths in the United States peaked at 41,669 in 1995, and has been rapidly declining ever since then (fig. 8.1). As of 2007, over 500,000 people have died of AIDS in the United States since the start of the epidemic, and over a million people live with HIV infection. Although there is still no cure, people like "Magic" Johnson and journalist/author Andrew Sullivan showed that it is possible to live a long time (both are still going strong more than twenty years after they became infected) if you take good care of your health and take the cocktail of antiretroviral drugs that keep HIV from attacking your immune system.

Nonetheless, HIV remains a serious threat in the United States, especially when "AIDS fatigue" develops and people stop following safer-sex practices. Increases in many other sexually transmitted diseases are disturbing signs that safer-sex practices are not as widespread as they should be.

THE SCOURGE OF AFRICA

Society's fittest, not its frailest, are the ones who die—adults spirited away, leaving the old and the children behind. You cannot define risk groups: everyone who is sexually active is at risk. Babies too, [are] unwittingly infected by mothers. Barely a single family remains untouched. Most do not know how or when they caught the virus, many never know they have it, many who do know don't tell anyone as they lie dying.

Johanna Mcgeary[12]

Although the spread of HIV and AIDS is declining in the developed world, the opposite is true of sub-Saharan Africa, where it originated. This tropical-subtropical region is a hotbed for some of the world's deadliest diseases, from the fatal infection caused by the ebola virus, to sickle-cell anemia, cholera, and tuberculosis, to the many mosquito-borne infections such as malaria, yellow fever, sleeping sickness, and trypanosomiasis. Anyone who has traveled to Africa knows about all the health warnings you must read, and extra vaccinations you must take for diseases not common in the developed world.

But of all these diseases, HIV/AIDS is more deadly and prevalent in central and southern Africa than they are in any other part of the world.[13] Africa has less than 15% of the world's population, but it has about 88% of the world's AIDS cases (Africa had roughly 25 million in 2005) and 92% of the world's deaths due to AIDS (over 2 million died in 2005 alone). Overall, roughly 6% of all adults sixteen to fifty years of age in sub-Saharan Africa are infected with HIV, compared to only 1% worldwide, only 0.55% in North America, and 0.3% in western Europe. If that seems bad, the numbers in specific countries are even more alarming: 16% of adults in Cameroon and over 20% in many southern African nations, where life expectancy has dropped dramatically (fig. 8.2). South Africa may have as many as 30% of its adults infected. Although the infection rate is only 2.5% in Nigeria, it is such a large country that it still has 3.6

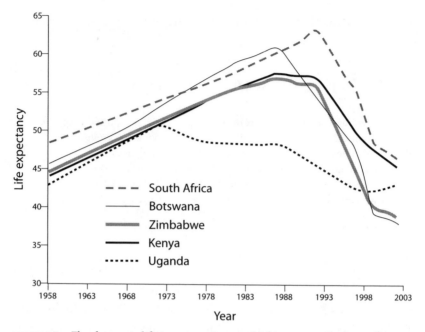

FIGURE 8.2. The changes in life expectancy in several African countries due to the spreading AIDS epidemic. *Redrawn from image created by xcd, "Life Expectancy in some Southern African Countries, 1958 to 2003," Wikimedia Commons, en.wikipedia.org/wiki/ File:Life_expectancy_in_some_Southern_African_countries_1958_to_2003.png.*

million infected people. The infection rate in the small country of Swaziland is 26%, one of the highest in the world; about 50% of adults in their twenties are infected. Consequently, the life expectancy there is only thirty-two years, the lowest value in the world (the second lowest, thirty-eight years, is in Angola, which had both a civil war and AIDS). AIDS-related illnesses cause over 60% of the deaths in Swaziland each year. In other words, about 30 people per 1,000, or 2% of the total population, die each year. The epidemic is so severe in Swaziland, and is wiping out much of the adult working population, that "the longer term existence of Swaziland as a country will be seriously threatened" according to the United Nations Development Program.[14] Much of the deadliness of the epidemic comes from the fact that most HIV-infected people die of an easily spread disease such as tuberculosis or pneumonia, which swiftly attacks people with immune systems ravaged by HIV.

The numbers are appalling enough, but the effects on African societies are even more alarming. Because it tends to strike younger adults who would be the main workforce in these countries, it has a disproportionately damaging effect on the economies of these already desperately poor nations. Not only is the overall economy slowed down, but each of these nations must spend a relatively high percentage of their meager budgets on health care and other expenses to care for all the dead and dying people and their dependents. The most striking effect is that the epidemic has produced over 12 million orphans, many of whom have no relatives to care for them, and so must fend for themselves in desperate poverty. This means that a high percentage of the next generation of Africans is growing up without parental care or role models, wretched, poor, and malnourished. Thousands of them wander the streets, desperately looking for work, and they are often swept into lives of crime.

These orphans are also deprived of the chance for education that would help lift their nations out of poverty. The shortage of teachers in these countries compared to the huge number of children means class ratios of 1 teacher per 50 students to as high as 1 teacher per 120 students, making learning all but impossible.[15] The Catch-22 of the situation is that education is essential to teaching the next generation about safer-sex practices to finally slow down the epidemic, yet those who cannot get education are doomed to make the same fatal mistakes as their parents.

Many writers have described the horrors of the AIDS epidemic and how it changes nearly every city and village. Johanna Mcgeary wrote,

> Imagine your life this way. You get up in the morning and breakfast with your three kids. One is already doomed to die in infancy. Your husband works 200 miles away, comes home twice a year and sleeps around in between. You risk your life in every act of sexual intercourse. You go to work past a house where a teenager lives alone tending young siblings without any source of income. At another house, the wife was branded a whore when she asked her husband to use a condom, beaten silly and thrown into the streets. Over there lies a man desperately sick without access to a doctor or clinic or medicine or food or blankets or even a kind word. At work you eat with colleagues, and every third one is already fatally ill. You whisper about a friend who admitted she had the plague and whose neighbors stoned her to death. Your leisure is occupied by the funerals you attend every Saturday. You go to bed fearing adults your age will not live into their 40s. You and your neighbors and your political and popular leaders act as if nothing is happening.

Across the southern quadrant of Africa, this nightmare is real. The word not spoken is AIDS, and here at ground zero of humanity's deadliest cataclysm, the ultimate tragedy is that so many people don't know—or don't want to know—what is happening.[16]

The first question that comes to mind is Why Africa? Even though HIV/AIDS was a serious threat all over the world during the early stages of the epidemic, most non-African parts of the world have managed to contain the spread, and their (relatively small) rates of infection are slowly declining. The reasons, of course, are complex. Certainly at the top of the list is the extreme poverty of these nations, their low rate of education that might inform people about AIDS and safer-sex practices, and little or no real health care for most of their people, which limits the amount of AIDS medicines and care that might be available. In addition, the poor infrastructure, rampant corruption in many countries, and lack of coordination between foreign donors and the local agencies also contribute to the problem. Nevertheless, a number of non-African countries are donating billions to supply Africa with antiretroviral drugs (ARV) that suppress HIV everywhere else in the world.[17] According to Peter Piot, executive director of UNAIDS, "Treatment is technically feasible in every part of the world. Even the lack of infrastructure is not an excuse—I don't know a single place in the world where the real reason AIDS treatment is unavailable is that the health infrastructure has exhausted its capacity to deliver it. It's not knowledge that's the barrier. It's political will."[18]

Some of the problems are cultural and behavioral. The biggest problem in Africa is the overwhelming ignorance of AIDS and its causes. This ignorance is fostered by myths that spread through villages, and the lack of strong government programs to educate people about the realities of AIDS and how to avoid it. There are also stigmas attached to admitting one has AIDS. Those who are infected cannot discuss it until the symptoms are undeniable, and thus they infect others before they are clearly sick. Yet another problem is the resistance to using condoms, particularly among African men—who view it as an affront to their masculinity, or something that might decrease their potency, or an insult that implies they might be infected. In many African societies, wives are expected to be monogamous and stay at home to raise the kids, while the men are allowed to roam freely and have as many affairs as they want. The cultural problems are often exacerbated by economic forces, especially when the men have to leave villages to find work wherever they can. For example, much of the spread of AIDS in southern Africa can be traced to truck drivers moving from city to city and country to country, having sex with infected prostitutes, and then returning home to infect their wives. In the drought conditions that recently hit Kenya, more women have turned to prostitution, and more of them are unable to insist that their clients wear condoms.

IGNORANCE THAT KILLS: AIDS DENIALISM

It [South Africa] is the only country in Africa ... whose government is still obtuse, dilatory and negligent about rolling out treatment ... It is the only country in Africa whose government continues to promote theories more worthy of a lunatic fringe than of a concerned and compassionate state.[19]

Stephen Lewis, UN Special Envoy for AIDS in Africa

The strangest of all the factors that contributes to the rapid spread and high death rates of Africa's AIDS epidemic is totally unnecessary. In nearly all countries of the world, the political leadership listens to the medical community and helps when they can. Not so in Africa. The brutal dictator Robert Mugabe, who has ruled Zimbabwe with an iron fist for decades and driven his country into poverty and bankruptcy, has done little to fight the rampant epidemic in his country.[20] Even more shocking were the actions of the former president of South Africa, Thabo Mbeki, who ruled from 1999 to 2008. In 2000, when the International AIDS Conference was held in Durban, South Africa, Mbeki convened a presidential advisory panel that featured several prominent AIDS deniers. One of them, David Rasnick, argued AIDS testing should be legally banned, and that there wasn't "any evidence" of an AIDS catastrophe in South Africa (even though the disease was killing thousands at the time). Another prominent AIDS denier, Peter Duesberg, "gave a presentation so removed from African medical reality that it left several local doctors shaking their heads."[21] Mbeki himself addressed the conference and claimed that poverty, not HIV, was responsible for AIDS, and hundreds of delegates walked out of his talk outraged.[22] Because of his bizarre statements, activists and scientists at the Durban Conference prepared a document affirming that science has conclusively shown that HIV is the sole cause of AIDS, and more than 5,000 scientists and doctors signed it.[23]

Nevertheless, Mbeki installed a health minister, Manto Tshabalala-Msimang, who was a committed AIDS denier. She used her government agency post to call antiretroviral drugs poison and curb their use, and prescribed old witch-doctor potions like garlic, beetroot, olive oil, and lemons instead. (She got the nickname Dr. Beetroot for her strange ideas.) The South African medical community fought hard against her dangerous policies and called for her removal, even though Mbeki kept his hard-line stance throughout most of his term. He even fired Deputy Health Minister Nozizwe Madlala-Routledge for speaking out against the pseudoscience that his government promoted. Only after a group called the Treatment Action Campaign brought suit in South African courts over the rights of sick people to receive ARV medication did his government gradually relent, late in 2004. Mbeki called Malegapuru Makgoba, one of South Africa's leading scientists, a racist defender of Western science for opposing AIDS denialism.[24] Nonetheless, in 2002 it became a political issue, as Mbeki's followers savaged the father of their country, revered former president Nelson Mandela for challenging the government's AIDS policy. When Mandela told the world that his son had died of AIDS in 2005, his public statement served both as an effort to combat the stigma associated with AIDS, and as a "political statement designed to . . . force the President [Mbeki] out of his denial."[25]

How could something as weird as the pseudoscience of AIDS denialism become the prevailing policy in a country with hundreds of thousands of AIDS patients? Some of it may be the individual quirks of Mbeki and his political followers, but a lot has to do with the fact that only recently have the majority of nonwhite people in South Africa become familiar with Western medicine. Deprived of modern medical techniques by years of oppression by the white minority and apartheid, they have come to mistrust anything that originates from the so-called white devil, and think of Western medicine as racist and a tool of oppression. Even today, in most villages there are local witch doctors, and most people still practice folk medicine. Another factor appears to be the false hope that AIDS denial promises to people who have suffered for a long time with the disease, and do not want to face the harsh reality that (so far) AIDS is 100% fatal and there is no cure. As reporter Mark Schoofs wrote in the *Village Voice* during the 2000 International AIDS Conference in Durban,

> Winstone Zulu, a prominent HIV-positive activist on the panel, welcomed the discussion because it gave him hope. "For 10 years I've lived with HIV," he told the *Voice*, "and for 10 years I've preached the main line. To hear that I could be wrong is good news. If you were in my shoes, you could understand." Indeed, Zulu comes from Zambia, where people are so poor that the costly drugs that have reduced the AIDS death rate in rich nations amount to a cruel mockery. Yes, people must "face reality," Zulu said, but added, "Ideas from the other side will find very fertile ground in Africa because the conventional thinking hasn't been of much use."[26]

Providing people with hope where there is none is a powerful incentive for people to believe in false cures and snake oil, but unfortunately, such cures do not stop the problem. Much worse is the fact that during the period from 2000 to 2008, when Mbeki and his minions were in power, there were about 340,000 AIDS deaths—and about 171,000 new AIDS infections that might have been prevented if the government had promoted the real methods of avoiding infection, such as safer sex. Even sadder, 35,000 infants were infected with HIV during that time, innocent victims of political endorsement of pseudoscience and stupidity.[27] If there is any happy ending to this sad tale, Mbeki was ousted from office in 2008 by Kgalema Motlanthe. In his first day in office, the new prime minister fired the ignorant health minister Dr. Beetroot. His new health minister, Barbara Hogan, declared, "The era of denialism is over completely in South Africa."[28]

DENYING DEATH

People are focusing on the wrong thing. They're focusing on conspiracies rather than protecting themselves, rather than getting tested and seeking out appropriate care and treatment.

Dr. Stephen Thomas, Director, University of Pittsburgh Center for Minority Health[29]

It is one thing to try to understand why poorly educated Africans who have no exposure to Western science or medicine, or might even have a hostile attitude toward Westerners, might question the consensus

that HIV causes AIDS. But what excuse is there for educated Westerners? How could people raised in cultures like that of the United States
or Europe or other developed countries, where the causes and effects
of AIDS have been well known for over twenty-five years, still deny an
obvious reality?

The answers are similar to those we witnessed in Africa. Some deniers
are people who cannot face the inevitability of death by AIDS and then
go to extraordinary lengths to try to negate the science behind it. The
second group is the tiny minority of dissenting scientists and researchers (very few with the appropriate background to evaluate the science of
AIDS) who insist on challenging the mainstream accepted view, no matter how much evidence proves them wrong. And finally, there is a third
group of fringe crazies with no medical credentials who adopt the AIDS
denier agenda because it fits their antiscience view of life.

In the first category, there are many examples of people and groups
driven to denialism by the scariness of the AIDS death sentence. During
the early phases of the AIDS epidemic in the United States, a number of
radical gay activist groups were inclined to deny the evidence that HIV
causes AIDS, or that it cannot be cured. For example, the AIDS Coalition
To Unleash Power (ACT UP) in San Francisco is a hotbed of denialism,
even though the chapters in Los Angeles, New York, the East Bay Area,
Philadelphia, and nationally are not. ACT UP San Francisco's website
claims, "HIV does not cause AIDS. . . . HIV antibody tests are flawed
and dangerous. . . . AIDS drugs are poison."[30] Its online statements read
like classic paranoid fantasies with the government, business, and mainstream medicine cast as the bogeyman, blaming the deaths not on AIDS
but on the drugs used to treat it:

> The fact is that there is no plague of contagious AIDS. Every year of the so-called
> AIDS "epidemic" in the United States more people died from car accidents than
> from AIDS. Government estimates of the number of HIV positive Americans has
> been continually revised downward from 1.5 million in the mid-1980s to between
> 400,000 to 600,000 today. In addition, the life span of HIV positives that refuse
> toxic AIDS treatments is over twenty years—as long as HIV has supposedly been
> around.
>
> The AIDS scam affects us all—men, women, gay, straight, Americans and
> global citizens alike. As this tragic mistake is elevated to the level of an "economic
> threat to national security" we are facing a death toll higher than the hundreds

of thousands already poisoned to death by dangerous experimental AIDS treatments. Future generations are poised to fall victim to the AIDS propaganda campaign that demonizes sex, treats adults like children and victimizes us all with the fear of death. Investigate the U.S. AIDS lie and stop the terror.[31]

Of course, the reason HIV cases and AIDS deaths have been declining is already well established: the efforts to educate people and encourage safer sex has turned the corner on the once-exploding epidemic (fig. 8.1). Their claim that there are lots of HIV-positive people who refused treatment yet are still healthy is completely unsubstantiated by any hard data. Even if there were a few such cases, the overwhelming number of known HIV-positive people who refused ARV drugs have already died of AIDS-related illnesses.

Given the history of persecution of gays, and especially the government indifference to AIDS in the gay community in the Reagan years, some of this suspicion might be understandable. But as the rest of the ACT UP organizations nationwide have shown, one need not deny well-established reality to be an AIDS activist. On the contrary, such intransigence is probably making the lives of HIV-positive people who refused drugs much shorter and more painful. Just look at how well HIV-positive public figures like Andrew Sullivan and "Magic" Johnson are doing when they have followed ARV treatments. If ACT UP San Francisco continues to deny that AIDS is caused by a sexually transmitted virus and if they fail to advocate safer-sex practices, they are enabling the spread of the disease and condemning many more to this awful fate.

Another prominent AIDS denier was Christine Maggiore, who became wealthy from the Alessi International clothing company in the late 1980s. In 1992, during a routine physical exam, she discovered she had HIV, and her boyfriend at the time also tested positive. At first she was active in the various AIDS charities and groups promoting medical research (AIDS Project Los Angeles, L.A. Shanti, and Women at Risk), but after meeting prominent denier Peter Duesberg, she became a denier and formed her own group, Alive & Well AIDS Alternatives, which denied not only that HIV is the cause of AIDS, but recommended that HIV-positive women should avoid ARV medications for themselves and their infected children. She and her husband, Robert Scovill, continued to have unprotected sex and bragged to the media about her continued health despite no ARV treatments.[32] They had two children, Charles and Eliza Jane. Neither child was tested for HIV, and both were breastfed by their infected mother, which is one of the surest ways to transmit the virus. Eliza Jane contracted AIDS at age two and died at age three in September 2005 of HIV-related pneumonia, the most common opportunistic infection in pediatric AIDS cases. The autopsy revealed all the classic symptoms of a childhood AIDS death, and HIV was found in numerous tissues and organs. Yet Maggiore refused to believe the coroner and had an unqualified *veterinary animal* pathologist (and AIDS denier) Mohammed Al-Bayati perform another autopsy. Naturally, he did not agree that AIDS was responsible, but instead he blamed it on an allergic reaction to amoxicillin. In 2006, the district attorney's office considered bringing charges of child endangerment against Maggiore but did not prosecute. The Medical Board of California did revoke the license of Dr. Paul Fleiss, a noted anti-vaxxer, anti-circumcision activist, and AIDS denier (and also a convicted felon for having abetted the prostitution ring of his daughter, the "Hollywood Madam" Heidi Fleiss). He was thrown out of medicine because he refused to counsel Maggiore about taking ARV medicines and avoiding breast-feeding.[33]

Other medical authorities denounced Maggiore as someone who was misleading people into dangerous and fatal practices, all because of her antiscientific denialism. At the Sixteenth International AIDS Conference, Dr. John Moore said,

> Infants whose HIV infected mothers listen to AIDS deniers never got the chance to make their own decisions. The Maggiore case received wide publicity. Christine Maggiore is a person who's proselytized against the United Statese of antiretrovirals to prevent HIV/AIDS. She's a classic AIDS denier, and she gave birth to a child who died at age three late last year of an AIDS-related infection. The coroner's report clearly reports that the child died of AIDS. That was another unnecessary death.[34]

Maggiore continued her campaign of denying her daughter's death was AIDS related and urging women to avoid testing and shun ARV drugs. She continued to treat herself with "holistic cleansing," "alternative medicines," and other quack remedies. But ultimately reality caught up with her denialism, as it does with every case of untreated AIDS. She died on December 27, 2008, of AIDS-related pneumonia, although her supporters naturally blame it on a toxic alternative medicine, or stress, or a cold, or the flu.[35] Her denier website, www.aliveandwell.org, continues on without her, as virulently antiscientific as ever.

There is a long rogues' gallery of AIDS deniers (all profiled on the website www.aidstruth.org) who are not trained in medicine but continue to promote their cause. Journalist Celia Farber has written numerous articles and a book sympathetic to the denier cause, spouting all sorts of scientific mumbo-jumbo that clearly shows she has no understanding or training in medicine, immunology, or even basic science. There is Nate Mendel, bassist for the rock band Foo Fighters, who has written songs supporting the cause and organized concerts as fundraisers for the AIDS deniers. German vitamin salesman Matthias Rath has promoted his vitamins as a treatment for AIDS, and claims that the entire AIDS scare was cooked up by the drug companies to boost their profits. Ironically, his AIDS denialism helps sell his worthless vitamins to desperately sick people looking for easy answers from quacks (a clear conflict of interest).

What is truly revealing is how few people with actual medical training have bought into the entire AIDS denier movement. We have already discussed Dr. Mohammed Al-Bayati, a veterinary pathologist with no human pathology experience or even an MD. He claims that AIDS is due to so-called toxins and drug use, and he profits (to the tune of $100 per hour) for consultations "on health issues related to AIDS, adverse reactions to vaccines and medications, exposure to chemicals in the home, environment or workplace."[36] Dr. Rebecca Culshaw is a mathematical modeler with no medical training, yet people use her name on denier literature over and over again. Maverick scientist Kary Mullis received a Nobel Prize for his development of the polymerase chain reaction, which revolutionized molecular biology—but his training is not in immunology or medicine, so he has no qualifications of relevance to HIV and AIDS research. (Consistent with his embrace of one kind of denialism, Mullis embraces a wide range of kook notions. He is also a global warming denier, denies that CFCs caused the ozone hole, says he encountered aliens that looked like fluorescent raccoons, takes LSD, and also believes in astrology.) Some medical scientists who were at first part of the AIDS denier community, such as Robert Root-Bernstein and Joseph Sonnabend, have now rejected their former positions after seeing the ability of ARV drugs to suppress HIV. Despite their changed positions, they have had to castigate AIDS deniers for continuing to use their names in their literature.

About the only significant denier who has relevant credentials is Dr. Peter Duesberg of the University of California at Berkeley. Early in his career, in the 1970s, he did research on cancer-causing viruses in birds, and won numerous awards, tenure at age thirty-six, and election to the National Academy of Sciences. But he was always known as a contrarian, and in 1987 he began to publish articles disputing the role of HIV in AIDS. Instead, he claimed that AIDS was caused by recreational drug use and ARV drugs used to suppress HIV, which (in his view) was only a harmless passenger virus.[37] As we have already seen, his ideas were picked up by Christine Maggiore, Thabo Mbeki, and many others— making him one of the central figures in the AIDS denialism movement. However, his research is clearly contrary to a huge number of experiments proving HIV causes AIDS.[38] According to most scientists who know the subject better than Duesberg, his papers are classic examples of misinterpreted and cherry-picked data that violate all rules of scientific inquiry, quotations out of context, and disregard for the huge mountain of literature that doesn't agree with his ideas.[39] When one of his papers was rejected by the respected and peer-reviewed *Journal of Acquired Immune Deficiency Syndrome* (JAIDS), he sent it to a non–peer reviewed journal, *Medical Hypotheses,* and got it published in only two days. The uproar over this maneuver caused the journal to retract the paper. According to one reviewer who saw the paper rejected by JAIDS and then saw the resubmitted version, unchanged despite all the criticisms and suggestions,

> "They select only those publications that (allegedly) support their arguments, but ignore all those (many, many more) that do not. Worse, they take quotations and statements out of their original context, to create a message that is the opposite from the one provided by the original publication. This conduct is so egregious," the reviewer states, "that if the present article were to be published, Duesberg et al. could well find themselves answering scientific misconduct charges."[40]

As a member of the National Academy of Sciences, Duesberg demanded the privilege of publishing one of his papers in the *Proceedings of the National Academy of Sciences* without peer review, even though it caused a scandal because of the poor quality and bad interpretation of data. As of this writing, he is under investigation not only for his unethical and poor scholarly practices, but also for his failure to disclose conflict of interest, since his co-author, David Rasnick, was an employee of vitamin huckster Matthias Rath. The case has not been resolved yet, but the penalties could include expulsion from his university position.

The final set of players in the AIDS denial rogues' gallery should not surprise us: right-wing conservative deniers of evolution and global warming. We have already heard about the wild ideas of Kary Mullis in this regard. Duesberg's work is published by the arch-conservative Heritage Foundation and Regnery Press. Tom Bethell, a journalist known for his attacks on evolutionary biology and support of ID creationism, as well as attacks on global warming science, sides with the AIDS deniers in *The Politically Incorrect Guide to Science*.[41] The founder of the ID creationism movement, lawyer Phillip Johnson, wrote five papers denying that HIV causes AIDS and accused the CDC of fraud in relation to HIV/AIDS.[42] Moonie creationist Jonathan Wells is also a rabid HIV denier, as are several other creationists.[43] The right-wing libertarian blog www.LewRockwell.com also features articles in support of Duesberg and denying that HIV causes AIDS.

Given the religiously motivated hatred of gays common among right-wing movements, the alliance of conservative Christians and gay activists would seem to make strange bedfellows. One might think that they would be more like the Reagan administration and regard the gay plague as just punishment for the sin of homosexuality, and not find any problems with the fact that HIV is the cause. But the fact that Big Government and Big Science have come to agreement on the causes of AIDS and have worked hard to set up health programs to prevent it is, apparently, more important. Such an alliance trumps their antigay instincts, since they apparently hate the government and the scientists even worse than they hate gays. As Steven Epstein wrote, "the appeal of Duesberg's views to conservatives—certainly including those with little sympathy for the gay movement—cannot be denied."[44]

And so we find ourselves back where we were at the beginning. Most of these antiscience denial movements have a strong connection, from antievolution, to anti–climate science, to anti-HIV (and in the case of Kary Mullis, even astrology and UFOs). In some cases, we see the exact same political and debating strategies: quoting out of context, attacking Big Science as a religious movement or a conspiracy with Big Government, using fringe scientists to challenge well-established facts, making lists of scientists who dispute the orthodoxy (and improperly including the names of scientists who do not agree with them, or who have changed their minds), pointing to the fact that their researchers cannot publish in mainstream peer-reviewed journals as evidence of a conspiracy to suppress their revolutionary ideas, and so on. Anything in science (evolution, global warming, HIV/AIDS) that threatens the right-wing view of government, business, and evangelical religion is challenged by a pseudoscientific movement funded by their right-wing think tanks and conservative millionaires.

However, there are some important differences. Unlike the smoke-screen the deniers have created which has caused great public confusion and doubt about evolution and global warming, the AIDS deniers are still a tiny, shrill minority who have a limited effect on a handful of individuals who foolishly stop taking their ARV medications. The exception is when the leadership of a foolish regime, like that of Thabo Mbeki, follows AIDS denial as a national policy and kills thousands of people.

And there is one other important difference—this kind of denial is truly deadly to the individuals in it, and the movement itself is losing members to death by AIDS, not gaining them. The editors of AIDS denier magazine *Continuum* all died of AIDS, and the magazine died with them. Christine Maggiore's crusade against reality eventually led not only to her own death but also to the death of her innocent daughter. The website www.aidstruth.org maintains a list of deniers who have died of AIDS, and that list keeps growing.[45] Creationism never seems to go away, and global warming denialism will only vanish when the climate is so bad that no one can ignore it, but the AIDS denier movement should vanish long before then. Let us hope they never again have the power to cause millions of innocent people to die, as they once did in South Africa.

FOR FURTHER READING

Epstein, S. 1996. *Impure Science: AIDS, Activism, and the Politics of Knowledge*. Berkeley: University of California Press.

Kalichman, S. C. 2009. *Denying AIDS: Conspiracy Theories, Pseudoscience, and Human Tragedy*. Berlin: Springer.

Shilts, R. 1987. *And the Band Played On: Politics, People and the AIDS Epidemic*. New York: St. Martin's.

Smith, T. C., and S. P. Novella. 2007. HIV Denial in the Internet Era. *PloS Med* 4 (8): e256. doi/10.1371/journal.pmed.0040256, www.plosmedicine.org/article/info:doi/10.1371/journal.pmed.0040256.

9

If It Quacks like a Quack: Snake-Oil Con Artists in an Era of Medical Science

All drug doctors are quacks.

Benjamin Franklin

Trust not the physician; his antidotes are poison.

Shakespeare, Timon of Athens

Nearly all men died of their remedies and not of their illnesses.

Molière, Le Malade imaginaire

MODERN SNAKE-OIL SALESMEN

Before the era of modern medicine and its antibiotics, anesthesia, and antiseptics (which were developed in the late nineteenth and early twentieth centuries), medical practices were crude, primitive, and often more dangerous to the patient than going untreated. Doctors were often called leeches because they used those bloodsuckers to cure their patients, based on the false notion that bloodletting got rid of the humors or evil spirits causing the illness. Frequently, patients died from blood loss instead of recovering.[1] Most of the medical kit contained various types of remedies, elixirs, and patent medicines to foster the impression that the doctor was doing something positive. In most cases, these chemicals were largely harmless mixtures of alcohol and other common substances for flavor (so at least the patient got drunk and did not feel as much pain), but sometimes they contained toxic substances (such as arsenic, iodine, or mercury) that sickened or killed the patient instead.

Prior to the development of modern medicine, the main job of most doctors was to spend time with the patient, listen to their complaints, and give them something to make them think they were going to get better (even if it was a placebo). In cases where the patient had a strong immune system, they might recover with a bit of positive thinking, but more often the patient got worse and died. Either way, the doctor was helpless in doing any real curing. As Voltaire wrote, "Doctors are men who prescribe medicines of which they know little, to cure diseases of which they know less, in human beings of whom they know nothing."[2] Even the prominent physician and writer Oliver Wendell Holmes complained of his profession, "If all the medicine in the world were thrown into the sea, it would be bad for the fish and good for humanity."[3]

None of the readers of this book would be old enough to remember this, but before the 1920s, it was common (especially in the more rural parts of the United States) for a variety of peddlers to travel town to town, selling their various types of miracle cures, remedies, elixirs, and patent medicines.[4] Their victims were mostly people who were desperate for relief from a serious illness and had not been warned of the past victims of these people. The major newspapers and magazines ran small ads for their miracle cures, which were sold by mail order. In the United States, these traveling salesmen would do a heavy-duty sales pitch to crowds all along their routes (often portrayed in movies about life in small towns in the nineteenth century), using all the rapid patter and extravagant claims and other rhetorical techniques of a carnival barker, or the fire and brimstone tactics of a preacher trying to terrify his flock. These salesmen became known as snake-oil peddlers in the United States, because they claimed that many of their products had exotic ingredients with magical curative powers. The general term quack came to be applied to any type of fraudulent or ignorant medical practice. It comes from the Dutch *quacksalver*, or "boaster who applies a salve." In the Middle Ages, "to quack" meant "to shout," as these peddlers did when they hawked their wares.

Finally, the examples of fraud and people dying from impure medicines distributed by unlicensed peddlers with outrageous unsubstantiated claims prompted the reformers to act. In 1880, the Dutch Society against Quackery was founded, and began its crusade against fraudulent and dangerous medicines. In the United States, action against quacks and snake-oil medicines did not take place until pushed by activists of the Progressive Era of the early twentieth century. Muckraking authors such as Upton Sinclair and Samuel Hopkins Adams wrote exposés of the dangers of patent medicines, which raised public awareness. Finally, thanks to political pressure from Theodore Roosevelt, Congress passed the Pure Food and Drug Act on June 30, 1906. This law mandated standards for medicines and drugs, and required accurate labeling of what was in a product, and soon led to similar laws being passed in Britain and elsewhere in the world. In 1938, as part of Franklin D. Roosevelt's New Deal, the Food and Drug Administration (FDA) was created to further regulate the testing and marketing of food and drugs, and insure that foods we buy are safe and free of contaminants. Food and drug producers also recognized the value of keeping their products pure, and regulated themselves through the NBBB (National Better Business Bureau). Today, we take it for granted that food in the supermarket or drugs in the pharmacy are safe and effective, but it would not be so without the hundreds of scientists who constantly monitor them. In fact, an outbreak of bad food or drugs is now so rare that it makes the news when it happens today.

We think of quacks and patent medicines as a thing of the prescientific past, but there are surprising anomalies. For example, Beecham's Pills (which contained only aloes, ginger, and soap, yet claimed to cure thirty-one medical conditions) were first sold in 1842, and were still on sale in 1997 (the company that once sold them became the pharmaceutical giant GlaxoSmithKline).[5] But today the most common forms of medical fraud concern our modern health issues, especially overhyped products that claim to cause easy weight loss, or penis enlargement, or greater sexual stamina. We hear and see their ads every day in a variety of media, and almost everyone's spam folder is filled with dozens of internet messages for various quack medicines. Most of us ignore these ads and know that their claims are exaggerated or false, but still they sell millions of dollars of their products to enough gullible people to rake in the money and enjoy thriving business.

Likewise, we are bombarded with the modern equivalent of pat-ent medicines, the huge market in herbal remedies, organic or natural medicines, and health food products. Millions of dollars in these pills and liquids are sold everywhere today, yet few have met the strict FDA standards of safety and efficacy proven in scientific studies, so they can-not call themselves licensed drugs. When they advertise, they cannot claim to cure certain illnesses or prevent certain medical conditions, because they have not met the FDA standards. Consequently, they use unrestricted weasel words such as "promotes sexual health" or "supports a healthy immune system." Despite their lack of any proven value what-soever, they have a huge market share, and large numbers of people buy them thinking that they somehow feel better after taking them. This is probably due to wishful thinking and the placebo effect—and also the fact that our immune systems will overcome most mild illnesses like warts or the cold or flu sooner or later, no matter what medicines we take.

Given all the miracles of modern medicine discussed in chapter 4, you would think that people would only use what is scientifically proven to work, and not fall back on prescientific quackery. Yet there are still an alarming number of nonscientific practices out there claiming to cure people, and making trillions of dollars from millions of patients. In analyzing all these supposed cures, we will use the practices of *evidence-based medicine*. It is not enough to claim that Patient X was cured by your methods. This is anecdotal evidence, which may be impressive to people who listen to their friends and family (or celebrities on TV), but this evidence does not constitute a rigorous study. To be scientifically valid, there must be a number of careful scientific studies, usually comparing the patients who received the cure with patients who received only a placebo, so we can rule out things like simple positive psychological ef-fects on the immune system.[6] The studies must be controlled so there are no other variables to account for. Ideally, these studies are conducted in a double-blind fashion so that neither the investigators nor the patients know who received the medicine and who received a placebo until the results are in. This prevents the possibility of investigators biasing the results. For the medical community (and informed consumer) to take a medicine seriously, there must be dozens of such studies, with hundreds of patients over an extended period of time, all done by different labs, so that the results can be replicated and cross-checked, and any symptoms or side effects identified. One favorable scientific study is not sufficient. (This is why FDA approval of real drugs often takes years, since many such clinical trials are slow and expensive.)

There are many such questionable medical practices out there; they are constantly monitored by sites such as www.Quackwatch.com, www.Sciencebasedmedicine.com, and others. You can also consult the books listed at the end of this chapter, or the peer-reviewed journal *The Scientific Review of Alternative Medicine and Aberrant Medical Practices*. Space considertations prevent a complete debunking of every fraudulent form of alternative medicine, and the resources just mentioned do a complete job of this anyway. Instead we will focus on just the largest and most clearly bizarre practices out there today.

HOMEOPATHY: THE "WATER CURE" REVISITED

It's a miracle! Take physics and bin it!
Water has memory!
And while its memory of a long lost drop of onion juice is Infinite
It somehow forgets all the poo it's had in it!

Tim Minchin, Storm

Charles Darwin spent most of his adult years wracked by a mysterious debilitating illness that caused him to have nausea, headaches, stomach-aches, indigestion, and vomiting almost every day from 1838 (shortly after he returned from the voyage of the HMS *Beagle* a strong and vigorous man who had climbed the highest mountains in the Andes) until he died in 1882. During that time, Darwin sought one cure after another. One of the strangest was Dr. Gully's Water Cure, which Darwin followed from 1849 to 1866. It involved drinking lots of water, along with being heated by hot lamps until he sweated profusely, then being doused with buckets of cold water, followed by vigorous rubbing with cold wet towels and cold foot baths. The cure also mandated a strict diet and taking long walks.

No one knows for sure what caused Darwin's mysterious illness. Scholars think it was probably stress and anxiety from constant work on his dangerous and controversial ideas, or possibly Chagas disease contracted when he was in South America. Whatever the reason, the water cure apparently got his mind off his stressful work, and gave him a good diet and regular exercise, which would help anyone, sick or not. The plunging between sweating sessions and cold baths probably did not hurt him, although there was apparently no medicinal value, either. Later, Darwin brought his dying ten-year-old daughter Annie to Dr. Gully, but the water cure could do no good for her scarlet fever, and by the 1860s, Charles Darwin was getting no benefits as well and stopped traveling the long distance to Dr. Gully's clinic in Malvern.

Gully's water cure was popular not only with Darwin, but also with many other prominent Victorians, including Archbishop Wilberforce (an early foe of evolution) and Alfred, Lord Tennyson, but by the 1870s it was out of fashion. But another nontraditional cure practiced by Dr. Gully and many others of the time was *homeopathic medicine*. It was first developed by the German doctor Samuel Hahnemann in 1796, and based on the old Greek and medieval Principle of Similars, or the idea that like cures like. For example, Hahnemann observed that cinchona bark helped treat malaria in sick people, and caused similar symptoms in healthy people, so he reversed the logic and reasoned that whatever causes similar symptoms can be used to treat it. (Cinchona bark does contain natural quinine, a cure for malaria.) He argued that if poison ivy causes skin rash, then diluted poison ivy is a cure for skin diseases. But his theory of likes cure likes was pure medieval alchemy and mumbo-jumbo, completely invalidated when modern chemistry developed in the nineteenth and twentieth centuries.

This archaic version of premodern medicine has persisted virtually unchanged until today. The ingredients used in homeopathy sound like those of a witch doctor or medieval apothecary (or something out of Harry Potter's potions class): snake venom, ground honeybee, crushed bedbugs, live eels, wolf milk, arsenic, poison ivy, diseased tissues (including pus, tumors, feces, urinary discharges, blood, and tissues from sick individuals), quartz, gold, oyster shell, and common salt. Their other ingredients, known as imponderables, supposedly capture electromagnetic energy by exposing alcohol or lactose to sunlight, X-rays, or lightning.

What makes homeopathy different from other alternative medicines that use bizarre ingredients is its key method: dilution. All these substances are then ground into fine powder to make a tincture, diluted in water, and then diluted over and over again. The hitting, or shaking, technique used by the homeopath is very important because of the supposed kinetic energy input and its effect on the medicine. (This is just a fancy way of saying that more shaking probably mixes a solution better, but it is given a mystical mumbo-jumbo meaning in homeopathy.) A typical homeopathic remedy has a strength of "30C" in their terminology. This means that the original agent has been diluted thirty times by a factor of one hundred each time. A simple calculation shows that this material has been diluted by a factor of 1×10^{60}, or 1 followed by sixty zeroes (in common terms, one part tincture in one million trillion trillion trillion trillion trillion trillion parts of water). There are only 3×10^{25} molecules of water in a liter, so for all intents and purposes, there are no molecules of the so-called active ingredients left and the *homeopathic remedy is just water and nothing more.* At this concentration, a patient would need to swallow 10^{41} pills (a billion times the mass of the earth) or 10^{34} gallons of elixir (a billion times the volume of the earth) to consume even a *single molecule* of the substance. Even a stronger homeopathic solution of 12C has only a 60% chance that one molecule of the tincture is present. Yet according to homeopathic theory, *the medicine is stronger the more it is diluted.* Most homeopathic remedies have been diluted so much that they are just small bottles of water. In effect, homeopathy has become the modern water cure.

Of course, drinking lots of clean water is good for you, but this is something different. Homeopaths are claiming that a medicine made of nothing but water somehow has acquired magic properties, or a memory of the substance that was once diluted in it, and somehow this water memory transfers the therapeutic value of the original material. Anyone with common sense, and especially anyone who has taken even an introductory high school or college chemistry class, can immediately tell that something is wrong with this theory. First of all, in chemisty lab you learn by repeated experiments that dilution only makes a solution weaker and less reactive. There is absolutely no known instance where a *less* concentrated substance has a *stronger* effect. Second, you learn early in chemistry class that water is a simple molecule—just two atoms of hydrogen and one of oxygen. It does have some remarkable properties, like its high heat capacity and its ability to become less dense after it freezes (which is why ice floats). But water (and even the crystalline form of ice) is still an extremely simple molecular structure, with no potential for memory. There are complex organic chemicals such as DNA, or the complex sheet structure of a clay mineral, which do have the complexity to record changes in their code, and thus have a memory. But water (even in the case of the most complex ice crystals or clathrate structures) has no such ability. The Australian Council against Health Fraud wrote, "Strangely, the water offered as treatment does not remember the bladders it has been stored in, or the other contents of the sewers it may have been in, or the cosmic radiation which has been blasted through it."[7]

But could homeopathic remedies work like a vaccine, where we introduce tiny amounts of some germ to a patient to stimulate immunity? Superficially, they sound like the same thing, but in reality there is no comparison. A typical vaccine has billions of killed viruses or virus fragments in a few milliliters of solution, and these work by immediately stimulating the immune system to kick into high gear and recognize their chemical signal, so that the white blood cells quickly attack when a full-strength virus invades at some later time. By contrast, a homeopathic remedy is pure water, or at best one or two molecules of an active ingredient, and there is no reason to believe these things stimulate the immune system or anything else. The analogy is so ridiculous that the nineteenth-century doctor Oliver Wendell Holmes wrote that this comparison was like "arguing that a pebble may produce a mountain, because an acorn can become a forest."[8]

Reading the basic theory of homeopathy (like cures like, tinctures, dilution increases potency, water has memory, and the long list of peculiar substances used in cures), it is clear that it was based on medieval notions of alchemy (as homeopathy was when it was founded in 1796). It bears no resemblance to what chemists have discovered in the past two hundred years about the nature of matter and chemical compounds and their behavior. In fact, homeopathy would have probably vanished like many other forms of quack medicine in the early nineteenth century, except for the fact that the application of conventional medicine at that time (with leeches, many types of toxic medications, and no sanitation, antiseptics, antibiotics, or anesthesia) was often deadlier to the patient than leaving them alone. Thus, early homeopathic hospitals in England had greater success than conventional hospitals of the time because all they did was give their patients a water placebo and a lot of attention, and did not make the patients sicker. Those who had stronger immune systems survived, while similar individuals died after being treated in conventional hospitals and being bled and infected by doctors who did not wash their hands.

This may explain why homeopathy was popular over 150 years ago, but modern medicine has made enormous advances since then, and has demonstrated real success in curing diseases all over the world. How, then, could an archaic theory like homeopathy based on obsolete and prescientific notions of chemistry, still survive? Amazingly, Singh and Ernst show that it is widespread in many countries worldwide.[9] India has over 300,000 homeopaths, 182 colleges of homeopathy, and 300 homeopathic hospitals. The British Faculty of Homeopathy has over 1,400 doctors on its register. In Belgium, over half the population relies on homeopathic remedies, while in France, the use of homeopathy increased from 16% to 36% between 1982 and 1992. Even in the United States, where the medical profession is much more Westernized than in India, sales and profits of homeopathic products increased from $300 million in 1987 to $1.5 billion in 2000.

The reasons for its continued popularity are complex, but has largely to do with historical accidents and the nature of medicine in each country. India has had a long history of nontraditional folk medicine, so homeopathy fit right in when it was introduced by Transylvanian Dr. Martin Honigberger in 1829. It became popular because conventional Westernized medicine was associated with the hated British colonial masters. Although homeopathy had fallen out of favor in Britain by the early twentieth century, King George VI was a firm believer, and helped get homeopathic hospitals under the coverage of the National Health Service when it was formed. In the United States, senator Royal Copeland (also a homeopathic doctor) persuaded the other senators to include homeopathy in the 1938 Federal Food, Drug, and Cosmetic Act. Homeopathy's more recent revival in the United States, ironically, came in the 1960s and 1970s, when hippies and New Agers began to import not only the philosophies and religions of India, but also some of their unconventional medical practices.

These cultural factors go a long way toward explaining the popularity of homeopathic quackery, but the real question is this: Does it work?

Does it pass the muster of scientific scrutiny? Is it based on evidence of effectiveness, or just anecdote? Of course, the practitioners and patients argue that they have evidence that it works. First, like most alternative medicines, homeopathy relies largely on the testimony of individuals who believe that the treatment worked for them. But as we outlined above, this is anecdotal evidence, and we have no way of controlling other factors (such as natural immune reactions, or the placebo effect) that might also explain why the patient got well.

Underground videos and press visits to a homeopathic hospital show how the process works.[10] The homeopaths all dress in white lab coats and act like real doctors, so they provide the psychological reassurance that the patient is in a legitimate medical institution. The homeopaths typically take a long time in their consultations (thirty to sixty minutes or more), listening to all of their patient's complaints, as the old country doctors did in the eighteenth and nineteenth centuries. These listening therapies are shown to have a beneficial effect on patients whose problems might be psychosomatic, and a caring, attentive doctor's time and attention is what they really need. (Here is one place where homeopathy has a slight advantage over most modern medicine. In today's modern medical system, bottom-line profit motives and HMO pressures force doctors to see as many patients as possible, and reduce interviews to just a few minutes.) Then the doctor gives them a vial of the magic water, which to the true believer, is an effective medicine, even though it is actually plain water and is only a placebo. Nevertheless, a lot of patients get better by taking placebos simply thanks to belief and the comfort that a medicine provides, and their immune systems kick in and spur recovery without any real drugs involved. Studies show that people who do not know they are receiving a placebo but think they have the real medicine overestimate its effectiveness by 17%, and if the experimenters do not randomize their samples and patients but administer them in a biased way, they can overestimate the efficacy by 30–40%.[11] These numbers alone, plus the psychological benefits of listening therapies, explain most of the anecdotal success stories of homeopathy.

But what about rigorous clinical studies in which many patients are involved—which all *real* medicines are required to pass in order to meet FDA approval? For years, nearly all the evidence favoring the efficacy of homeopathy were individual anecdotal cases, or examples like the fact that early-nineteenth-century homeopathic hospitals were less deadly than conventional hospitals (as we pointed out already, only because conventional medicine of that time was worse than no treatment at all). Every other study that had favored homeopathy suffered from one deficiency or another: insufficient sample size, bias by the investigator, improper statistical techniques, inability to replicate the results in any other lab, and many other flaws.

Then in 1988, Jacques Benveniste, a former French racing driver who had taken up medicine after suffering a back injury, submitted a paper to *Nature* that appeared to be a rigorous study supporting homeopathy.[12] *Nature* editor John Maddox, was skeptical, but after it passed peer review because there were no obvious flaws in the reported methods, he consented to publish it—but with a disclaimer that *Nature* would send its own panel to investigate Benveniste's lab and see if the results were credible. The team consisted of Maddox himself (a trained chemist and physicist), chemist Walter Stewart, and magician, skeptic, and mythbuster James "The Amazing" Randi. They watched the laboratory closely as they repeated the experiment, and at first the lab kept getting results consistent with their original report. But then the panel realized that the investigators might be biased in their sampling, so they set up a procedure where the investigators could not know which test tube was which. While the results were being analyzed, Randi performed some sleight-of-hand magic and card tricks to amuse the people in the lab while they waited. Sure enough, when the new results were reported and the panel revealed the actual identities of each test tube, there was no positive effect of the homeopathic medicine.[13] The results were no different from random chance reactions.

After *Nature* reported the debunking by the panel, and the probable reasons for the experimenter bias, they discovered that Benveniste had been partially funded by a French homeopathic company, so there was a clear conflict of interest. Benveniste was not an impartial scientist, but stood to gain if he found results that improved the profits of his sponsor. Since Benveniste's study, three different independent labs failed to replicate his results, and Benveniste lost all credibility in the scientific community. Subsequently, he left the conventional medical community and went private. With big funding from homeopathic medicine producers, he set up his own private lab without scientific oversight where he continued to make amazing pronouncements about his miraculous discoveries. Fancying himself a scientific martyr à la Galileo, he felt he deserved a Nobel Prize, but instead he has twice (in 1991 and 1998) won the satirical Ig Nobel Prize for the worst pseudoscientist of the year.[14]

The claims of homeopathy have been tested over and over again, and none has passed muster.[15] In 2005, the premier British medical journal *Lancet* published an analysis of 220 studies about homeopathy, half of them conventional studies of medicines, and half of them of controlled homeopathic experiments.[16] They found no evidence that homeopathy had any real value except as a placebo. In 2006, the *European Journal of Cancer* surveyed six studies, and found homeopathy had no effect.[17] Even studies by homeopaths themselves often show that their products have no significant effect, despite their biases to prove otherwise.[18] As Singh and Ernst document, the trials continue, but not once has any study shown a statistically meaningful evidence that homeopathy works.[19]

In January 2010, a group of British skeptics decided to stage an event to show the uselessness of homeopathic remedies. They planned a homeopathic overdose day on Jan. 30, where they would take hundreds of times the recommended dose of homeopathic remedies to commit suicide.[20] If homeopathic remedies were real drugs, such overdoses would indeed have made these people sick, or killed them. Nothing adverse happened to them—except that some of them had to go to the bathroom more often from consuming so much water. Their protest was an effort to expose the fraudulent nature of homeopathic remedies sold in British drugstores, to the tune of £12 million, between 2005 and 2008. Since the original stunt, groups in America and Australia have staged similar events to publicize the worthlessness of homeopathy.

With all this evidence, it seems clear that homeopathy is just a well-established form of quackery, and patients would be well advised to stay away from it, especially since modern medicine is proven to be truly effective. But it is taking a long time for the nations of the world to realize this and avoid supporting it, especially because millions of people will believe in it, no matter what.[21] The national medical systems of India, Mexico, France, the United Kingdom, Denmark, and Luxembourg still reimburse homeopathic treatments, but those of Germany, Belgium, Austria, and Switzerland no longer do so.

In the United Kingdom, the pressure to stop wasting taxpayer money on homeopathy has increased. The number of homeopathic remedies prescribed by doctors declined 40% between 2005 and 2006, and homeopathy accounted for a miniscule 0.006% of the prescription budget.[22] One homeopathic hospital has closed due to a lack of referrals and a bad review by a medical association. In British medical schools, courses on homeopathy are vanishing because they rightly suggest that these schools are teaching pseudoscience.[23]

In February 2010, the UK House of Commons Science and Technology Center concluded that

> the NHS should cease funding homeopathy. It also concludes that the Medicines and Healthcare products Regulatory Agency (MHRA) should not allow homeopathic product labels to make medical claims without evidence of efficacy. As they are not medicines, homeopathic products should no longer be licensed by the MHRA.
>
> The Committee concurred with the Government that the evidence base shows that homeopathy is not efficacious (that is, it does not work beyond the placebo effect) and that explanations for why homeopathy would work are scientifically implausible.
>
> The Committee concluded—given that the existing scientific literature showed no good evidence of efficacy—that further clinical trials of homeopathy could not be justified.
>
> In the Committee's view, homeopathy is a placebo treatment and the Government should have a policy on prescribing placebos. The Government is reluctant to address the appropriateness and ethics of prescribing placebos to patients, which usually relies on some degree of patient deception. Prescribing of placebos is not consistent with informed patient choice—which the Government claims is very important-as it means patients do not have all the information needed to make choice meaningful.
>
> Beyond ethical issues and the integrity of the doctor-patient relationship, prescribing pure placebos is bad medicine. Their effect is unreliable and unpredictable and cannot form the sole basis of any treatment on the NHS.[24]

Unfortunately, this strong recommendation by the British Labour government was overturned by the elections of July 2010, when the Tories and Liberal Democrats took over in a coalition government. The new secretary of state for health decided to defer to the National Health Service (NHS) on what to do about homeopathy. Since homeopathy is already well entrenched in the NHS, the response was unsurprising: leave it alone. In their report, the authors write,

[O]ur continued position on the use of homeopathy within the NHS is that the local NHS and clinicians, rather than Whitehall, are best placed to make decisions on what treatment is appropriate for their patients—including complementary or alternative treatments such as homeopathy—and provide accordingly for those treatments.[25]

No one knows how events will play out in the United Kingdom and other countries, but efforts to deny public funds to these quacks are still in the early stages.

In Canada, homeopathy is allowed but regulated as a private health technique; it is up to the individual provinces as to whether to support it with public funds. So far, none have done so.

In the United States, of course, there is no national medical system fully in place yet, but homeopathy is not covered by most private medical insurance providers. About $3.1 billion were spent on homeopathic medicines in 2007, and about 2% of the United States population seeks homeopathic treatment each year.[26] Homeopathic remedies are still regulated as drugs for purity by the FDA, although the FDA does not endorse their medicinal qualities. However, the FDA considers most of homeopathic drugs harmless, because they are so diluted that they are just water. (The FDA is not empowered to tell consumers whether the drugs are worthless or a waste of money, just whether they are safe.)

Slowly, the quackery of homeopathy is retreating from the position it once held among nontraditional medicines. Still, we are a long way from saving people from this worthless form of medicine that gives them false hope and costs an enormous amount of money.

BACK-CRACKING: THE CHIROPRACTIC CON GAME

Injured by a chiropractor? Call 860-529-8826

Bus ad for Chiropractic Stroke Victims Awareness Group

When I was young, my mother was a regular patient of a particular chiropractor. He was a handsome, friendly man with a long handlebar mustache, a gentle manner, and very strong hands and arms. She got me to try a few treatments when I was a teenager, and since I had real back problems (scoliosis, or curvature of the spine), I felt better after he had "adjusted" me by pushing hard on my back until the muscle tension relaxed, and "cracking" my neck. I figured that chiropractic was simply an elaborate, more physical form of massage, and never looked it up or thought about it further. My parents also sent me to real doctors (orthopedic specialists and physical therapists) who diagnosed my scoliosis correctly. They had me doing a hard regime of exercise (pull-ups and sit-ups, mostly) that eventually straightened my spine and strengthened my back muscles, so my spine is normal now and I no longer suffer from back pain.

Chiropractic is immensely popular in many countries of the world as a form of healing, and even some conventional doctors refer patients to chiropractors. According to Singh and Ernst, roughly $3 billion a year is spent on chiropractic treatment in the United States.[27] Between 1970 and 1990, their numbers tripled and in 2002 there were about 60,000 chiropractors practicing in North America. This figure doubled by 2010, while the number of MDs increased by only 16%. Chiropractors are licensed in all fifty states, and a few insurance plans cover their services. As this chapter was being written, Kaiser Permanente (one of the largest HMOs in the United States) rejected chiropractic as a valid form of medicine, and refused to provide any reimbursements.[28] In the United Kingdom, they are regulated as are true doctors and nurses, and they are equally well established in the health care system.

Thus it came as a real surprise to me (as it does to most Americans) that chiropractic is not just a form of massage or vigorous physical therapy recognized as safe and effective by the AMA. Instead, the AMA regards it as a bizarre cultish form of medical quackery whose belief system—based on dogmas that have been entirely debunked by the last century of medical research—and roots go back to the days before modern medicine.[29]

Chiropractic was founded in the 1890s by a well-known faith healer and bone-setter quack, D. D. Palmer, and carried on and expanded by his son, B. J. Palmer. Most of the Palmers' ideas date back to a point just before the development of modern neurobiology allowed us to understand the spine and nerves and muscles that surround it. D. D. Palmer was a notorious believer in weird therapies of the period, including magnetic healing and bone-setting, and his ideas competed with homeopathy at that time. According to chiropractic theory of Palmer, the spine is the source of the vitalistic nerve energy or life force he called "innate intelligence" (based on medieval notions discredited long before 1890) representing God's presence in man. Early chiropractic was full of religious imagery and faith healing. In the early days as they were getting their methods accepted, the Palmers actually considered filing for protection as a form of religion, as did Mary Baker Eddy's Christian Science movement. This mystical life force supposedly flows from our spine to the rest of the body, including organs like the heart, lungs and liver (which in reality have only limited connection to the spine through nerves) and affect our health. According to chiropractic theory, when the spine is misaligned (a subluxation, in their terminology), it can cause illness, so it needs to be manipulated and straightened so these life forces can flow properly and the patient gets well. Originally, chiropractic also denied that bacteria and viruses cause many diseases, and argued that every ailment could be cured by spinal manipulation. Most modern chiropractors no longer deny the germ theory of disease, but there are some who are vaccination deniers (see chapter 7). Despite all the advances of modern medicine and biology, there are some who still hold to many of these archaic false notions of the Palmers.

Boiled down to this bare minimum, the theory behind chiropractic sounds truly bizarre and antique, like something out of the Middle Ages. Even to the layperson familiar with basic medicine, it should be obvious that there are no mysterious life forces running through the body, let alone that manipulating the spine would not cure illnesses of the heart or lungs or digestive tract.[30] About the only way in which a spinal realignment could conceivably help people is if there are pinched nerves, damaged disks, or other problems in the spine. Treating these problems is a tricky and sometimes dangerous business, and chiropractors have been sued for permanently damaging the backs of people with nerve damage, fused vertebrae, slipped disks, and so on.[31] It is difficult even for orthopedists, who are real doctors trained in the latest research in spinal problems, to solve every back problem, so it is absurd to think that quacks using century-old, outdated theory could do much better.

If you have never been to a chiropractor (please, don't waste your time and money if you haven't), it typically starts out just like a conventional medical appointment (waiting room, receptionist, annoying easy-listening music, doctor in white lab coat, fill out the questionnaire about your health issues, X-rays to see your bones). As practiced by the chiropractor I saw as a teenager, when you are called in, it seems to be a conventional massage: you take off your clothes in private, and don some combination of gowns or towels to preserve modesty while exposing your back. You lie face down on a special table with multiple hinges that allow any part of it be elevated or dropped, with heat lamps to warm and relax your back and relaxing music (again) in a quiet, dimly lit room to minimize distractions. Sometimes the chiropractor applies one of those salves that deep-heat your back muscles while you lie on your belly. Sometimes they used one of those vigorous plug-in vibrating massagers to shake and pound your muscles. When you are suitably warmed up, the chiropractor then begins a conventional massage, working hard to relax your back muscles and loosen everything up. Eventually, he tries to crack your neck (by gently rocking your head back and forth to relax the neck muscles, then giving a sudden twist that definitely stretches the muscles around your neck vertebrae) and adjust your back (by gently twisting your torso on a special table until the back muscles are looser, then giving a sudden sharp twist to really stretch the muscles that hold your vertebral column in line). It actually feels pretty good if the chiropractor is doing it right and is experienced, but there is a big difference between stretching muscles and bones (which is the job of properly trained physical therapists and orthopedists, who employ modern medical theory to do the same thing) or relaxing you with massage (which is relatively harmless) and claiming to solve all problems of the body by cracking your neck and spine.

This was my experience with a modern chiropractor, but there are many kinds of chiropractors out there. The so-called straights are those who adhere to the original vitalistic Palmer theories of innate intelligence. But even as the Palmers were still alive and practicing, some of their followers began to use the methods of real medicine: X-rays, massage, exercise, good diet, nutritional supplements, and other developments that *do* have some proven medical value. B. J. Palmer scornfully called these methods "mixed," but they appear to be common in most modern chiropractors' offices. Thus, it is not surprising that the average patient, after having spinal X-rays taken, thinks that chiropractic is just a form of normal modern medicine.

Leaving aside the weird vitalistic theory of the straight followers of the Palmers, the crucial question in everyone's mind is this: Does it really work? Once again, we have to put aside any anecdote-based arguments from individual patients who claimed they got better due to chiropractic. Heck, after the vigorous backrubs and stretches from my mom's chiropractor that I experienced when I was a teenager, I could not help but feel great afterwards. The real question is whether chiropractic methods have a success rate greater than just a placebo effect of a good massage, and whether they can really cure anything besides spinal problems. On that regard, the jury is in: manipulating the spine has little or no effect on the health of other systems such as the heart, liver, kidneys, and other internal organs.[32] In 2006, Ernst and Canter summarized ten systematic reviews of seventy experimental trials claiming that spinal manipulation helped with headaches, infantile colic, asthma, allergies, and menstrual pain.[33] Their conclusion was this: chiropractic could alleviate none of these symptoms. The only possible beneficial effect in this regard is the relaxation of tension that it might provide, and (if the chiropractor is a mixed practitioner advocating good diet and exercise) possibly the indirect benefit of the same advice that any MD would give.

But what about back and neck pain, whiplash, and other possible medical benefits that logically chiropractic might help? What does our insistence on evidence-based medicine say about chiropractic? Here again, the evidence is pretty clear: many different controlled studies have been done comparing chiropractic with conventional medicine, and so far there are no demonstrable beneficial effects of chiropractic compared to controls or placebos.[34] Stephen Barrett, a psychiatrist who has conducted much research into chiropractic fraud, once ran a simple experiment to test for the shady practices of chiropractic therapy. He took the same healthy twenty-nine-year old woman to four different chiropractors to see how their diagnoses differed. In his words,

> The first diagnosed "atlas subluxation" and predicted "paralysis" in fifteen years if this problem was not treated. The second found many vertebrae "out of alignment" and one hip "higher" than the other. The third said that the woman's neck was "tight." The fourth said that misaligned vertebrae indicated the presence of "stomach problems." All four recommended spinal adjustment on a regular basis, beginning with a frequency of twice a week. Three gave adjustments without warning—one of which was so forceful that it produced dizziness and a headache that lasted several hours.[35]

If a group of four MDs gave such radically different diagnoses to the same healthy patient, and demanded that you come in for lots of expensive treatment, would you trust any of them?

Given this evidence, it seems clear that chiropractic is just another form of quackery, albeit one with a much better image and stronger lobbying and political influence than the others. The AMA tried more than once to have chiropractic officially treated as quackery, and prevent it from receiving the same support and access to health plans that real evidence-based medicine has. For many years, chiropractors were routinely jailed for practicing medicine without a license, but the deep pockets of their trade association always paid their legal bills. Although the AMA's battle with the chiropractic profession failed to force them out of business (mostly because the courts saw it as an issue of "restraint of trade" and really did not consider the scientific evidence for the failure of chiropractic), the exposure of the reality of chiropractic was that the medical profession now only rarely takes chiropractic seriously, or refers patients to chiropractors.[36] In addition, the exposure of the quack beliefs of the straights means that most chiropractors have tried to incorporate more mixed traditional methods to get away from their dodgy past.

So the two professions have reached an uneasy truce for now. But what is the consumer to do? If the patient wants to follow practices that are based on solid scientific evidence, there is no question that it is a waste of time and money to go to a chiropractor. More importantly, one could be taking unnecessary medical risks from such a visit. Singh and Ernst (2008) document a long list of reasons which chiropractic can be risky or dangerous, leading to conditions that modern medicine could treat but a chiropractor cannot. These include damage to the spine or neck, internal bleeding, exposure to unnecessary X-rays, and especially death from stroke. Chiropractic manipulation is especially problematic when it comes to manipulating the soft and still developing bones of young children, as well as those of older (mostly female) patients whose bones are fragile due to osteoporosis. There are numerous cases of death and permanent disability from inappropriate actions by chiropractors, when a simple visit to a conventional doctor could have cured the patients.[37] Risk-benefit analyses all show that the risks of chiropractic treatment far outweigh any possible benefits for patients.[38]

Consequently, as the word gets out, the popularity of chiropractors is gradually declining in the United States. In a 2006 Gallup poll, chiropractors ranked dead last in reputation for honesty and ethics, with only a 36% favorability rating, whereas most other professions had higher ratings (from 62% for dentists to 84% for nurses).[39] Since most health care plans do not recognize chiropractic as a legitimate treatment, it has become harder for them to get patients with HMO coverage. Consequently, in the United States chiropractic use declined from 9.9% of adults to only 7.1% between 1997 and 2002, the largest decline among the alternative medicine providers of homeopathy, naturopathy, acupuncture, and the like.[40]

Chiropractic also has a nasty, litigious side as well. The muckraking journalist, writer, and filmmaker Simon Singh found this out when the British Chiropractic Association sued him for libel in 2008 for writing a column in the *Guardian* that exposed their weird unscientific beliefs and dangerous practices.[41] Libel laws in the United Kingdom are very different than they are in the United States. In the United States legal system, it is not possible to win a libel case if the person writing the exposé is telling the truth. English libel laws, on the other hand, allow any plaintiff to sue when they feel they have been damaged, even if the accusations are true. The British chiropractors' attempt to intimidate and silence Singh backfired, however, when a huge backlash against the chiropractors forced them to drop the lawsuit (luckily, the newspaper paid his legal costs so Singh did not have to spend untold pounds on his defense). In the backlash, formal complaints of false advertising were filed against five hundred chiropractors within twenty-four hours, and 25% of the British chiropractors found themselves under investigation.[42] They were advised to remove claims that they could cure whiplash or colic from their leaflets, to be wary of telephone calls and new patients, and to take their websites down immediately for fear of exposure. The damage to their field appears to be severe and long lasting, and the attempt by British chiropractors to operate under the radar as a semi-legitimate form of modern medicine has been exposed for good. One chiropractor was quoted as saying, "Suing Simon was worse than any Streisand effect and chiropractors know it and can do nothing about it."[43] (The "Streisand effect" is a common shorthand based on Barbra Streisand's 2003 attempt to suppress photos of her; the attempt got her far more, and far worse, publicity than she wanted.) One can only hope that homeopathy and other quack practices make similar publicity blunders, so that they, too, are exposed for what they are.

WHERE IS THE EVIDENCE?

First, do no harm.

Hippocrates

There is not enough space in this short chapter to discuss every other form of quack medicine out there, of which there are dozens. Singh and Ernst (2008) provide an excellent review of the scientific evidence of each of them, and more can be found in the resources listed at the end of the chapter.

There is no question that individuals believe they have been cured by homeopathy, chiropractic, acupuncture, naturopathy, and many other alternative medicines that are promoted all over the world. As we saw earlier, however, these are anecdotal accounts, and are not legitimate when we want to evaluate whether a given treatment is truly effective based on scientific and statistical analysis. Many versions of alternative medicine are happy to escape rigorous evaluation and scientific scrutiny, in which case they are just folk nostrums that you can forget. Unfortunately, many of them take on the trappings of medicine and science, but cannot (or will not) subject their treatments to rigorous testing. If a form of alternative medicine wants to be taken seriously as evidence based, it must be subjected to the rigorous scrutiny of the scientific method, especially in ruling out placebo effects and other instances of individuals recovering on their own.

In addition to the issue of whether alternative medicines have any demonstrable scientifically evaluated benefits is the even larger issue of their costs and their dangers. Every dollar spent on quack medicine that does not cure a patient's complaint is a dollar wasted, when it could have been spent getting legitimate medical care. And the staggering amounts of money spent on alternative medicines in the United States alone shows that a whole lot of Americans are getting swindled buying useless snake oil.

The second danger is the restatement of fundamental phrase of the Hippocratic oath: First, do no harm. As we saw with homeopathy and chiropractic, the greatest risk of all is that alternative medicine providers (especially unlicensed ones with no medical training to recognize danger signs in a patient) can harm or kill the patient because they do just the wrong things to alleviate the patient's complaints. By contrast, if the same patient had only gone to a real doctor, the chances are much better that they would have had a favorable outcome.

Thus, our warning from the beginning of the book applies here: *Caveat emptor.* Before spending money on questionable treatments, *do your own reading and research,* and examine whether the alternative medicine provider has any real scientific basis for their claims. *Do not* rely on their own literature, but search the internet for studies that are unbiased and that critically evaluate their claims. These are easy enough to find; I found them when I researched this chapter. Remember that personal testimonials are just anecdotal evidence, and that only rigorous controlled studies of hundreds of patients over time are really good criteria to decide if an alternative medicine is worth considering. This is not just an abstract issue about science and pseudoscience; this is your own life and health at risk when you gamble on treatments that do not pass scientific muster, and could harm you more than help you. It is your choice: be smart about it!

FOR FURTHER READING

Barrett, S., and W. T. Jarvis. 1993. *The Health Robbers: A Close Look at Quackery in America*. Buffalo, N.Y.: Prometheus.

Bausell, R. B. 2008. *Snake Oil Science: The Truth about Complementary and Alternative Medicine*. New York: Oxford University Press.

Benedetti, P., and W. McPhail. 2002. *Spin Doctors: The Chiropractic Industry under Examination*. Toronto and Tonawanda, N.Y.: Dundern.

Jonas, W. B., and J. Levin, eds. 1999. *Essentials of Complementary and Alternative Medicine*. Philadelphia: Lippincott, Williams and Wilkins.

Maddox, J., J. Randi, and W. W. Stewart. 1988. High-Dilution Experiment a Delusion. *Nature* 334: 287–291.

Singh, S., and E. Ernst. 2008. *Trick or Treatment: The Undeniable Facts about Alternative Medicine*. New York: W. W. Norton.

Shapiro, R. 2009. *Suckers: How Alternative Medicine Makes Fools of Us All*. New York: Random House.

Shelton, J. W. 2004. *Homeopathy: How It Really Works*. Buffalo, N.Y.: Prometheus.

10

What's Your Sign?
The Ancient Pseudoscience
of Astrology

What utter madness in these astrologers, in considering the
effect of the vast, slow movements and change in the heavens,
to assume that wind and rain have no effect at birth!

Marcus Tullius Cicero

THE SILLIEST SUPERSTITION OF ALL

Among the weird examples of pseudoscience that are still widely followed by modern, educated rational people, perhaps the silliest of all is astrology. Its concepts date back to some of the earliest recorded history. The Babylonians were the first known culture to practice a form of astrology over five thousand years ago. Various forms of astrology were followed by nearly every ancient culture, including the Egyptians, Chinese, the Hindus, Muslims, the Persians, the Mayans, the Greeks, the Romans, and many others. During the Middle Ages, astrology fell out of favor in the Christian world due to its pagan implications, but was maintained in the Arab and Muslim world and eventually reintroduced to Europe during the Crusades. Even founders of science and reason—including Galen, Copernicus, Paracelsus, Tycho Brahe, Galileo, Kepler, and Newton—followed it, as did most people until the rise of modern astronomy and of critical thinking in the nineteenth and twentieth centuries. Astronomy (a true science with testable data and results) arose largely from the astrological tradition of studying the sky and the apparent motion of celestial objects, but diverged in the eighteenth and nineteenth centuries as the scientific conclusions of astronomers began to undermine the myths of astrologers. In 1698, the great Dutch scientist Christiaan Huygens wrote, "and as for the Judicial Astrology, that pretends to foretell what is to come, it is such a ridiculous, and oftentimes mischievous Folly, that I do not think it fit to be so much as named."[1] After the critiques of astrology and its lack of scientific testability and a plausible physical mechanism, scientists and philosophers have considered it a pseudoscience for more than a century.

Yet this scientific rejection of astrology has had relatively little impact on the general public, thanks to the general scientific illiteracy of the American people. Despite the increased level of overall level of education in the past century, astrology experienced a new surge in popularity in the early twentieth century, when newspapers began to run daily horoscopes. A high percentage of people *do not even know the difference between astrology and astronomy,* something that drives real scientists (astronomers) crazy. Various polls estimate that about 25–30% of Americans, Canadians, and British believe in astrology, or at least read their daily horoscopes in the newspaper or online.[2] There are roughly ten to twenty thousand astrologers practicing in the United States alone.[3] Popular astrologers such as Sydney Omarr, Jeane Dixon, and Joan Quigley wrote (and others still write) daily columns in the newspapers for decades, and sold (and still sell) hundreds of thousands of copies of books on astrology. A surprising number of powerful people—including modern world leaders, as well as businessmen, athletes, and entertainers—make crucial decisions based on the junk that astrologers tell them.

It is still common to go to a social event and be asked (sometimes as a pickup line), "What's your sign?" The stranger then instantly makes snap judgments about you based on phony generalities about the nature of each astrological sign, and often ends up accepting or rejecting you based on this fantasy rather than who you really are. (It is bad enough that people make snap judgments of you based on other irrational and unfair criteria, but astrology should not be one of them.) Astrologers have their own cable T v shows, websites, and a presence in nearly every medium. They have long been common guests on talk shows like the old *Merv Griffin Show, The Tonight Show, Larry King Live,* or *Oprah,* as well as in the conventional news media.[4] With this kind of pervasiveness and acceptance across the entire culture, it is not surprising that most people have not heard why astrology is bunk, or that they consider it a legitimate form of understanding life and the future.

SO WHAT IS WRONG WITH ASTROLOGY?

As Neil deGrasse Tyson, author and director of the Hayden Planetarium, has observed, astrology was discredited six hundred years ago with the birth of modern science: "To teach it as though you are contributing to the fundamental knowledge of an informed electorate is astonishing in this, the 21st century." Education should be about knowing how to think, he notes, "[a]nd part of knowing how to think is knowing how the laws of nature shape the world around us. Without that knowledge, without that capacity to think, you can easily become a victim of people who seek to take advantage of you."[5]

My own experience is probably typical for many people. I vividly remember in eighth grade expressing my doubt about horoscopes to a boy who was deeply into astrology. He challenged me to let him do a detailed horoscope and prove me wrong. I gave him all the detailed information of the exact place, time, and date of my birth, and he came back the next day with a horoscope. As is typically the case, it was the usual mix of incorrect or inaccurate descriptions of my personality, with only a few minor hits that seemed close to reality.

This is how all horoscopes work. For the general audience of a newspaper or the internet, horoscopes are deliberately very vague, so that they can fit just about anyone who reads them. Astrologers know that the credulous mind will interpret whatever happens in life to fit the horoscope, no matter how off base it is. If you have someone cast your personal horoscope, it may be a bit more specific in its predictions, but one thing is still true: it will be wrong more often than it is right. All horoscopes work on the same principle that psychics and fortune-tellers use to trick people (known as cold reading): tell them lots of generalities that apply to most people, and their minds will do the rest. As psychologists have long shown, we are prone to notice the handful of times that the random guesses of a psychic or horoscope get it right, and ignore or undercount the number of times they were wrong. This habit of noticing the hits and ignoring the misses is a familiar psychological effect known as *confirmation bias.* Our minds tend to notice things that favor or support what we already believe, and discount or ignore observations that do not fit our preconceptions. If you are doing the fortune-telling or psychic reading in the presence of the cold reader, they will use not only your verbal answers, but your body language to tell if they are getting close to your own experience, even if you do not say anything. And the cold reader often says things that prompt you to respond, and then lets you tell him or her stuff so they can further fine-tune the reading.

Another relevant psychological phenomenon, known as the *Forer effect,* is the tendency of people to rate as highly accurate any personality profile that seems to describe them, when in fact the profile is so general that it could apply to a great many people. Astronomer Phil Plait puts it this way:

> I have talked to many people who claim their horoscopes are accurate. These people routinely say that it predicted something that came true. But there are several possible logical missteps here! First, was the prediction really that accurate? Did it say something like "you will come into money today" and you found a quarter on the ground? Or was it something specific, like "you will find a quarter on the ground"? The difference is that a specific prediction is rarely right, while a vague one is rarely wrong.
>
> Second, was that horoscope right in everything it said? Did an old friend contact you? Were you able to resolve a thorny issue today? Did you really find love today? In other words, how many predictions were accurate, and how many were not? People tend to remember the hits and forget the misses.[6]

There are many ways to show the problems with astrology by looking at individual horoscopes and counting how often they got something right by random guessing, and how most of the predictions are misses. However, magician and skeptic James "The Amazing" Randi provides a very simple demonstration of this effect (available as a short YouTube video).[7] Take any group of people (Randi uses a classroom of teenagers), give each of them a sealed horoscope, and tell them that the horoscopes were personally prepared for each of them, using the exact times of their births. Each person then reads the horoscope, and the group leader then polls them as to how close to correct each individual horoscope is on a scale of 1 to 5 (1 being least accurate; 5 being most accurate). No matter how many times you run the same experiment, most of the subjects will agree that the horoscope was a very accurate description of (ranking it 4 or 5). Then have them pass their personal horoscopes to the next person to read. After a few seconds, they all break out in nervous laughter, because *all of the horoscopes were identical!* Our confirmation bias is so strong that each of us will read into *any* horoscope (especially one that is suitably generalized and vague) exactly how it fits into our lives and futures, and will ignore the parts that do not fit us.

If you think about it, the entire idea that there are unique predictions about you based on the date and time of your birth is obvious nonsense. Why should it be your moment of birth that matters so much? Many babies are born prematurely or late, and some are born ahead of their natural gestation period by medical intervention. If a surgeon removes a baby a week early by C-section, does that change the baby's fate? The time of birth is arbitrary and influenced by factors such as the baby's and mother's health. If astrology made predictions based on the moment of your conception (a truly unique event), it might be slightly more plausible. But there is a good reason they do not use it for predictions: few people know the exact moment they were conceived, whereas the date and time of birth is usually well recorded.

On average, any horoscope should be applicable to about 500 million people born at the same time of the year. Simple common sense tells us that these 500 million people are not identical in most aspects (especially in different cultures from our own). Many studies of these so-called time twins (two individuals born within minutes of each other) show conclusively that there are no real statistically significant similarities, no matter who casts the horoscope. As Dean and Kelly said in the abstract to their study of this phenomenon,

> Many tests of astrologers have been made since the 1950s but only recently has a coherent review been possible. A large-scale test of persons born less than five minutes apart found no hint of the similarities predicted by astrology. Meta-analysis of more than forty controlled studies suggests that astrologers are unable to perform significantly better than chance even on the more basic tasks such as predicting extraversion [sociability]. More specifically, astrologers who claim to use psychic ability perform no better than those who do not.[8]

A larger study in 2006 used the personality profiles of fifteen thousand individuals, crunched them all in a huge computer database, and attempted to see if there was any correlation based on their birthdate and time and the position of the planets. The study found no correlation beyond that which would be expected by random chance.[9]

So what about the big predictions that make the news? These are the examples when we hear that an astrologer or psychic predicted a natural disaster, or predicted that a world leader should not take what became a disastrous course of action. Before you believe the account in the newspaper or on the internet, do some detective work. More often than not, the prediction was never recorded at the time it was made in a way that could be confirmed by a neutral third party, so we have it only on the astrologer's word that he or she "predicted" the event before it happened—and there is a good chance he or she is lying to get some publicity and fame. Also, check the wording of the prediction carefully. If it is vague, as are most astrological forecasts, then it proves nothing about the astrologer's ability to divine the future.

This was the strategy of many of the oracles in ancient Greece. They used vague predictions that could be read several ways, and when a general or king followed one course of action to disaster, the oracles claimed that he had misinterpreted them. Even if a modern astrologer's prediction was highly specific and reliably recorded by some third party not in on the con game, you should check the entire history of this astrologer's predictions. If they were recorded in any detail, it will become apparent that this prediction was just a lucky guess following of hundreds of incorrect predictions; the pattern is the same for most psychics and fortune-tellers. Once again, the apparent success of the prediction is just an example of confirmation bias, although here the media are reporting one lucky hit and shirking their journalistic duties by failing to do the digging required to report all the misses.

A rigorous way to evaluate this issue is to conduct a double-blind test of the claims of astrologers and show whether they match reality with better than a random probability. A 1985 study by Shawn Carlson, a researcher in the University of California, Berkeley, astronomy program, tested this in detail.[10] He sent natal charts of real people to thirty prominent American and European astrologers considered to be among the best in the field, but the astrologers knew only the exact time and date of birth. They had no face-to-face contact with the person whose chart they were reading, so they could not watch the body language cues of the mark to refine their hits and misses as psychics and astrologers do when they tell your fortune in your presence. The astrologers were given three personality profiles that might match the natal chart, one from the client and two others chosen at random, and had to decide which profile matched the chart they were reading. The experiment was double-blind, so that neither the astrologers nor the experimenters knew which profile was which until the results were all in. Sure enough, the astrologers were able to select the right profile only one in three times, exactly the probability to be expected if they were randomly guessing. The astrologers had claimed they could do better than one in two, and some who were most confident of their matches were the most wrong. As Carlson wrote in the published study,

> We are now in a position to argue a surprisingly strong case against natal astrology as practiced by reputable astrologers. Great pains were taken to insure that the experiment was unbiased and to make sure that astrology was given every reasonable chance to succeed. It failed. Despite the fact that we worked with some of the best astrologers in the country, recommended by the advising astrologers for their expertise in astrology and in their ability to use the CPI, despite the fact that every reasonable suggestion made by advising astrologers was worked into the experiment, despite the fact that the astrologers approved the design and predicted 50% as the "minimum" effect they would expect to see, astrology failed to perform at a level better than chance. Tested using double-blind methods, the astrologers' predictions proved wrong. Their predicted connection between the positions of the planets and other astronomical objects at the time of birth and the personalities of test subjects did not exist. The experiment clearly refutes the astrological hypothesis.[11]

In addition to these tests, every other rigorous examination of astrology has shown that it is incapable of making predictions any better than would be expected by chance. In short, astrology is bunk, baloney, bullshit. When the horoscope is cast in the presence of the victim, however, the astrologer (like a typical psychic) uses nonverbal body language cues as well as the victim's own encouraging remarks to improve the cold reading. With this feedback, the astrologer can make the predictions sound more accurate. Even in these cases, however, if you listened closely and counted the hits and misses, you would find plenty of mistakes (as any cold reader would make). Only because humans are preprogrammed to notice the hits and ignore the misses does the method seem to work.

DO THE STARS AND PLANETS HAVE
INFLUENCE OVER US?

The fault, dear Brutus, is not in our stars, but in ourselves.

William Shakespeare, Julius Caesar, *1.2.135*

When various ancient cultures first dreamed up astrology over five thousand years ago, they had a very different view of the cosmos than we have today (fig. 10.1). The stars were fixed balls of fire that represented divine beings, or objects of the gods' intent, and their position in the

FIGURE 10.1. Typical woodcut illustrating the premodern view of the cosmos. The flat earth was covered by a dome on which were located the fixed stars. All the planets plus the sun and moon were thought to move on huge wheels or gears outside the dome of the stars. *From a nineteenth-century illustration by Flammarion.*

sky was a divinely placed message to us. These stars were thought to be very close to the surface of a flat earth, and have strong magical effects on individual people and their fates. The position of the moon was also thought to influence people, and when a so-called bad moon influenced people, they became lunatics (*luna* is "moon" in Latin). Early astrologers made maps of the stars and lumped certain clusters of stars into constellations, which appeared to be natural shapes or divine objects in their religious perspective. These maps of the stars were originally used strictly for astrological forecasts to divine the will of the gods, but the careful observations of stars, planets, and their positions in time and space laid the foundation for the science of astronomy.

Ironically, the modern fields of astronomy and physics completely destroyed the ancient view of stars as divine messengers. By the time of Newton, the stars and planets were viewed as we view them now: huge balls of burning gases millions and billions of miles away from us. Newton used this modern understanding of stars and planets to calculate his physics of motion and gravitation. These scientific observations and the basic physics of astronomy, in turn, made it impossible to imagine that the stars were divine messengers or could influence human events. Thus, the astrological roots of astronomy eventually brought about the destruction of astrology as a serious interpretation of the cosmos.

Here are just a few of the physical absurdities of astrology, and why there is no basis in modern physics and astronomy for taking their claims seriously:

1. The ancient astrologers had the idea that stars might influence the fates of people through some sort of magical force applied at a distance. Today, thanks to Newton and subsequent physicists, we know how extremely far away these objects are, and how much gravitational and other forces they exert on the earth. You can do the simple calculation yourself using Newton's law of gravitational attraction ($F = \text{constant} \times m_1 \times m_2 / d^2$), which states that the gravitational force is a product of the masses of the two objects (m_1 and m_2), and diminishes rapidly as the square of the distance between them (d^2). Simple calculations show that even the closest stars (none of which are in the constella-

tions of the zodiac), such as Alpha Centauri, are so far away that they have the same attraction as a cell phone more than 20 feet away from you (that is, so small it is not even measurable). Even the planet Mars is so far away from you that it has the same gravitational attraction as a truck placed 45 feet away from you (again, too weak to feel or even measure with the most sensitive gravimeter). Only the moon and sun are close enough and massive enough to have a gravitational pull on earthly objects, and they do so in well-understood ways: the sun holds the earth in orbit with its immense gravity, and the moon causes the bulging of water on the earth's surface that we know as tides. If astrology were about the *actual gravitational attraction of different celestial bodies,* the moon and sun would be the only ones of importance and the rest of the zodiac would be irrelevant. But astrology is all about the alleged effects of distant stars and planets, with only limited use of the sun and moon, so there is no correspondence between astrological ideas and any real physics of gravity. Likewise, the only other known force acting at distance, electromagnetism, is too weak to operate at these immense distances, and—more to the point—most celestial objects have no electrical charge! Nor do electromagnetic changes corresponding to known astrological patterns to have anything to do with stars and planets influencing our lives. If there is any other force out there that astrologers are using, it has never been detected by science. Thus, there is no physical basis for the "power of the stars" on our fate.

2. The constellations and the basic view of the stars as patterns are completely arbitrary and culture bound. You can demonstrate this by going out to see the sky on any dark night with good visibility (away from the light pollution of the cities, on a night with no clouds and no moon). Most people look up at the sky and see huge numbers of stars (some brighter than others), but the untrained eye discerns no obvious patterns or constellations. It actually takes a lot of practice and using star charts to and find the North Star (Polaris) and force your imagination to see the Big Dipper or any other constellation—and many people still have trouble allowing their imaginations to make an arbitrary number of points of light into a meaningful shape. Those constellations are actually even more arbitrary, because we have inher-

ited them from the astrological traditions of the Arabs of the Middle Ages, or from the ancient Greeks. Babylonian or Chinese or Mayan or Hindu astrology used entirely different star maps and made entirely different astrological predictions—*based on the exact same view of the sky.* Thus, the entire concept of constellations and signs of the zodiac is based purely on human imagination, and has no basis in reality.

3. Even more to the point, we see a constellation as a series of stars arranged in a pattern on a flat plane from our perspective on earth. But, in fact, the stars in any given constellation are not equidistant from us. They actually form a complex three-dimensional array, and only from our earth perspective do they form a pattern we might call the Big Dipper. If we viewed them from out in space so we were on the "side" of the cluster, the Big Dipper pattern would disappear entirely.

4. Another glitch in the astrological view of the universe is the fact that the earth wobbles around its axis like a top (this is known as *precession*). Thus, its rotational north polar axis (currently pointed at Polaris, the North Star) has pointed in different direction as the wobble goes through a cycle of roughly 21,000–23,000 years. Five thousand years ago, when the Babylonians first developed astrology, it would not have pointed at Polaris at all. If a human looked at the sky 10,000 years ago, halfway through the precession cycle, it would have pointed to a completely different "north star," Vega (one of the brightest stars in the northern sky). Due to precession, the phases of the moon and the positions of the sun have changed with respect to the constellations that were studied by the Babylonians over 5,000 years ago. For example, the astrological calendar originally began with the day of the vernal equinox (the first day of spring in the Northern Hemisphere—March 21 in our current calendar), when the sun was supposedly in the constellation Aries. Due to precession, however, the position of the sun during the timeframe of Aries (March 21–April 19) has "slipped" in the past 2,200 years, so that today it is actually "in" the constellation Pisces (February 20–March 20). Thus, the zodiac signs and the predictions made by astrologers based on patterns from 2,200 years ago no longer actually match the night sky, or the sun's so-called position with respect to the zodiac signs.

5. In addition to archaic notions of the stars and the sky, astronomers have since discovered several planets (Uranus, Neptune—and maybe Pluto, if you consider it a planet) that were unknown when astrology originated—yet they have been out there for millions of years. If planets have an effect on us, why did astrologers fail to predict the effects of as-yet-unseen planets like Uranus or Neptune, or the dozens of new planets in other solar systems that have been found by astronomers in the past ten years?

This partial list should clearly demonstrate that the notions on which astrology was founded were based on a conception of the universe that dates back to ancient times, and is completely contradicted by what we now know about the universe. The stars have no effects on us on earth; the constellations are not real phenomena; the position of the sun in the sky no longer matches what astrologers first observed 2,200 years ago, so your astrological sign is not your *real* sign; and so on. In short, astrology has nothing to do with the real universe, but only with a prescientific concept of the universe that visualized the earth as flat, the stars mounted on a fixed sphere around the earth, and the sun, moon, and planets moving on great wheels around the flat earth. There is no reason to take astrological concepts of the universe seriously in the twenty-first century, any more than we take Aristotelian physics to be real. This is on top of the fact that there is no reason to take astrology seriously because it has failed every rigorous scientific test of its predictive powers.

SO WHAT IS THE HARM?

Skepticism is not in evidence, and is in fact discouraged.... The way astrologers treat researchers and skeptics is just the way they treat other astrologers who disagree with them—continuing on as if they and their disagreements never existed.... The thing that I find least comfortable about astrology discussions (and not just on the internet) is their immateriality, their lack of grounding. Astrologers are less literate than average; they write badly and they read badly; there is almost no critical response; errors are ignored, corrections are not acknowledged. They answer off the top of their heads, quote from memory, claim that anything published anywhere at any time is general knowledge, and then get sidetracked into arguing about who's a liar instead of sorting out the facts of the original question. There is nothing resembling peer review, except in regard to political correctness. The fact is they don't look stuff up, not even when they disagree with you! Most astrologers would rather have an iffy quotation from Dane Rudhyar or C. G. Jung to support their opinions than some good research.

Joanna Ashmun[12]

We have established that astrology is bunk, baloney, bullshit. Predictions based on it are useless for knowing the future, and completely based on guesswork (cold reading as used by psychics) aided by quirks of human psychology such as confirmation bias and the Foret effect. When subjected to actual rigorous testing, astrologers do no better than random chance. Likewise, there is no basis in modern physics for the alleged effects of the stars upon our lives, and the entire astrological concept of the universe dates back to the Middle Ages and ancient times, with no correspondence to what we know about the universe today.

As numerous authors have documented, however, astrologers do not pay any attention to what the outside world thinks of them or their predictions, and do not even agree among themselves.[13] As long as there is a steady market of gullible people who are willing to listen to their garbage, they make money—like any other con artist who has no one to challenge them. As long as people buy their books, come to their offices to have their horoscopes cast, and read the horoscopes in the newspaper or on the internet, they keep on making big money selling bullshit. So what is the harm if 25% of the population believes this garbage?

As astronomer and skeptic Phil Plait indicates, there is serious harm.[14] First of all, enormous amounts of time and money are wasted on pure garbage. This is money spent on something that does not work, and which could be spent on more important and real needs. Even worse, astrology promotes uncritical pseudoscientific thinking, and leads people into making all sorts of bad, irrational decisions not only about their futures, but also about other topics. In Plait's words,

> The more we teach people to simply accept anecdotal stories, hearsay, cherry-picked data (picking out what supports your claims but ignoring what doesn't), and, frankly, out-and-out lies, the harder it gets for people to think clearly. If you cannot think clearly, you cannot function as a human being. I cannot stress this enough. **Uncritical thinking is tearing this world to pieces,** and while astrology may not be at the heart of that, it has its role.[15]

Astrology takes away from the grandeur of our view of the real universe, and puts us back in the bad old days of ancient astronomy, which was demonstrably wrong. As Plait puts it, "Astrology dims the beauty of nature, cheapens it." Added to this problem is the fact that the media (TV, the internet and newspapers) give astrology a big place in their coverage, but not the real world of science. Again, to quote Phil Plait,

> Hey, you might say, sure it's in the newspapers, but they put it next to comics, right? How seriously do newspapers take it then? My answer is, if newspapers don't take horoscopes seriously, then they shouldn't publish them in the first place. People know that comics aren't real, but not everyone understands astrology has as much legitimacy as "Blondie and Dagwood." Saying their location indicates their rationality is a cop out. Most newspapers in this country don't even have a science section, and science is critical to our daily lives (you're reading this on a computer, right? Do you wear glasses, or clothes, do you brush your teeth, take medicine, invest in tech stocks, drive a car? Thank science for all of those things then). They don't have a science section, but they'll publish horoscopes.[16]

Finally, believing astrological bullshit could have serious political and economic ramifacations. During the Reagan years, it was revealed that Nancy Reagan consulted astrologer Joan Quigley for advice regarding her husband and his presidential actions.[17] This meant that actions of the U.S. president (assuming Ronnie took Nancy's advice) were subject to the whims of a pseudoscientist. It is chilling to think that from 1981 to 1989 we may have experienced unnecessary presidential disasters and mistakes, all due to the instigation of a charlatan who happened to have the ear of the First Lady. Once this was revealed to the press, the astrologic consultations ended, but as the back cover of Quigley's 1990 *What Does Joan Say?* boasts, "not since the days of the Roman emperors—and never in the history of the United States Presidency—has an astrologer played such a significant role in the nation's affairs of State." This should be cause for alarm and consternation, not just another amusing anecdote about a president with a limited grasp of reality.

FOR FURTHER READING

Abell, G. O., and B. Singer, eds. 1981. *Science and the Paranormal: Probing the Existence of the Supernatural.* New York: Scribner's.

Carlson, S. 1985. A Double-Blind Test of Astrology *Nature* 318 (December 5): 419.

Cohen, D. 1967. *Myths of the Space Age.* New York: Dodd, Mead.

Culver, R. B., and P. A. Ianna. 1988. *Astrology: True or False? A Scientific Investigation.* Amherst, N.Y.: Prometheus.

Dean, G. 1987a. Does Astrology Need to Be True? Part I: A Look at the Real Thing. *The Skeptical Inquirer* 11 (Winter 1986–87): 166–184.

———. 1987b. Does Astrology Need to Be True? Part II: The Answer Is No. *The Skeptical Inquirer* 11 (Spring 1987): 257–273.

Gauquelin, M. 1979. *Dreams and Illusions of Astrology.* Amherst, N.Y.: Prometheus.

Hyman, R., 1977. Cold Reading: How to Convince Strangers That You Know all About Them. *The Zetetic* 1 (2): 18–37.

Lindsay, J. 1971. *The Origins of Astrology.* New York: Barnes and Noble.

Randi, J. 1982. Into the Air, Junior Birdmen! In *Flim-Flam! Psychics, ESP, Unicorns, and Other Delusions,* 55–92. Amherst, N.Y.: Prometheus.

11

Down the Slope of Hubbert's Curve: The End of Cheap Oil and Other Natural Resources

Dear Future Generations: Please accept our apologies. We were roaring drunk on petroleum. Love, 2006 A.D.

Kurt Vonnegut

THE NEVER-ENDING OIL CRISIS

Item: In response the United States resupplying Israel's military after their victory over their Arab neighbors in the Yom Kippur (or Ramadan) War in October 1973, the OPEC cartel decides to punish the Western nations by slowing down their petroleum sales and production.[1] The price of oil quadruples (fig. 11.1) in the United States to nearly $12 a barrel (about $40 a barrel in 2011 dollars). The Nixon administration is forced to impose price controls to curb inflation, which further raises the price of oil. Soon there are long lines at gas stations, rationing of gas, and social disruption and anger. The entire world economy is sent into a tailspin by the "oil shock," inflation rises severely, and many nations made rapid adjustments after decades of complacency built on the false assumption that oil would always be cheap and abundant. After months of diplomatic negotiations and some small concessions by Israel, in March 1974 the OPEC nations decide to end the embargo and allow the price of oil to return to pre-embargo levels.

Meanwhile, automakers are rapidly shifting from gas-guzzlers to fuel-efficient cars. For the people of the United States, this first message that oil will not be cheap and abundant forever begins to sink in, especially

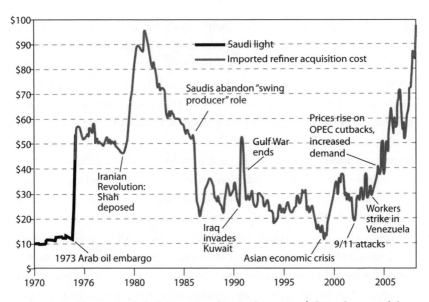

FIGURE 11.1. Fluctuations of oil prices over the past forty years (adjusted to 2008 dollars), showing the price spikes during the 1973 and 1979 oil embargoes, and the steady climb to new record levels in the past five years due to growth of demand from booming economies like India and China.

among people with interests in environmental issues. The most vulnerable countries, such as Japan and many European nations (which have no oil of their own), quickly make adjustments: they tax oil to reflect its real value in terms of the costs to society (its "externalized" cost) and conserve it, forcing cars to become smaller and more fuel efficient; they focus on nuclear, solar, and wind energy to reduce their dependence on imported oil; and they institute energy policies that will minimize the oil shock effect again. In the United States, however, once oil drops in price the people go right back to their wasteful ways and drive gas-guzzlers again, and Detroit loses money on all the new fuel-efficient cars that no one seems to want.

Item: It's déjà vu all over again. After the Iranian Revolution in early 1979, the Shah flees to the United States, leading to the Iranian hostage crisis. Iran is in chaos, greatly reduces its oil production, and causes another shortage of oil. Then in 1980, Iraq invaded Iran, disrupting supplies from both countries. Oil prices jump from about $16 a barrel to almost $40 a barrel (fig. 11.1), or about a 250% increase in a few months. In April 1979, the Carter administration begins a phased deregulation of the Nixon-era price controls, which had been put into place to restrain inflation but made the price of oil rise instead. Even as other OPEC nations pump more to offset the supply disruption, events in Iran, Iraq, and elsewhere keep oil prices rising. Americans find themselves in gas lines again, burning 150,000 barrels a day as their cars idle in line.[2] In Detroit, automakers satisfied the post-1973 American demand for bigger cars once the first oil shock ended, and are caught flat-footed for a second time. Japanese, German, and other foreign automakers are ready with well-built, fuel-efficient vehicles. Companies such as Honda, Toyota, and Audi gain a foothold in the American auto market and soon dominate it.

Jimmy Carter gives his famous "malaise" speech and calls the energy crisis "the moral equivalent of war." He urges the American people to recognize that our wasteful ways cannot continue forever, and that we need to seriously address our energy problems and over-consumption. Many Americans tune out or reject his message, and his popularity plunges—he is viewed as a pessimist and spoilsport, not as a prophet. A year later, he loses the election to the optimistic Ronald Reagan—largely due the effects of the energy crisis and his inability to resolve the Iranian hostage crisis. To spite Carter, the Iranians promptly release their American hostages the day Reagan is inaugurated. Reagan, in turn, rejects most conservation measures put in place by Carter, and even goes so far as to remove the solar panels and the wood-burning stoves that Carter put in the White House as a symbol of national energy conservation efforts. Americans go right back to driving gas-guzzlers, and by the 1990s the sales of monster SUVs take off, making Detroit wealthy and inclined to abandon their fuel-efficient cars a second time.

Item: Oil prices rise steadily since the late 1990s, and the price of a barrel of oil rises from $30 in mid-2003 to $60 by August 2005, and then climbs steeply to an all-time record of $147 by July 2008 (fig. 11.1). This oil crisis has no simple cause such as the 1973 OPEC embargo, or the Iran-Iraq turmoil of 1979. Instead, the real effects of declining petroleum reserves are being felt as Asian giants such as China and India demand more and more for their rapidly growing economies. Short-term events such as the 2006 Israel-Lebanon conflict, the Iranian nuclear buildup,

Hurricane Katrina's disruption of oil production in the Gulf of Mexico in 2005, and other factors cause spikes in the prices, but even after these events end, prices keep climbing. The effect is so gradual we do not see the gas station lines in the United States (but they do occur in China). Yet as the price at the pump reaches painful levels (above $4 a gallon), people start to conserve again. The fad for oversized, overpriced gas-guzzling s u vs finally goes out of fashion. Hummer sales are so poor that the non-military models are discontinued. Only a global recession, triggered by excessive speculation on overpriced real estate and too many people (who could not afford them) being given mortgages on these overpriced houses, pushes down oil demand and the price drops—but nowhere near 1973 or 1979 levels.

Item: The April 20, 2010, explosion of the oil platform *Deepwater Horizon* in the Gulf of Mexico triggers the worst oil spill in U.S. history, dwarfing the famous 1989 Exxon-Valdez oil spill or the 1979 Santa Barbara oil spill. Before it is over, the deep gusher releases almost 5 million barrels of oil into the Gulf. It is not capped until July 15, eighty-six days after it starts. The effects of the spill on Gulf ecology and on the fishing industries have not yet been fully evaluated. Even after most of the surface oil seems to have been collected or evaporated or eaten by microbes, huge amounts are found to have sunk to the bottom and coated the seafloor with a thick layer that kills all marine life.[3] While the cleanup is well underway, another platform to the west of the *Deepwater Horizon* caught fire and shut down in September 2010. The world is forced to realize what an enormous price (both economic and environmental) our insatiable demand for oil exacts from us, and what huge risks we take drilling in more and more dangerous and inaccessible places. At any given time, there are a half dozen or more oil spills going on all over the world, and there have been two to five major ones nearly every year for the past four decades.

All of these incidents point to a larger picture: oil is a very precious, but very hazardous, commodity, and we are extremely dependent upon it. Yet if we go back a century, there was almost no use for petroleum (except to lubricate axles) prior to 1900, when the modern age of oil exploration began. The invention of the internal combustion engine

and the gas-powered automobile changed society forever, and tied our
economy to oil for over a century. In the prosperous post–World War II
era, when servicemen came home, got married, and moved to houses in
the suburbs, the need for oil exploded as more and more people owned a
car and most had to drive it to work or any number of other destinations
not reachable by public transit. Breakthroughs in chemistry made oil
indispensable to the synthetics industry, since nearly all plastics, synthetic fabrics, pesticides, and fertilizers are petroleum based. The 1940s
and succeeding decades were an oil-fueled boom in the growth and expansion of American economy, on a scale never seen before—and, in
retrospect, unsustainable over the long run. Similar dependence on oil
occurred in every other developed or developing nation, until the entire
global economy was easily disrupted by any change in oil prices.

Ironically, in 1956 an oil geologist by the name of M. King Hubbert
studied and understood this pattern—and also predicted the future with
remarkable accuracy. An incredibly brilliant and innovative geologist
and geophysicist, he was born in San Saba in central Texas, and trained
at the University of Chicago. He taught at Columbia University (my
graduate alma mater) before going to work for Shell in 1946 and retiring
in 1964. But he was not done. After leaving the oil business, he then went
to work for the United States Geological Survey (USGS), and retired for a
second time in 1976. While at the USGS, he spent several years teaching
at Stanford and the University of California, Berkeley, before passing
away at age eighty-three in 1986. Late in his career he received nearly all
the awards a geologist can earn: the Arthur Day Award, and the Pen-
rose Medal of the Geological Society of America (which he served as
president in 1962), the Vetlesen Prize, and the Cressen Medal; and election to the prestigious National Academy of Sciences and the American
Academy of Arts and Sciences. (There is no Nobel Prize for geology.)
These are a mark of how much respect the geological profession holds
for him and his ideas.

Many geologists and geophysicists know Hubbert's name from his
early mathematical demonstrations that the rocks of the earth's crust
flow at depth, and for the correct formulation of Darcy's Law of flow of
fluids such as oil and groundwater through a porous medium, such as a
porous reservoir sandstone or permeable aquifer. Hubbert made many

other contributions toward understanding the flow of both ground-water and oil in the subsurface, which led to much greater success in finding oil. Based on this track record, geologists had good reason to take his ideas seriously. There was one idea that neither Shell nor the USGS wanted to accept: Hubbert's predictions about the future of oil resources, which were much more limited than the wild overestimates of the oil companies and the USGS at that time. They had heard too many false alarms before, and they did not want to believe that the party would soon be over, so they rejected Hubbert's prediction largely on emotional grounds, rather than on sound science.

What was Hubbert's prediction, and why was it so startling? While working for Shell, Hubbert gave a presentation at the 1956 American Association of Petroleum Geologists meeting (a conference I have attended many times in my own research career), which suggested that oil production should follow a bell-shaped curve, from slow growth at the beginning, to exponential growth to a peak, then a steady decline afterward (fig. 11.2A). Back in 1956, he made the astonishing prediction that U.S. oil resources would peak between 1965 and 1971, depending upon which figure you used for oil reserves. Hubbert lived long enough to see that U.S. oil production clearly peaked in 1970, and has been steadily declining since then (fig. 11.2B).[4] Thus, although his ideas were long rejected in oil company circles, they have long been accepted by academic geologists with no vested interest to defend, and are now considered more and more realistic even by the oil companies. Nothing convinces like a successful prediction!

How did Hubbert get the idea that oil production should follow a bell-shaped curve? Economic geologists had long known that this is the normal pattern seen in just about any nonrenewable mineral resource—such as oil, gas, coal, uranium, or any of the metals. Early in its history, a resource is consumed slowly, since it does not have a lot of established markets. But then a market develops, and suddenly the resource will be produced and consumed at a rapidly increasing rate as the deposits easiest to reach are quickly mined or drilled. This growth curve cannot last forever, and production slows down as most of the easily reached deposits are exhausted. The market for the resource, however, continues to put high demand and even higher prices on the resource, although

A

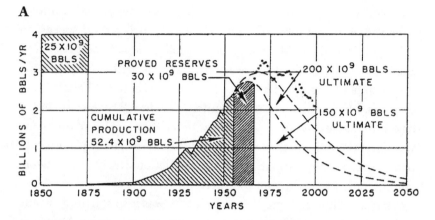

FIGURE 11.2. (A) The original Hubbert curve, published in 1956, with its bell shape and its predictions for when the peak would occur in the United States based on different assumptions of total reserves (redrawn from M. K. Hubbert, "Nuclear Energy and the Fossil Fuels," Coming Global Oil Crisis website, March 7–9, 1956, www.hubbertpeak .com/hubbert/1956/1956.pdf). (B) The actual history of oil production in the United States, which followed Hubbert's prediction with remarkable convergence. The upward trend in the last few years is due to the extraordinary push to find more American oil by going to more unconventional (more dangerous and more expensive) sources, such as secondary recovery, deep-ocean drilling, tar sands, and other previously untapped sources and it is nowhere near the peak production of the 1970s. (Modified from several sources, including "Hope for U.S. Oil: Where It Could Come From," Failing Gracefully website, February 10, 2012, failinggracefully.com/?p=3118.)

only very low-quality deposits remain. Exploration for less and less desirable deposits occurs in an attempt to extract even the most expensively obtained, lowest-grade resources that are left. Sooner or later, however, even exploitation of low-grade deposits cannot keep up with demand, and production rapidly declines as there are no more new discoveries. The supply eventually runs out, no matter how much prices rise and demand increases. Then there is an abrupt economic adjustment as the mineral resource can no longer be found at any price, and people learn to do without it.

We can see this with the production history of anthracite coal from Pennsylvania (fig. 11.3) or many other resources. The demand for coal increased rapidly in the late nineteenth century as the Industrial Revolu-

B

U.S. oil production in mbpd 1940-2011

tion and the development of steel created demand for coal-fired furnaces in western Pennsylvania. But the dangerous job of mining these coal deposits in deep, collapse-prone shafts hundreds of feet below mountains meant that the good deposits were exhausted by the 1920s, and production in Pennsylvania has declined ever since, as mining companies shifted to states with more abundant and more easily mined coal. There has been no significant coal production in Pennsylvania in over thirty-five years, even though the legends of Pennsylvania coalfields are still part of our culture.

Some people ask this: What about supply and demand? It is true that in the short term, there are price fluctuations due to changes in the supply-demand balance. In the case of coal (fig. 11.3), there are major

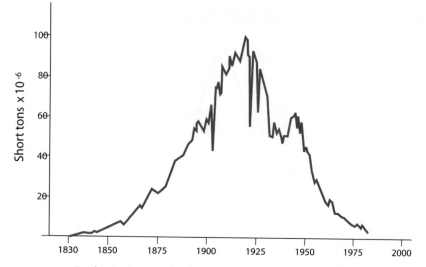

FIGURE 11.3. Production history of anthracite coal in Pennsylvania, showing the classic bell-shaped curve of nonrenewable resources between 1820 and 1980. There are
some short-term wiggles (such as during the Great Depression, 1929–1938) and spikes
(during World War II, 1941–1945), but overall the shape is strongly bell shaped. *From
M. T. Halbouty, ed., "Future Petroleum Provinces of the World,"* AAPG *Memoir 40 (1986):
177–205; used by permission.*

downward spikes during World Wars I and II and the Great Depression, as wars and economic slowdown decrease or disrupt demand or
production. There were also peaks during the postwar boom times of the
1920s, 1940s, and 1950s. But these are just short-term blips on a long-term
trend that is not affected by simplistic economics. As any Economics 101
course will teach you, supply and demand only apply when the supply
is *elastic*—that is, you can always make or grow more of it if the price is
high enough. But nonrenewable mineral resources are *inelastic:* they cannot, over the long term, increase their supply. There was only a limited
amount of them in the earth's crust to begin with, the minerals are no
longer being generated at any significant rate by the earth, and when they
are exhausted no more can be made.

Let us consider another example: silver production in Nevada, the
"Silver State" (fig. 11.4A). Shortly after the California Gold Rush ended,
bonanzas of silver at Ophir, Gould, and Curry mining districts drew
hundreds of miners to the Nevada Territory in the 1860s. The first pros-
pectors literally found crusts of pure silver right on the ground that were

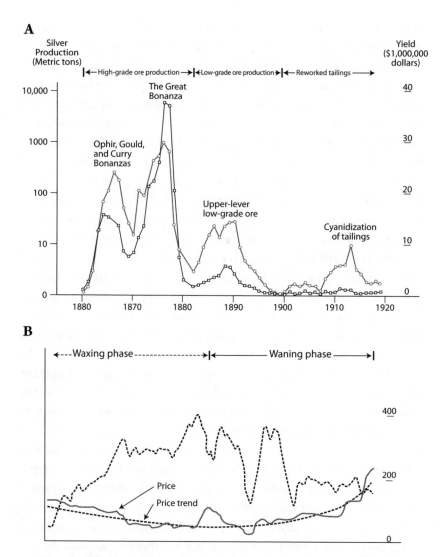

FIGURE 11.4. (A) Silver production history of the Comstock Lode, Nevada. The curve with the darker line represents yield (in millions of dollars). The light gray curve represents production (in millions metric tons). (B) U.S. production history of silver. The solid curve represents price in troy ounces. The dashed line represents production (in millions metric tons). *Redrawn from Cook, 1965.*

thick and crumbly and could be removed with a shovel. They got rich overnight. Other miners flocked to the region, and during the Civil War, Nevada's population had grown so much that it was admitted to the Union as a state. But those incredible surface deposits were actually rare.

Even the silver veins just below the surface were soon mined out, and the early bonanzas ended as there was no more easy silver to mine.

A few years later, the Great Bonanza was found, turning bare desert into some of the largest cities west of the Mississippi—such as Virginia City, Silver City, and Gold Hill. Virginia City once had ten thousand residents, and was considered one of the richest cities in America. The wild untamed mining town had the usual share of saloons, gambling halls, and whorehouses, but it also had first-class hotels with the finest amenities. The young Samuel Clemens spent several years in Virginia City as a writer and editor of the local newspaper, the *Territorial Enterprise,* and described the bizarre events that took place this rugged mining town in his classic *Roughing It.* As a newspaper reporter, he used the pseudonym Mark Twain, and that is how most people know him today. But by the 1880s, the rich veins of the Comstock Lode were running out fast, and it was becoming too dangerous and too expensive to mine any deeper. During the 1890s, miners were reduced to crushing and milling millions of tons of extremely low grade ore (containing only a few grams of precious metal per ton of waste rock), and that final boom (fig. 11.4A) petered out by 1900. By that time, Virginia City and nearly all the other mining towns were abandoned, and most have reverted to weathered old ghost towns or vanished entirely. Before the boom was over, at least $400 million was taken out of the ground.

The Comstock Lode died, but there were always desperate miners who sought ways of extracting silver the early miners had missed. In the 1900s, the technique of running crushed mine tailings through cyanide was discovered. The cyanide pulled the silver away from the waste rock and extracted the material that other methods of low-grade processing could not recover. However, that mini-boom lasted only another decade (fig. 11.4A), and it left huge environmental problems behind as the mining companies went bust and abandoned giant pits of cyanide to leach into the streams and groundwater. Since that time, Nevada has yielded small silver deposits in other parts of the state, but nothing on the immense scale or value of the Comstock Lode.

The Comstock Lode demonstrates in a microcosm the bell curve of production from boom to bust, with spikes when particularly rich deposits are found and then exhausted. What about U.S. silver production? It, too, shows a roughly "bell-shaped curve" which rises in the 1860s due to the Comstock Lode, peaks through the 1920s to 1940s, and has been in steady decline since the 1950s (fig. 11.4B). There are large short-term wiggles, of course, due to proximate economic conditions (such as the big drop in the middle of the peak due to the Great Depression), but overall it shows a clear waxing (increasing) phase and a waning (declining) phase. Today, the United States imports 75–85% of its silver, so we are no longer a net producer but an importer of silver. The prices have also climbed significantly as silver has grown rare in more and more countries. During the 1980s and 1990s, there was a serious fear that the rise of silver prices would make film for cameras prohibitively expensive, since all camera and movie films used a silver-based emulsion to react to the light exposure and create images. Fortunately, this problem was averted when digital images reduced the need for silver in photography. Other factors continue to cause a rise in demand, since world supply is not keeping up. World silver production is still climbing, so we have not reached the global peak of the bell curve yet, but in many countries, such as the United States, the peak has long since passed.

THE WEALTH OF NATIONS

No country is wealthy whose economy is based on fishing, forestry, or agriculture. Wealthy nations are those whose economies are based on the exploitation of fossil fuels, metals, and construction minerals. Technology applied to these resources is the basis of the high material levels of living enjoyed by the industrialized portion of the world's people. Therefore, if the resources prove to be finite and nonrenewable, the wealth of nations not only faces severe resource constraints, but is ephemeral.

Earl Cook[5]

This is the challenge that the United States and many other countries face. As the great economic geologist Earl Cook put it above, the wealth of nations lies in mineral commodities more than in any other single factor. We can see that played out all over the world today. The United States reached its economic supremacy largely on the strength of its mineral resources (many of which are depleted now), and today Russia, China, South Africa, and the oil-rich OPEC nations are the richest

Table 11.1. U.S. dependence on foreign sources for more than 35% of our non-fuel minerals

Mineral resource	Percent imported	Foreign sources	Uses
Aluminum	100	Jamaica, Suriname, Australia, Brazil, Guinea	Aluminum metal, abrasives
Columbium	100	Brazil, Thailand, Nigeria, Germany	Steelmaking
Graphite	100	Mexico, Canada, China, Brazil	Brake linings, lubricants, pencils
Arsenic	100	China, Japan, Mexico	Preservative, pesticides
Thorium	100	France	Ceramics, welding, electronics
Fluorine	100	China, South Africa, Mexico	Steelmaking
Manganese	100	South Africa, Australia, Gabon, France	Batteries, agricultural chemicals
Thallium	100	Belgium, Canada, Mexico	Superconducting, electronics, glass
Yttrium	100	China, Japan	TV, fluorescent lights, temp sensors
Mica	100	India, Brazil, Belgium, China	Electronics and electrical equipment
Strontium	100	Mexico, United Kingdom, Spain	TV picture tubes, ferrite magnets
Vanadium	100	Korea, Canada, Austria, Czechia	Steel alloys, high-speed tool steel
Cobalt	100	Democratic Republic of Congo, Belgium, Finland, Canada	Steel alloys, pigments and dyes
Rare earth elements	100	China, France, Russia	Alloys, electronics, spacecraft
Tantalum	100	Thailand, Canada, Australia, Brazil	Electronics, high-temperature alloys
Bismuth	99	Peru, Japan, Mexico, United Kingdom	Replaces lead in cosmetics, medicines
Antimony	93	South Africa, China, Bolivia, Mexico	Alloys for electronics, fire retardants
Chromium	91	South Africa, Russia, Turkey, Zimbabwe	Steelmaking, alloys, dyes, tanning
Asbestos	86	Canada, South Africa	High-temperature insulation
Platinum group metals	89	South Africa, United Kingdom, Russia	Precious metals, electronics, alloys
Tin	80	Malaysia, Thailand, Bolivia	Anti-corrosion alloys, solder, pewter

Barium	80	Ireland, Peru, Mexico	Oil drilling mud, X-rays, vacuum tubes
Diamonds	78	South Africa, Russia, India, Australia, Canada	Gems, drill bits, cutting tools
Mercury	73	Canada, Algeria, Mexico, Spain	Thermometers, medicine, lighting
Nickel	71	Canada, Norway	Stainless steel, coins, magnets, batteries
Zinc	76	Canada, Mexico, Australia, Honduras, Peru	Galvanizing, deodorant, shampoo, paint
Tungsten	63	Canada, Bolivia, Thailand, Peru	Lightbulbs, superalloys, catalysts
Tellurium	59	Peru, Canada	Alloys, solar panels, semiconductors
Selenium	58	Canada, Japan, Mexico	Glassmaking, pigments, electronics
Cadmium	50	Mexico, Canada, Australia, Belgium	Batteries, solar panels, pigments
Potassium	49	Canada	Gunpowder, fertilizer, ceramics, dyes
Gold	45	South Africa, Canada, Russia	Precious metal, electronics, dentistry
Gypsum	39	Canada, Mexico, Jamaica	Drywall, plaster of Paris, fertilizers

Sources: Concressional Budget Office, "Strategic and Critical Nonfuel Minerals: Problems and Policy Alternatives," www.cbo.gov/ftpdocs/50xx/doc5043/doc15-Entire.pdf; *US Bureau of Mines Bull. 667* (1975): 12; U.S. Geological Survey, minerals.usgs.gov/minerals/pubs/mcs/2010/mcs2010.pdf.

in terms of mineral wealth—and also the major players on the world economic stage. Before World War II, the United States was completely sufficient in most of its mineral commodities, but after the huge postwar expansion of our economy and the rapid depletion of many of our mineral resources, we no longer have enough of most key commodities and are now net importers (which also makes us weaker in the balance of trade). The United States is now on the depleted tail of Hubbert's curve for many natural resources, not just silver. For the past thirty years, the United States has become dependent on foreign sources for twenty-one of the twenty-four major nonfuel minerals.[6] If you also consider oil, we have only about 20 billion barrels left, less than 3% of the world's reserves, and must import 60% of our oil.

A quick run through the list (table 11.1) is sobering. The top fifteen minerals in the list are 100% imported; the United States has no significant production of these minerals left. The next eight minerals on the list are at least 75% imported, so we have only minimal reserves and import most of them. Most people know little about industrial production of minerals in the world and what they are used for, so I provided some notes in the right-hand column of table 11.1. Most surprising of all is aluminum, which we use in a huge number of ways—from drink cans and foil and many other containers, to siding on houses, to entire buildings and many other structures that need to be light but strong (such as the frame, chassis, and even engine blocks of many cars). The United States has no significant aluminum stores left, since what little we had was mined out long ago. Aluminum ore is produced by the deep weathering of tropical soils to produce an aluminum- and iron oxide–rich clayey soil known as bauxite. Not surprisingly, it is imported mostly from tropical countries such as Jamaica, Suriname, Guinea, and Brazil, as well as the tropical northern region of Australia. That is why I am a fanatic about recycling every aluminum can that I find, *because we do not produce aluminum here,* and recycling is crucial to reducing our dependence on other foreign suppliers. (If you save aluminum cans at home, you will be surprised how well they pay back if you take them to a recycling yard.) Tragically, many states have no incentives to recycle aluminum yet. It is painful for me to travel and have no choice but to throw aluminum away. Even most airlines and airport food providers have not figured out how to separate the huge volume of aluminum cans from their regular garbage and make money from recycling.

Most people understand the importance of aluminum in their lives (even if they throw it away), but there are many minerals at the top of the list that are important in surprising ways. Take graphite, for instance. You use it all the time not only in the "lead" of your pencil (which is not real lead, but graphite), but it is critical to the brake linings of your car and a graphite lubricant works in high-temperature settings, such as engines, in which fluid lubricants fail. Columbium, cobalt, and fluorine are essential to steelmaking, which is still a big business in the United States. Arsenic and manganese are critical to the manufacture of agricultural chemicals. Yttrium, thallium, mica, strontium, tantalum, and the platinum group metals are essential to all sorts of electronics, computers, TV, and many other high-tech devices. Many of us have never heard of these minerals before, but ask any metallurgist or electrical engineer, and they recognize the importance of these rare minerals. The Rare earth elements (REEs) consist of seventeen elements near the bottom of the periodic table that are found in only a few places on earth, and are used to make everything from the phosphors on flat-screen TVs, lasers, and fluorescent lamps (ytterbium); permanent magnets for high-efficiency motors (neodymium); and many other products. They are produced almost exclusively in China due to their large supply and lax environmental and safety laws. In 2010, China held Japanese electronics manufacturing hostage by embargoing their REE shipments to Japan, all over a dispute over territory in the South China Sea.[7] We all know of chromium from car bumpers and wheels, but it is critical to making high-quality steel that is harder and more corrosion resistant, so chromium is used in most steel alloys, as well as in dyes and pigments ("chrome yellow") and tanning leather. Tin, zinc, and nickel are familiar metals that are required in making certain alloys and products, and for which there are no substitutes.

I have skipped one other imported mineral that everyone knows: diamonds. There are a few mines in the United States, but more than half of the world's diamonds come from South Africa, Botswana, Angola, and a few other African countries—followed by mines in Russia, Australia,

and India. Most of think of diamonds in terms of jewelry—as in engage-
ment rings and "Diamonds are girl's best friend." Indeed, diamonds are a
very valuable commodity in that regard, controlled almost entirely by a
single monopoly, the DeBeers Corporation, which has ruled the African
diamond mines for decades. But they have many other uses that most
people do not appreciate. As the hardest natural substance known, diamonds are essential for drilling, cutting tools, abrasives, and many other
uses. Some are also used in industrial lasers that require their optical
properties. Their biggest use besides jewelry (especially for diamonds
that are not jewelry-grade quality) is for drill bits to cut rocks in drilling
for oil. We could survive the loss of diamonds in jewelry, but there would
be no drilling for oil, and our worldwide oil industry would collapse. That
brings us back to the political dimension of these resources. We have all
heard of "blood diamonds" or "conflict diamonds," which were extracted
by childhood slave labor and used by powerful warlords in Angola, Sierra
Leone, Liberia, Ivory Coast, the Republic of Congo (Congo-Brazzaville),
the Democratic Republic of Congo (formerly Zaire), and Zimbabwe
to finance their civil wars—or by oppressive governments to support
their activities against rebels. The tragedies that these diamonds have
financed have made them pariahs in the international market (as well as
the subject of several movies).

Here is another political dimension to consider: since the United
States is dependent on imports of so many mineral commodities, we are
also vulnerable to political events that might cut off our supply of these
commodities. We have twice gone to war in the Persian Gulf region
since 1991, spending trillions of dollars and thousands of American lives.
These wars were ostensibly to push back Saddam Hussein's invasion of
Kuwait, and then to displace him altogether because he allegedly had
"weapons of mass destruction" (which he did not). But it was no secret
that we paid far more attention to this region than any other tyrannical
regime around the world, since these Middle Eastern countries are the
chokepoint for the world's supply of oil. We also support the oppressive
regimes in Saudi Arabia and other Persian Gulf monarchies because they
control the oil supply and the votes of OPEC. We did not attack or even

reprimand Saudi Arabia, even though most of the 9/11 terrorists came from there, and were motivated by our presence in their country.

Even more damning were accounts uncovered by a number of reporters that George W. Bush and Dick Cheney (both former oil men themselves) consulted with oil industry executives in the earliest days of their administration to plan their policies.[8] They then planned the invasion of Iraq in the first weeks after they took over the White House in January 2001 to secure the world's third-largest reserve of oil. They secretly had Iraq in their sights from the very beginning. Saddam Hussein's control of that oil, not terrorism, was the main reason for their invasion plans. After all, this was many months before the 9/11 terrorist attacks gave Bush political cover to invade the Middle East. Ironically, we have spent trillions there and yet not benefited from the Iraqi oil deposits anywhere approaching the amount we spent—and we leave a country with a weak divided government that may still deny us their oil.

Those of us old enough to remember might think of the days in the 1970s and 1980s when the racist practice of apartheid made South Africa an international pariah. If it were any other nation, we would have cut off trade long ago and destabilized their regime. But look at the countries listed on table 11.1. South Africa is the top supplier of the world's diamonds, produces over 80% of the world's gold, and is also the richest source of many essential rare minerals—from chromium to manganese, fluorine, and all the platinum group metals. Without these substances, we would have a worldwide diamond and gold crisis, and many of our critical high-tech industries would be starved of essential minerals. That is why the developed nations handled South Africa with kid gloves for a long time, and only gradually ratcheted up the pressure until the apartheid regime finally fell. We could not put a hard embargo on South African diamonds, gold, or precious metals without risking their cutting us off and causing global economic meltdown. Look at the list again, and you will see another big mineral-rich country, Russia (and the former Soviet Union). They control much of the world's non-OPEC oil, plus diamonds, chromium, gold, platinum group metals, and many other precious resources that give them enormous global leverage. As the collapse of the Soviet Union in the 1990s and the subsequent instability of the region show, we cannot count on their being friendly and willing to trade with us forever. We were locked in a Cold War with them once, and there are no guarantees as to what Russia might do in the future.

Thus, oil is part of a bigger world mineral resources picture. Nations that are dependent on foreign sources of commodities are vulnerable to any political or economic event that threatens that supply, and geopolitics thus becomes a game between the haves and the have-nots. And this raises an even larger question: If the United States is already at the exhausted tail end of Hubbert's curve, how much does the world have left? Can we apply the bell-curve dynamic to global supplies, especially of oil?

THE END OF CHEAP OIL

We've embarked on the beginning of the last days of the age of oil. Embrace the future and recognize the growing demand for a wide range of fuels or ignore reality and slowly—but surely—be left behind.

Mike Bowlin, chairman and CEO of ARCO, speech in Houston, February 9, 1999

Energy will be one of the defining issues of this century, and one thing is clear: the era of easy oil is over.

Chevron

While major new finds cannot be ruled out, recent statistics do provide worrisome signals.... Discoveries only replaced some 45% of production since 1999. In addition, the number of discoveries is increasing but discoveries are getting smeller in size. The 25 biggest fields hold some 33% of discovered reserves and the top 100 fields 53%; all but two of the giant fields were discovered before 1970.

USGS WPA 2000 Part 1: A Look at Expected Oil Discoveries

All the easy oil and gas in the world has pretty much been found. Now comes the harder work in finding and producing oil from more challenging environments and work areas.

William J. Cummings, Exxon-Mobil company spokesman, December 2005[9]

It is pretty clear that there is not much chance of finding any significant quantity of new cheap oil. Any new or unconventional oil is going to be expensive.

Lord Ron Oxburgh, former chairman of Shell, October 2008[10]

So far, we have examined the individual bell curves of the oil resources of limited regions such as the United States or other countries, where a resource has been exhausted and the country depends almost entirely on imports. But the world is a big place, with lots of different countries that might have a lot of a given resource. Assuming a best-case scenario in which international geopolitics does not restrict our ability to trade for what we need, how long will some of these resources last? In particular, how long will oil last?

Many different approaches have been taken to estimate how much oil is left, and when it will run out. At one extreme are the "cornucopians" (they might also be described as "corn-utopians") such as Julian Simon, Bjørn Lomborg, and Matt Ridley, who assume that market forces and supply and demand will always find a way to give us whatever we need, no matter how much we increase demand, and how much the supply is depleted. They argue that technology and free markets create a substitute for something once it becomes scarce and expensive. But there are several problems with this school of thought. First of all, many of these scarce resources are not replaceable by another resource, let alone by a technological fix. As discussed below, there are some alternatives to oil, for example, but in many applications we cannot replace oil with anything else. In addition, this idea that we can replace one scarce resource with another misses the key point: nearly all resources are becoming scarce on worldwide basis, so you cannot find a cheap substitute for one resource when almost everything becomes scarce and expensive.

If you examine the cornucopians' writings closely you will find they are based on economic models that do not take into account the inelasticity of nonrenewable resources, or the demonstrated fact that with a limited resource such as oil or silver, there is typically a bell-curve distribution, and eventually the resource runs out completely. Sometimes you will see these economists draw a curve showing the global increase in oil demand and production, and then a plateau at which the world can somehow maintain its level of production indefinitely. There may be economic models that describe such behavior, but they are not based on any real-world economic data or the past behavior of a commodity; they are pure fantasy. It is hard to get this point across, so let us repeat it one more time: where nonrenewable resources are concerned, supply and demand only applies is the short term. Eventually, that resource is exhausted, and the earth makes no more of it. When that happens, *you cannot have more of it, no matter how much you raise the price in response to demand!*

Opposing the cornucopians are nearly all the geologists (including many oil company geologists) who deal with the realities of natural resources and oil supply as part of their daily experience, and have no fantasies of oil suddenly becoming abundant again. As the many quotations above illustrate, the leaders of nearly all the oil companies are fully aware that oil is becoming scarcer, and that there are fewer and fewer new oilfields found, and they are nowhere close to keeping up with demand on a worldwide basis. Figure 11.5 provides a very sobering reality check: a plot of the discovery dates of American oil fields. Notice that there was a bell curve with a peak in the 1930s. Even though U.S. oil companies have spent billions and made huge technological advances since the 1930s and 1940s, the rate of discovery has continued to drop, and large oilfields no longer can be found in the lower forty-eight states. Even those in Alaska are near exhaustion, since they peaked in 1988 and are nearly dry now.[11] The right-wing "Drill, baby, drill" slogans are just fantasies. U.S. oil companies have indeed been drilling as fast as they can everywhere in the United States—and, as figure 11.5 shows, getting very little no matter where they look. Now they are spending most of their time and money on increasingly risky operations such as offshore oil platforms—and the 2010 Gulf oil disaster (along with previous oil disasters on platforms around the world) shows just how dangerous it is to drill so far offshore. As I discovered when I worked for them, most U.S. oil companies are largely invested in foreign operations, with little exploration in the lower-48 any more. I have heard this over and over again from my friends in the oil business, and you will see it if you consult with oil companies (as I have) and find that their rooms and corridors are lined with maps and seismic profiles from all over the world—except the United States.

One of the favorite themes of the oil companies and the right-wingers is to drill more in Alaska, especially in the ecologically sensitive Arctic National Wildlife Reserve (ANWR) on the North Slope of Alaska. The entire issue became a political hot button in the 2008 presidential election, as environmentalists pointed out how much habitat would be de-

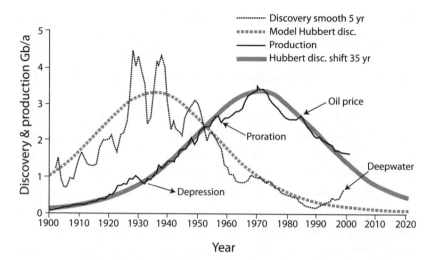

FIGURE 11.5. Discovery dates of American oil fields, and the later bell-shaped curve of production from those fields as their discovery rate drops off. *J. Laherrère, "Future of Oil Supplies" (paper presented at the Seminar at the Center of Energy Conversion, Zurich, May 7, 2003), reprinted in T. Rehri and R. Friedrich, "Modelling Long-Term Oil Price and Extraction with a Hubbert Approach: The LOPEX Model,"* Energy Policy *34, no. 15 (October 2006): 2413–2428, www.sciencedirect.com/science/article/pii/S0301421505000984.*

stroyed in the short-term search for oil. But the answer is clear, no matter what your politics: such exploration would be futile. In 1998, the United States Geological Survey estimated that there were at best only 16 billion barrels of oil in the ANWR and most of these reserves are prospective resources, not proven resources.[12] Sixteen billion barrels sounds like a lot until you realize that it is less than 1% of the total world oil consumption each year. *The United States alone consumes over 20 million barrels of oil per day,* so even if every drop of oil were actually in the ANWR, it would at best provide two to three years' worth of oil for the United States—and then it would be exhausted, and what would remain would be an ecological disaster.

Despite the Republican attacks on President Obama for not allowing oil companies to drill wherever they want, it turns out that American oil companies have hugely ratched up their production, and at the moment there is a short-term increase in oil being produced in the U.S. (fig. 11.2B). This is because of the discovery of the Bakken field in North Dakota and Montana, the big increase in natural gas supplies due to fracking, the push to exploit the Athabasca tar sands of Canada, and lots of risky deep-water drilling. However, this apparent increase in our oil production is due to exploiting these ever more risky and expensive oil sources; they do not return us to the days when large oilfields were producing lots of cheap oil.

Late in 2012, there was a big fuss made over a report that claimed that North American production would exceed that of Saudi Arabia or Russia by 2020.[13] This makes it sound like the United States and Canada will soon become another pool of oil and OPEC powerhouses. What most reporters and commentators missed in reading the story, however, is based on the fact that both Saudi Arabia and Russia are experiencing rapid declines in production from their oilfields as they are tapped out, while North American nations are increasing production only by pushing to explore increasingly expensive and risky prospects.

Also missed by these overly optimistic commentators is the fact that even if the North American nations did manage to surpass Saudi Arabian production in 2020, we would not be able to keep all this domestic oil in the United States and Canada and have cheap gas again. No, the oil we produce here still goes into the overall worldwide supply of oil, and most will end up in China or India. It may help our trade imbalance a bit, but we will not be able have cheap oil again when the worldwide demand is so large, and other traditional sources like Saudi Arabia and Russia are in decline.

So what about the world discovery rate? The answer has been known for a long time. World discovery rate peaked in 1965 (fig. 11.6), and has been declining ever since, even though more and more exploration is conducted in the farthest reaches of the globe in the past forty-five years.[14] The peak-oil effect has already occurred, and we are clearly on the decline in discoveries of oil.[15] Knowing that there are likely no more huge fields in our future, the next step is to calculate how much oil is left. Estimates of the ultimate recovery have fluctuated all over the place in the past few decades, but in recent years most of the estimates place the total volume of ultimately recoverable oil in the range of 1.6 to 2.6 trillion barrels, with most estimates around 2.0 trillion barrels.[16] This was the number that Hubbert himself used when trying to determine the amount of oil left and when the peak would occur (he estimated

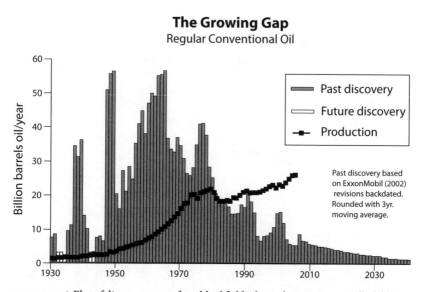

FIGURE 11.6. Plot of discovery rate of world oil fields through time. *From Kjell Aleklett, "International Energy Agency Accepts Peak Oil," Association for the Study of Peak Oil and Gas website, www.peakoil.net/uhdsg/weo2004/TheUppsalaCode.html.*

a window between 1995 and 2000). Depending upon how the model is run, most scientists predicted that the peak of world oil production would occur between 2005 and 2010, with most estimates around 2006 to 2007. Although it is still too early to tell if the peak has fully passed, so far that prediction has proven accurate.[17] According to the International Energy Agency (IEA), the peak of global oil production was in 2006, and declined by 6.7% in 2007.[18] As figure 11.7 shows, the peak occurred in 2005 and 2006, and has hit a plateau since then, despite increasing prices and pressure to produce more oil. In an interview the head of the Brazilian oil conglomerate Petrobras pointed out that the decline in supply was so severe that *we would need one new discovery the size of the entire Saudi Arabian oil reserve every two years to keep up with increasing demand!*[19]

Of course, there are skeptics who refuse to believe peak oil will ever occur, or doubt that it will occur anytime soon. To counter this illusion, oil analysts give a long list of reasons the peak has already happened, and will happen in the next few years:

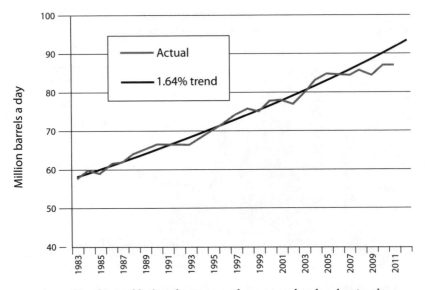

FIGURE 11.7. Trend in world oil production over the past two decades, showing the peaking and plateau of production since 2006, even though prices have been climbing steadily and rapidly. *Redrawn from "Graph of the Day: Growth in World Oil Supplies, 1983–2011," Desdemona Despair blog, April 12, 2012, www.desdemonadespair.net/2012/04/ graph-of-day-growth-in-world-oil.html.*

- USGS is the only organization private/public/government that has an estimate that high [peak at 2015 or later]. They were caught off-guard when oil production peaked in the lower-48.
- Biggest growth in demand is coming from Middle East itself. Saudi Arabia was a country of 6 million people in 1970 and today, the population is 22 million. Production has never reached the 1970s high again. Saudi Arabia threatened to increase production and their production didn't budge even in the aftermath of hurricane Katrina.
- World's second largest oil field ever discovered, Kuwait's Burgan oil field, peaked last year.
- Estimates about the coming scarcity of crude oil come from numerous sources, not the least of which is Vice President Dick Cheney himself. In a 1999 speech he gave while still CEO of

Halliburton, Cheney stated: "By some estimates, there will be an average of two-percent annual growth in global oil demand over the years ahead, along with, conservatively, a three-percent natural decline in production from existing reserves. That means by 2010 we will need on the order of an additional 50 million barrels a day."

- U.S. government had done a study on the peaking of production, headed by a gentleman by name Robert Hirsch. Findings of this study are alarming. A copy of this report is cited here.[20]
- Europe's only major oil field in the North Sea has peaked and the production is in decline. Decline rates in North Sea has been dramatically high so far.
- World's second largest oil field in terms of daily output—the Cantarell field of Mexico—has peaked according to the Mexican National oil company Pemex. The field could experience steep declines in production up to 8–10% every year. According to this report, 2006 production is already 13% below the 2005 level.
- China has been transformed from being a net exporter of oil to an importer 10 years ago. China's major oil field, Daqing has probably peaked. The recent economic growth in that nation has added pressure to the demand side.
- Matt Simmons, a successful investment banker and CEO of Simmons and Company International, has put his whole career on the line on the peak oil theory. He has gone through several Society of Petroleum Engineers papers and has come to the conclusion that Saudi will not be able to hit their 1970s peak again.
- World's largest oil field, Ghawar, is dying.
- Russian oil production had seen a revival since the mid 1990s but could be close to entering precipitous declines very soon. Russian oil reserves are estimated at 48 billion barrels, a lot lower than the giants in the Middle East. Russian oil output peaked in 2010.
- World oil discovery peaked in mid 1960s. Since the oil production in the lower-48 of U.S. peaked about 40 years after the peaking of discovery, there are reasons to believe the same thing could happen to world oil production.

- Reputed oil industry insiders other than Dick Cheney and Matt Simmons throwing their weight behind peak oil include billionaires Richard Rainwater and T. Boone Pickens.
- Financial analyst Jeffery Rubin—chief economist for the respected CIBC World Markets—foresees a peaking in world oil production between now and the end of the decade. Eric Sprott, Sprott Asset Management, has over $1 billion of his firm's assets invested in areas that will benefit from peak oil.
- Sadad al Husseini, Saudi Aramco's former head of exploration and production, wrote in fall 2005 that world oil production would peak and plateau by 2015, at between 90 to 95 million barrels a day.
- PhD academics like Dr. Al Bartlett (University of Colorado-Boulder) plus Robert Kaufmann and Cutler Cleveland (Boston University) have for at least two decades been pointing to upcoming problems associated with peak oil.
- Former President Bill Clinton and Vice President Al Gore both referenced peak oil. In June 2006, Gore spent a minute talking it up on CNN's *Larry King Live*. Then in early July 2006, Clinton—in an interview with *Atlantic Monthly*—gave substantial credence to the peak oil concept. He also wondered why he had never received a peak oil briefing, given its strategic importance.
- Looking at the latest available data, OPEC production has been more or less flat since 2004.
- Many OPEC nations have manipulated oil reserve data. Under the quota regime of the OPEC cartel, the OPEC countries were allowed to produce oil in proportion to their reserves. During the oil crash of the 1980s, many OPEC nations found themselves short of hard currency and upwardly revised their oil reserve data to increase their quota. In 1984, Kuwait doubled their reserves on paper, without ever finding a single new field of oil and the other OPEC nations followed suit. All these nations have produced billions of barrels of oil since then and have not found any significant amount of oil, but surprisingly, their reserves reported has remained the same ever since 1990. They have never deducted what has already been produced from the data. When the seven sisters were running Saudi Aramco, the estimated oil reserves

in Saudi Arabia were 110 billion barrels. There has been no significant discovery in that country since the Saudis nationalized Aramco, but the oil reserves were revised up to 260 billion barrels. Same thing can be said of all other nations belonging to the OPEC cartel.

· To quote former head of National Iranian Oil Company, Dr. Ali Bakhtiari, "It goes without saying that when assaying Middle Eastern oil reserves, one should tread carefully. Because, on the one hand, oil reserves' estimation is both a science and an art; and on the other hand, seen from the point of view of most Middle Eastern countries, oil reserves are more political than geological. Thus, nonscientific views come to prime over science and further enhance the various types of shades that have led to an overall opaque situation in the Middle East."

· Dr. Bakhtiari wrote this in the context of a discussion in which he estimated total oil reserves in the Middle Eastern group of major oil-producing nations (Iran, Iraq, Kuwait, Saudi Arabia, and the United Arab Emirates) as about half, or even less, than what the respective national governments claim. Isn't it fitting that Kuwait is the first country to admit that it does not have as much oil as it has always claimed?

· Major oil companies are joining the alternative fuels bandwagon, for example, BP calling itself Beyond Petroleum, and making significant investments in alternative energy. Chevron is running a conservation website titled "Will you join us?"

· Half of the world's population residing in India and China were completely absent from the prosperity of the industrial world. But that has been changing over the last decade or two. Close to 40% of world's population live in those countries alone. . . . Crude oil consumption in those two enormous countries alone are on a tear. China is at least 10–15 years ahead of India on the path to prosperity.

· Now, if all goes well in India, then the demand for goods, services, and, hence, commodities will continue to increase very substantially for another 10–20 years. As can be seen from the chart [on the website], Indian oil consumption has just recently started to turn up. Should its demand now accelerate, which is very likely, then China's and India's oil demand could double in the next eight years (see chart). Demand from China and India for oil already doubled in the last 10 years.

· World oil production has been flat at around 84.5 million barrels per day (bpd) since late 2004, amidst record high oil prices. While it is too early to say if oil production peaked just under 85 million bpd, the longer the production stays flat to down, it will indicate the arrival of peak production.

· Reputed petroleum geologist Chris Skrewbowski has published a work known as Mega Projects 2004. Mr. Skrewbowski has looked at all new projects expected to produce more than 100,000 barrels per day in production. According to mega projects, declines from existing fields are balanced by oil coming from new projects until 2007, but not beyond 2007.

· Canada's oil sands will not prevent peak oil. Uppsala Hydrocarbon Depletion study Group, Uppsala University, Sweden, has made a study of a crash program scenario for the Canadian Oil Sand Industry. Even in a very optimistic scenario Canada's oil sands will not prevent Peak Oil. If a crash program were immediately implemented, it may only barely offset the combined declining conventional crude oil production in Canada and North Sea.

· Geologists Colin Campbell and Jean Laherrere published a work in 1998 *Scientific American* titled "End of Cheap Oil."[21]

· Industry analysts like PFC Energy; Groppe Long & Littell; and Petrie Parkman & Co. Last fall, Tom Petrie said he expected peak oil by around 2010 and that he would be "shocked" if world oil production didn't peak by 2015. In PFC Energy's 2004 presentation on peak oil, they show world oil production peaking in the 2014 time frame; their 2006 study, to be presented at the ASPO-USA conference next month, likely points to a slightly earlier date. Henry Groppe sees world petroleum liquids production peaking by 2010.

· U.S. Congressmen Roscoe Bartlett (R-MD) and Tom Udall (D-NM) sound seriously concerned about peak oil, have been speaking out and writing about the issue, and have enlisted over a dozen colleagues to join them in the House Peak Oil Caucus.[22]

Many oil companies see the writing on the wall. They are already spending some of their profits in research on and development of alternative energy sources, so their business does not die out when the oil becomes too scarce. In 2000, British Petroleum (BP) decided to market themselves as the environmentally friendly oil company. They changed their logo to a green and yellow shape resembling a flower or starburst, and launched a high-profile $200 million ad campaign touting their alternative energy efforts, with the tagline that BP stood for "Beyond Petroleum." It is quite remarkable to hear an oil company announce its own transition to non-oil energy sources (even if it was mostly PR hype, since in actuality BP spent only a tiny portion of its research and development budget on non-oil research[23]). Of course, with the BP Gulf oil disaster of summer 2010, the company has other bad publicity to deal with now. As soon as BP dropped its "Beyond Petroleum" campaign, Shell stepped up with their "Let's Go" ad campaign, touting their research and investment in alternative energy sources with the slogan, "Let's make the most of what we've got."

To summarize: the idea that cheap, abundant oil will soon become scarce is as well documented and supported by existing data as the idea that humans are causing global climate change or that evolution is occurring. The fact that Hubbert's hypothesis exactly predicted the United States oil peak, and seems to be predicting the global peak, should be strong enough evidence in and of itself. There is also the fact that the peak of discovery of major oilfields occurred forty-five years ago, and there have been no giant oilfields found in a long time, and most of the world's older oilfields are nearing their ends. There are no polls that show just how many qualified experts (geologists and geological engineers in and close to the oil industry) accept the concept of peak oil and the end of cheap abundant oil, but many oil experts are on the record as accepting the concept, including a number of oil geologists and executives. I have many friends and former students in the oil business, and almost all tell me that peak oil is widely accepted among their colleagues, and they have long been forced to work with extraordinarily difficult exploration problems because there are no easy oilfields any more. This message was more apparent than ever at the 2012 American Association of Petroleum Geologists meeting in Long Beach, California. Wherever I went—in plenary addresses by the big names in the oil business, and in casual chats with my friends in the industry—I heard over and over again that the future of oil will be into increasingly dangerous and expensive directions: deeper and deeper offshore drilling, drilling deeper on land in more exotic ways, pumping water into oilfields to enhance recovery, converting the Athabasca tar sands of Alberta to energy, relying more on abundant natural gas (especially that recovered by fracking), and so on. Oil geologists will continue to work hard to find oil, but it will be much harder and more expensive to find than ever before, and the days when we could use it freely and waste it without thinking are coming to an end.

One look at the roll of peak oil denialists reveals they are almost entirely economists and cornucopians with no direct experience or training in oil geology, and little or no firsthand familiarity with the subject, either. They talk about finding technological replacements when oil becomes scarce (more on that below), or give unrealistic projections that somehow oil will magically keep appearing and our supplies will be sufficient for the near future. The statement from the Petrobras CEO (cited above) that we would need another Saudi Arabia every two years just to catch up with demand shows how little these cornucopians know about the real situation with oil, or with many of the other nonrenewable mineral resources we discussed. In a few years, we will be facing the reality of scarce oil, and wonder how so many people did not see it, or refused to acknowledge it.

WHAT DO WE DO?

At the end of this decade, in the year 1980, the United States will not be dependent on any other country for the energy we need.

Richard Nixon, 1974

We must reduce oil imports by one million barrels per day by the end of this year and by two million barrels per day by the end of 1977.

Gerald Ford, 1977

Beginning this moment, this nation will never use more foreign oil than we did in 1977—never. Ours is the most wasteful nation on Earth. We waste more energy than we import. With about the same standard of living, we use twice as much energy per person as do other countries like Germany, Japan, and Sweden.

Jimmy Carter, 1977

While conservation is worthy in itself, the best answer is to try to make us independent of outside sources to the greatest extent possible for our energy.

Ronald Reagan, 1981

When our administration developed our national energy strategy, three principles guided our policy: reducing our dependence on foreign oil.

George H. W. Bush, 1992

A significant number of petroleum geologists have warned that the world could be nearing the peak in oil production.

William Jefferson Clinton, 1995

We almost certainly are at or near what they call peak oil.

Al Gore, 2006

We can't conserve our way to energy independence, nor can we conserve our way to having enough energy available. So we've got to do both.

George W. Bush, 2001

Breakthroughs . . . will help us reach another great goal: to replace more than 75 percent of our oil imports from the Middle East by 2025.

George W. Bush, 2006

Much of the world has accepted the fact that the end of cheap oil is coming. Every U.S. president from Richard Nixon onward has admitted that the United States was too dependent on foreign oil and urged the United States to work hard toward energy conservation and alternative forms of energy. This was driven home by the 1973 and 1979 oil crises, which were due to local political events and not the ultimate scarcity of oil. It is beginning to be recognized again as gasoline topped $4 a gallon in the United States during the 2005–2012 interval (fig. 11.1), and never dropped much in price compared to where it started—and this price increase *is* related to the ultimate scarcity of oil.

In fact, many Americans agree (at least in principle) with the idea that we need to end our dependence on foreign oil, so that issue is not really controversial. Every president since Nixon (both Republican and Democrat) has made this point, and it is a popular line in every political speech on this topic. It seems clear to most of us that we do not want our dollars going to support petro-dictatorships in the Middle East or Venezuela, and we do not want to have a huge trade deficit due to foreign oil, and we do not want to keep getting entangled in the morass of Middle East politics due to our oil dependency. A number of environmentally active Americans are already walking the walk and reducing their carbon footprints, reducing their driving or switching to cars that get better mileage, and conserving energy in their homes. But every large-scale attempt to change American energy consumption plan has failed at the government level because there are too many powerful special interests (primarily oil companies) that spend millions on their political puppets who then do their bidding.

There are some who say, "If cheap oil ends, we will just go to alternatives like natural gas, or coal, or nuclear power." This sounds fine in theory, but when you look hard at the evidence, it does not hold water (or oil). First of all, even if we stopped using oil for our cars tomorrow, there would still be huge demands in other areas. Most of our nation's power plants are oil or gas burning, and they account for a huge percentage of our consumption. For the moment, natural gas is more abundant than oil, but it still has a significant carbon footprint, and cannot be used to run automobiles or produce synthetics. If we got rid of oil- and gas-powered energy, we would have to go to nuclear power (which is still controversial here, thanks to the Three Mile Island disaster), or coal. We do have abundant supplies of coal in the United States, but as many people have shown, coal is one of the dirtiest and nastiest of energy alternatives. Most coal must be extracted either by dangerous shaft mining (which is expensive and produces relatively low quantities of coal) or by strip mining, which literally rips a landscape apart. As discussed in chapter

3, most coal is high in sulfur, so it has long been the major source of acid rain. Finally, coal produces far more greenhouse gases than does oil or natural gas, so coal does not solve our carbon footprint problems. And no one is even thinking of using coal to run cars any more (let alone going back to the coal-fired steam locomotives of the past).

People also forget (or do not realize) that we use oil for many other things besides energy. Nearly every synthetic substance you use, from the huge array of plastics in every product we own, to all the fabrics (nylon, rayon, Dacron, polyester) are produced from cheap oil. Just look around you and you will probably see dozens of plastics and synthetic fabrics in your clothes, and nearly every object in a typical room has plastic in it. When oil becomes too expensive for these things, what will we do? Suddenly, we will no longer be able to import thousands of cheap plastic toys for our kids' Happy Meals, or wear synthetic fabrics (even when we need our polyester or spandex), or use products made largely of plastic (such as the computer parts I am using right now), or waste plastic on water bottles which we throw away by the billions. (Ironically, the tap water of most American cities is as good or better than bottled water, but this is an industry that created a demand for an unnecessary product). When cheap oil becomes expensive, plastics will have to be recycled and rationed, and become much too precious for most ways we use and waste it today. And you cannot make plastics cheaply from anything but oil—not coal or natural gas or anything else.

Anyone who lives in the Farm Belt knows that there is another huge consumer of oil: agriculture. When I lived in the farming country of central Illinois for three years, it struck me that all the advertisers during the dinnertime news broadcasts (aimed at farmers when they were having dinner and watching the upcoming weather reports) were producers of fertilizers, herbicides, and pesticides. All of these products are derived from oil. Nearly every strain of corn we use today is genetically modified by Monsanto to be immune to their powerful herbicide called Roundup, which kills all plants except this modified strain of corn. Thus, Monsanto can sell both the corn and the poison, ensuring a large crop each year. (To top that off, Monsanto genetically engineered the corn to be infertile, so the farmers must buy new seed from them each year as well.) An acre of corn consumes 80 gallons of oil in the form of pesticides, fertilizers, and fuel for the tractors. We have replaced the human and animal labor of a century ago with machinery that requires lots of cheap oil. Our entire modern agricultural system of monoculture crops that have no resistance to pests, and which deplete the soil rapidly, can be sustained only by throwing oil at it in the form of herbicides, pesticides, and fertilizers. Without it, our food supply would collapse, and the world would be looking at a global famine. The end of cheap oil will force everyone to reexamine agricultural practices, since you cannot make most pesticides or fertilizers out of coal.

Many of the energy alternatives once touted in political campaigns turn out to be illusions. Take the example of biofuels. They have been hyped way beyond their actual worth because they are popular in the Farm Belt, where politicians must curry favor (especially in Iowa, which has huge power because it holds the first presidential caucus). When there was a surplus of corn in the early 2000s, everyone was talking about turning it into ethanol and using it for fuel. But the end result was a classic example of unintended consequences. The increased consumption of corn for biofuels helped contribute to a worldwide food shortage, so that now most corn goes to fuel ethanol or animal feed, and very little is intended for human consumption. Meanwhile, other countries saw the opportunities, and began to cut down pristine rainforests (with their valuable effect of pulling carbon dioxide out of the atmosphere, and of maintaining the highest diversity of land life) and replaced it with biofuel crops such as sawgrass. As reported in *Time*, 750,000 acres of Brazilian rainforests (an area the size of Rhode Island) were cut down in just six months in 2007, all to raise biofuel crops.[24] When you do the calculations, one person could be fed for an entire year on the corn required to produce one tank of gas from biofuels. One editorial cartoon lampooned this brilliantly. It shows a rich fat American in a nice suit pulling the ear of corn away from a starving African child and saying, "Excuse me. I'm going to need this to run my car."[25]

Yet there are also signs of hope. Each time oil prices rise abruptly, or cross some psychological barrier (say, $4 a gallon), people *do* conserve, cut down on unnecessary driving, get rid of their gas-guzzlers and invest in higher-mileage cars. We may not be able to get Americans to act by preaching at them or by trying to get our political system to work in our best interests, but economic pressures do seem to work. And there are models for an even better alternative to the roller coaster of oil prices. Nearly all the European and Asian countries that import all their oil have already adopted measures that greatly reduce consumption. Through taxation, most of these countries price their gas at a realistic rate that reflects its true externalized cost in terms infrastructure and environmental damage (usually $5–$10 a gallon), so people are strongly inclined to conserve gas and drive small cars only when their excellent systems of public transport are not sufficient. Those taxes on gas then go into the energy and transportation infrastructure, so the citizens get better mass transit, better roads, and they are invested heavily in energy alternatives—such as nuclear, wind, and solar power.

Paul Roberts, in his excellent *The End of Oil*, describes a model for other countries: Germany.[26] Before 1990, German politics were controlled by big industry (especially coal companies) and coal miners' unions. But the 1986 Chernobyl nuclear accident in nearby Ukraine galvanized the environmental awareness of the entire world, and by the 1990s, wind farms and other energy alternatives were rapidly emerging in Germany, spurred on by a law passed in 1990 to invest in carbon-free energy production. In addition, the Green Party became a significant force (not just a token party, as it is in U.S. politics). In 1999, the Greens won enough seats that they formed a coalition government with Gerhard Schroeder's moderate Social Democrats. Soon laws were passed, policies changed, and subsidies granted, and Germany was meeting a higher and higher percentage of its energy needs with wind and solar power, along with biomass facilities that burn crop waste to make energy. Now Germany leads Europe in its energy conservation efforts, reduced carbon footprint, and the research and development of alternative energy sources. German policies are closely emulated by the Scandinavian countries, France, and many other European countries. This was all achieved while Germany was continue to thrive economically; even now, they are the strongest economy in Europe, and hold up many of the weaker nations in the Eurozone such as Greece, Portugal, Spain, and Italy. And today Germany is less dependent on foreign oil than ever, and has an economy stronger than ours in many ways. Americans who despair of our getting out of our current addiction to foreign oil need only look to Germany, Scandinavia, France, and Japan to see that if there is enough economic pressure and political will, there is a way.

I have taught introductory geology for non–science majors at Vassar College, Knox College, Pierce College, and Occidental College for over thirty-three years. I teach almost half the student body, since my course is a popular way to fulfill their lab science requirement. Unlike those at many universities, my course is not Rocks for Jocks, but it as rigorous as our time and resources allow, with customized labs and several field trips. After covering all the standard intro geology topics, I always end the class with my final lecture on the topic of energy resources and world population growth. The students call it the "scare the shit out of you" lecture. But year after year I have heard from grateful alumni who come back and tell me how much that lecture changed their lives and attitudes, and how amazed they are to see my predictions from the 1980s and 1990s come true. But the subject of this chapter is no secret or surprise. Geologists have known about it for almost fifty years now, we mention it in our introductory textbooks, and many of us keep trying to make the public aware of these inconvenient truths. If we do not come to terms with the realization of the end of cheap oil and the shortage of most mineral resources, we will only suffer more as these predictions become realities.

FOR FURTHER READING

Bartlett, A. A. 1978. Forgotten Fundamentals of the Energy Crisis. *American Journal of Physics* 46 (9): 876–888.

Campbell, C. J., and J. H. Laherrere. 1998. The End of Cheap Oil. *Scientific American* 278 (3): 78–83.

Cook, E. 1975. The Depletion of Geologic Resources. *Technology Review* 77: 15–27.

Deffeyes, K. S. 2001. *Hubbert's Peak: The Impending World Oil Shortage*. Princeton, N.J.: Princeton University Press.

———. 2006. *Beyond Oil: The View from Hubbert's Peak*. New York: Hill and Wang.

———. 2010. *When Oil Peaked*. New York: Hill and Wang.

Downey, M. 2009. *Oil 101*. Norman, Okla.: Wooden Table Press.

Goodstein, D. 2005. *Out of Gas: The End of the Age of Oil*. New York: W. W. Norton.

Heinberg, R. 2005. *The Party's Over: Oil, War, and the Fate of Industrial Societies*. New York: New Society.

———. 2007. *Peak Everything: Waking Up to a Century of Declines*. New York: New Society.

Hyne, N. J. 2001. *Nontechnical Guide to Petroleum Geology, Exploration, Drilling and Production*. 2nd ed. Norman, Okla.: PennWell.

Klare, M. T. 2002. *Resources Wars: The New Landscape of Global Conflict.* New York: Holt.

———. 2005. *Blood and Oil: The Dangers and Consequences of America's Growing Dependence on Foreign Petroleum.* New York: Holt.

Leggett, J. 2005. *Half Gone: Oil, Gas, Hot Air and the Global Energy Crisis.* London: Portobello.

Maass, P. 2009. *Crude World: The Violent Twilight of Oil.* New York: Vintage.

Roberts, P. 2004. *The End of Oil.* New York: Houghton Mifflin.

Simmons, M. R. 2005. *Twilight in the Desert: The Coming Saudi Oil Shock and the World Economy.* New York: John Wiley.

Tertzakian, P. 2006. *A Thousand Barrels a Second: The Coming Oil Break Point and the Challenges Facing an Energy-Dependent World.* New York: McGraw-Hill.

12

Far from the Madding Crowd: Human Overpopulation and Its Consequences

Democracy cannot survive overpopulation. Human dignity cannot survive it. Convenience and decency cannot survive it. As you put more and more people into the world, the value of life not only declines, it disappears. It doesn't matter if someone dies. The more people there are, the less one individual matters.

Isaac Asimov

THE TICKING TIME BOMB

All of the issues of resource depletion discussed in the previous chapter boil down to one fundamental issue: human population size and growth. This topic has been controversial and highly politicized for several centuries. In 1729, Jonathan Swift wrote the wickedly satirical *A Modest Proposal,* suggesting that the British Isles deal with their exploding population by letting the rich eat the poor. In 1798, Thomas Malthus saw the huge explosion in the growth of poor people in early industrial Britain, and pessimistically predicted that human populations would continue to grow out of control, held in check only by the Four Horsemen of War, Disease, Famine, and Death. Since that time, scholars and pundits and scientists have looked at the issue and held a wide range of opinions of how serious it is and what could be done about it.

In the heyday of ecological consciousness of the late 1960s and 1970s, biologists such as Paul Ehrlich of Stanford University became famous for arguing that the population situation had become dire. His 1968 bestseller *The Population Bomb* was the first major popular work to bring the issue to a wider audience. Along with two other scientists, Ehrlich formed the organization Zero Population Growth in the same year. Ehrlich's alarming predictions caused many people to reassess the issue of how many people the planet could support, especially since the growth of human populations was accelerating at an exponential rate. Naturally, the predictions themselves were closely scrutinized, and some that were too extreme did not come true, largely because neither he or anyone else could anticipate the efforts of Norman Borlaug and other horticulturalists to generate more food in the "Green Revolution." The libertarians and "cornucopians" who hate government interference and believe that human populations can grow indefinitely taunted him about the failed prognostications, but do not mention that most of his predictions were on target. As Ehrlich pointed out in a 2004 interview,

> When I wrote *The Population Bomb* in 1968, there were 3.5 billion people. Since then we've added another 2.8 billion—many more than the total population (2 billion) when I was born in 1932. If that's not a population explosion, what is? My basic claims (and those of the many scientific colleagues who reviewed my work) were that population growth was a major problem. Fifty-eight academies of science said that same thing in 1994, as did the world scientists' warning to humanity in the same year. My view has become depressingly mainline![1]

In the same interview, Ehrlich also noted that about 600,000 people were on the verge of starvation at that time, billions more were undernourished, and most of his predictions were correct. In fact, in 2006 there were 36 million people who died of starvation worldwide (mostly in the underdeveloped world), and this number continues to climb.[2] Thanks to prophets such as Ehrlich during the 1970s, the majority of government organizations around the world took population growth seriously, and to some degree their actions to curb population growth have helped prevent Ehrlich's most dire predictions from coming to pass.

Before we go further into the evidence and arguments about human population growth, let us step back in time and provide some context.

ONCE UPON A TIME . . .

> Which is the greater danger—nuclear warfare or the population explosion? The latter absolutely! To bring about nuclear war, someone has to DO something; someone has to press a button. To bring about destruction by overcrowding, mass starvation, anarchy, the destruction of our most cherished

values-there is no need to do anything. We need only do nothing except
what comes naturally—and breed. And how easy it is to do nothing.

Isaac Asimov

For more than 99% of human history, our species was not very numerous or widespread. When anatomically modern *Homo sapiens* first
appeared in Africa 100,000–120,000 years ago (assuming the Klasies
Mouth Cave fossils are among the earliest members of *H. sapiens*), our
population was restricted to one continent, and probably fluctuated between 10,000 and 100,000 individuals.[3] Atkinson, Gray, and Drummond
used genetic evidence to estimate that the early human populations in
Africa generally numbered less than 100,000,[4] and diversified all over
the continent due to some major cultural innovation before they first left
Africa around 70,000 to 50,000 years ago, and slowly expanded around
the globe.[5]

For most of the first four million years of hominid history, humans
were small and weak and often prey to larger animals—such as lions,
hyenas, and leopards. Indeed, hominid fossils—with holes in the skull
that match a leopard's canines, showing that humans were the hunted as
well as the hunters—have been found in caves in South Africa.[6] They had
only the simplest stone tools, and not even fishhooks or other more advanced artifacts. Most of the anthropological evidence shows that early
humans lived nasty, brutish and short lives (twenty to thirty years, and
often much shorter) before they died of accidents or predators or warfare
with fellow humans. Given the harsh life of a hunter-gatherer (as we can
still see in some African tribes today, such as the Kalahari Bushmen),
human populations could never expand much on such meager resources
and under such dangers. Angela and Angela (1993) estimate that about
ten billion individual members of our family Hominidae had lived during the four million years of human existence. Only with the invention
of agriculture and other methods that would sustain larger numbers of
people could the world population expand significantly.

Another recent discovery has given us further reason to suspect that
worldwide human populations have long been at a very low sustainable level. A huge volcanic crater in Indonesia known as Toba erupted

several times about 74,000 years ago. It was such a huge explosion and released so much volcanic debris (at least 2,800 cubic kilometers) that it dwarfs any volcanic eruption known in historic times, including the catastrophic 1815 eruption of Tambora and the 1883 eruption of Krakatau, also Indonesia. These post-Toba historic eruptions caused so much dust in the stratosphere that it caused global cooling and a "year without a summer," which led to famine, disease, and starvation as crops failed to grow. But the Toba eruption was several hundred times larger than Tambora or Krakatau, and would have caused a global climatic catastrophe. Not only did it blanket all of South Asia with a layer of ash 15 cm thick, but the stratospheric dust clouds are also thought to have caused global temperature to drop 3–5°C and led to another glacial pulse around the globe, based on chemical evidence from deep-sea cores from the Indian Ocean and South China Sea, which also have a layer from this ashfall.[7]

As this geological evidence was being developed in the 1990s, geneticists found evidence of genetic bottleneck in human genomes, where the human population worldwide dropped to only 1,000–10,000 breeding pairs—and the molecular clock dates on this bottleneck coincide with the eruption of Toba.[8] These bottlenecks about 74,000 years ago occur not only in our own genes, but also in those of some of our parasites (such as the human blood louse[9]) and pathogens (such as the *Helicobacter pylori* bacterium that causes ulcers).[10] According to the model proposed by Ambrose, and by Rampino and Ambrose, the extreme nature of the global winter caused by the Toba eruptions would have caused severe global cooling and darkness for at least a decade, radically reducing the human food supply (whether plant or animal), and causing human populations to plummet from nearly a million to just a few thousand to tens of thousands.[11]

There are, of course, skeptics of this startling story, but there are also surprising lines of evidence that support it. As we decipher the detailed sequences of more and genes of different animals, we find that they too went through a population bottleneck around seventy thousand years ago. This has already been shown in chimpanzees,[12] orangutans,[13] macaques,[14] gorillas,[15] and tigers.[16] Thus, there might be different degrees of acceptance of the extreme Toba population crash scenario, but it clearly was a worldwide event that had effects on nearly every animal on land.

From a bottleneck about 70,000 years ago, human populations slowly recovered to about 1 million people around the globe at 50,000–40,000 years ago.[17] Jared Diamond calls this time the "Great Leap Forward," because this was when the archaic stone tools of most of human prehistory were replaced by advanced stone tools, needles, awls, fishhooks, spear throwers, plus jewelry and the cave paintings we associate with the Cro-Magnon culture.[18] Human population stayed stable around this number through most of the rest of human prehistory (fig. 12.1), since the limited resources of primitive agricultural societies and the problems of disease and accidents and warfare and infant mortality prevented human populations from expanding much further. This Great Leap Forward also coincides with the first time *Homo sapiens* spread from Africa and Eurasia and invaded other areas, such as Australia. By the end of this period, Neanderthals were extinct in Europe, and only modern *Homo sapiens* remained. Whether modern humans drove the Neanderthals to extinction is another, and very controversial, question.

As the great ancient civilizations of Egypt, Mesopotamia, China, and the Indus Valley arose six or seven thousand years ago, larger agricultural resource bases sustained many more people, and populations began to grow again. Most estimates place the world population between 5 and 20 million around seven thousand years ago.[19] By about 300 CE, there were about 55 million people in the Roman Empire alone, and about 160 million worldwide.[20] By 1340, there were about 70 million people in Europe, until the Black Plague reduced populations significantly.[21] Worldwide, the bubonic plague is estimated to have reduced the population from 450 million to less than 350 million by the year 1400.[22] It took another two hundred years for Europe's population to return to pre-1340 levels.

Websites maintained by the U.S. Census Bureau (www.census.gov/ipc/www/worldhis.html) and Scott Manning (www.digitalsurvivors.com/archives/worldpopulation.php) summarize the various scholarly estimates of how human populations have climbed in the last ten thousand years. The first time worldwide human populations surpassed a billion people was in 1804, as the effects of the Industrial Revolution and the spread of modern medicine, vaccination, and modern sanitation reduced

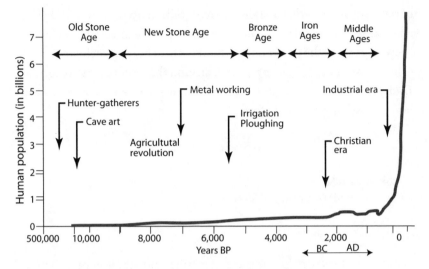

FIGURE 12.1. Human population over the past few thousand years.

the rates of dying from disease. As we saw in chapter 7, child mortality, even in civilized London, was a horrendous 75% in the 1730s and 1740s, but dropped to about 30% in the 1810s and 1820s.[23]

From that point forward, population growth around the world began to accelerate, first in the developed nations of Europe where the Industrial Revolution and the scientific revolution in medicine started, and then in the underdeveloped world in the twentieth century, when industrialization, modern agriculture, and modern medicine spread worldwide. The reasons for this are simple: as long as modern medicine reduces many forms of death (especially infant mortality), and industrialization and modern agriculture provide more food and economic opportunities, the population will quickly grow. Our first billion was reached in 1804, and by 1927, 123 years later, we had our second billion. Then the growth began to accelerate and population shot up like a rocket (fig. 12.1). Our third billion was achieved in 1960, only 33 years after we had 2 billion people. We reached 4 billion in 1974, less than half the time it took us to go from 2 to 3 billion. We reached 5 billion in 1987 (only 13 years later), and 6 billion in 1999. We passed 7 billion people on October 31, 2011 (not a Halloween prank, unfortunately). Based on previous rates of population growth, we should reach 8 billion by 2025, and 9 billion by 2045, at which time there will be as many people alive on the planet as there were in all the previous 4 million years of human prehistory. Put in other terms, 14 or 15 babies are born around the world every six seconds. Each hour there are nearly 8,800 more mouths to feed; each year, 77 million more than the year before.

DO THE MATH!

The greatest shortcoming of the human race is our inability
to understand the exponential function.

Albert A. Bartlett

If these statistics do not seem alarming enough, it is because most people do not comprehend the mathematics of exponential growth. Even in better educated, more developed nations, an alarming number of people have only the most basic grasp of math, so this is not surprising. Albert Bartlett, a physicist now retired from the University of Colorado, first pointed out the implications of this in 1978.[24] Responding to Jimmy Carter's pleas for energy conservation during the Iranian oil and hostage crisis, most critics of the president faulted him for not emphasizing increased domestic oil production. As discussed in the previous chapter, the idea that the United States could produce even a small part of the oil it needs domestically is pure fantasy, and this was true even in the 1970s. But Bartlett steps back from the immediate energy crisis to point to a larger issue: the issue is part of the overall problem of exponential growth of population and consumption.

When people think of "growth," they think about the slow incremental change of a growing flower or child. Such growth is usually constant and nearly linear, so that during the active growth phase the same amount of growth happens in the same increment of time (fig. 12.2). A good example of linear change is the sand in an hourglass: the amount of sand in the lower chamber accumulates with a constant rate, so that it is a good measure of time ticking away. Another example is the way a candle burns down the same amount of wax in a fixed amount of time,

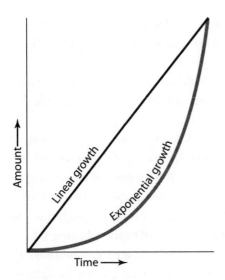

FIGURE 12.2. Examples of growth curves, contrasting linear growth with exponential growth.

which is why the Romans used candles with marked intervals as simple clocks (centuries before the modern clock, with its springs and gears, was invented).

In contrast to this slow, steady linear change, many systems in nature (bacterial growth, audio feedback, nuclear chain reactions, radioactive decay, viral epidemics) and in human systems (human populations, resource consumption, processing power in computers, internet traffic growth, compound interest, Ponzi schemes) have an *exponential* growth pattern (fig. 12.2). In exponential growth, the amount you have at Time One is a function of how much you had at Time Zero. Typically, it is expressed a fixed percentage of growth per year. One common exponential growth pattern is doubling whatever amount you start with at each time interval. If in each time interval you double the amount you had previously, then the total size of the population will grow slowly at first (fig. 12.2), in what is known as the lag phase. But as the standing crop gets larger and larger, doubling an already large starting number makes each step bigger and bigger, so the growth then takes off like a rocket (the log phase of growth). Once this happens, the numbers become astronomical, and soon they are greater than the total number of atoms in the entire universe. A quick way to estimate how long this doubling takes is to divide 70 by the percentage growth rate. Thus, if something grows at 7% per year, it will double in size in ten years.

Mathematically, this is a simple problem, but it is something that hard for most of us to grasp intuitively. Bartlett provides a vivid example that has been given in many different versions. Imagine you are a ruler who becomes devoted to playing chess, and you want to reward the wizard who introduced it to your court. The mathematically clever wizard tells the ruler that for his reward, he wants one coin (or grain of wheat, or what have you) on square 1 of the sixty-four-square chessboard, two on square 2, four on square 3, eight on square 4, and so on. In simple terms, the number is increasing as powers of 2, with the first square having 2^0 (= 1), the second 2^1 (= 2), the third 2^2 (= 4), and the final square should have 2^{63} coins. How many coins is 2^{63}? That comes to 9×10^{18}, or 9 with eighteen zeroes after it (9,000,000,000,000,000,000), or 9 trillion trillion! All on a single square (physically impossible, of course). And remember the rest of the chessboard. When you add up all the squares, you would have 1.8 $\times 10^{19}$ coins, or 18 trillion trillion coins—more than there are atoms in the solar system!

This example brilliant captures the counterintuitive nature of exponential growth. At first it is deceptively slow in the lag phase (1, 2, 4, 8, 16, 32 . . .). But as the standing crop gets larger and larger in the log phase, each time it doubles it becomes enormous, until it has reached astronomical numbers that are physically impossible. But other examples are just as instructive. For example, at a constant rate of inflation (12% inflation rate), in seventy years a 60-cent loaf of bread will be worth \$2,458! Take the example of compound interest: if you invest \$1 at 5% interest per year, it will be \$72 billion in only five hundred years!

This last example is a plot element in the Mike Judge movie *Idiocracy*, in which Joe Bowers, a man of average intelligence (played by Luke Wilson) wakes up from a suspended animation experiment to find himself centuries in the future, in which the whole world has been dumbed down and he is a relative genius. To bribe Frito (Dax Shepard), who claims to have access to a time machine, Joe promises to travel back in time, invest money in Frito's name, and then when Frito checks the bank account, he will find he is a millionaire. Five hundred years of compound interest would generate a LOT of money! The plan is doomed to fail since the "Time Masheen" is really an amusement park ride, but at least Joe understands compound interest.

Take another example: Jimmy Carter said in a speech, "In each of these decades (the 1950s and 1960s), more oil was consumed than in all of man's previous history combined."[25] People refused to believe this, but Carter was telling the truth. The growth rate of oil consumption averaged 7% per year for many decades. Thus, using the formula above, the doubling time is only ten years. It is indeed true that in the 1950s, we consumed as much oil as in all previous human history, and in the 1960s, we consumed as much as we consumed in the 1950s and all previous decades, and so on. There have been some fluctuations of that growth rate since the 1960s, but in general our demand for oil resources keeps expanding at almost the same absurdly high rate—and clearly this cannot continue. Recent estimates have suggested that we could double our current oil demand by 2050 or sooner. As we saw in the previous chapter, this will run into problems since the oil supply has peaked worldwide and is now diminishing.

Of course, it is obvious that nature does not allow things to reach such absurd astronomical quantities. Here we run up against something we discussed in the previous chapter: *the limits of growth and resources.* In natural systems, populations may expand exponentially in limited circumstances in which there are huge amounts of space or other resources. But as soon they go through a few doubling cycles, they run up against the limits of their environment and reach a stable state, or have a population crash and die out altogether if they exhaust all the resources. Ecologists talk about this steady plateau of no further growth as an equilibrium state, or the *carrying capacity* of the environment (fig. 12.3). Nature exhibits this in many different animal and plant populations, but humans seem to think they are exempt from the laws of nature, and that they can keep growing indefinitely.

Let us use another example to illustrate this folly. Imagine a petri dish in which you have placed a single bacterium, and this bacterium (and all its descendants) will split into two cells every minute in this

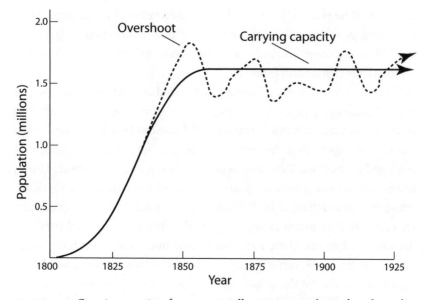

FIGURE 12.3. Carrying capacity of an exponentially growing population forced to stabilize growth at a particular level.

rich vacant patch of resources that is the petri dish. If there is one cell at the start, and the dish is full at sixty minutes, when will it be half full? Answer: at minute fifty-nine! Remember, the standing population at minute fifty-nine will double again in a single minute to fill the dish, so it must be half full exactly one minute before it is completely full. Now let us imagine that some intrepid bacterium pioneer wants to relieve the packed petri dish by exploring outside the dish and finding new petri dishes to populate. How long would it take this dish to fill another whole dish? Only one more minute! And it would fill two more dishes in two minutes, and so on. So much for the fantasies that we can solve our population and resource problems by going to outer space! Even if it were not so expensive, and the other planets were rich in resources (contrary to the scenario described in the movie *Avatar*—in which there is a rich lode of "unobtanium"—they are not), it will still not solve any problems when our standing population is huge and capable of doubling in very short order.

IN GROWTH WE TRUST

"Smart growth" destroys the environment. "Dumb growth" destroys
the environment. The only difference is that "smart growth"
does it with good taste. It's like booking passage on the *Titanic*.
Whether you go first-class or steerage, the result is the same.

Albert Bartlett

This illustrates another point: with exponential growth, a resource, a petri dish, or a planet can seem only half full and very spacious—and then the population doubles in a short time interval and suddenly it is completely full. Thus, even when it seemed like there was plenty of room for human population growth (as it did during the early twentieth century), the global population doubled and doubled again so fast that it was soon at capacity and then over capacity—and now we find cornucopians arguing that there is room and resources for another few billion people!

Nevertheless, these limits are very real, and there are no magical extra petri dishes that will give us a reprieve. Take, for example, the resources of energy and other minerals from the earth discussed in the previous chapter. As we saw in our summary of the opinions of nearly all the oil and mining geologists who really know how the system works (and not economists who work in a fantasy world of unlimited supply and demand), we are close to reaching the worldwide limits on many resources, and are on the downhill slope from the Hubbert peak of most of the important ones. As we discussed in chapter 11, with limited or inelastic resources, simple supply and demand do not work over the long run; no matter how much you raise the price, *you cannot have more if there is no more to be mined*. Yet worldwide population and demand continue to grow, and now we are seeing the price pressure on many of these resources as this demand has consequences. Our recent global recession has temporarily slowed demand in some parts of the world, but the global price of oil has not dropped that much despite the decreased demand, since the pressures from rapidly developing nations with huge populations, such as China and India, will not decrease over the long run. Once this recession is over, expect the pent-up demand to force those oil prices rapidly upward.

This raises an even larger point about resources and population: the global inequity in how they are produced and consumed. For most of the past few decades, there was a striking inequity operating in that the United States and other developed nations consumed roughly 70% of the world's resources, but produced only 30% of them. Conversely, 70% of the world's populations in the underdeveloped nations are left with only 30% of the resources to consume. To nobody's surprise, this is viewed as unfair, as people in China and India and elsewhere see our TVs and fancy cars and nice homes, and think it is their turn to enjoy their own wealth and amenities, and have what we have. Yet the cruel twist in this story is that we greedy Westerners consumed so much of the world's resources over the past few decades that we are now all in the same boat of fewer resources for an expanding global population. Naturally, this is no recipe for global stability, especially as the severe pinch of declining oil supplies and other resources begin to hurt all the global economies over the next few decades, and all people, rich and poor, are forced to cope with a declining standard of living.

As Bartlett points out, the real enemy here is the perverse psychology that *growth is good*. Our national motto is "In God we trust" but should be "In growth we trust." We measure our economic prosperity using such indicators as the GNP and GDP, which are indicators of how fast our economy is growing. Politicians of all persuasions constantly tell the voters that they will do what it takes to "grow the economy," which is always a popular message since the leaders of all the world's nations usually rise or fall with the economic fortunes of their people. When James Carville crystallized the 1992 Clinton campaign message as "It's the economy, stupid," he represented a common thread in the politics of nearly all nations: voters think about their own pocketbooks first, and seldom let global geopolitics or other issues take precedence unless there is a war, or unless the economy is growing robustly and they have the affluence to worry about other issues.

Yet not all growth is good. Cancer is a form of growth that we fear, and the faster it grows, the faster the patient dies. We should realize that growth that leads to bad consequences is much like a cancer. So the message we are receiving from the mathematics of exponential growth and limited resources is that we have very little time left, and we cannot afford to grow too much more or consume things too quickly. Otherwise, we will just set ourselves up for a crash when the world is much too over-crowded and there are fewer and fewer resources for everyone. Instead, we need to view the earth as a small lifeboat with limited supplies, and we should be trying to reach a stable state, not a growing demand that our lifeboat cannot support.

A 1978 article by Bartlett features numerous quotations from various leaders and media sources that demonstrate our complete misunder-standing of this fundamental dynamic. The people quoted therein ad-vocate more and more rapid growth and consumption even as the limits are already in view, just like bacteria that are multiplying in a nearly full dish, heedless of the consequences of their rapid growth and expansion. Most assume growth as a good thing, and argue that we have enough fuel resources such as coal to supply us for a long time. Missing from all these optimistic estimates is the reality that if we suddenly expand our mining and consumption of coal, it will not last very long as a resource. The quo-tations usually mention its production lifetime *using present rates* of min-ing and consumption, but if rates increase, then their projections are far too optimistic, and coal will be gone much sooner (none of these quota-tions mention the huge negative consequences of coal mining: huge strip mines, giant amounts of greenhouse gases, and additional acid rain).

Added to this fundamental ignorance of the mathematics of expo-nential growth, and the wrongheaded notion of pushing more growth as a solution to problems caused by too much growth, are additional fallacies. For example, as stated in *Time* magazine:[26] "Energy industries agree that to achieve some form of energy self-sufficiency the US must mine all the coal that it can." On August 31, 1977, in a three-hour CBS special on energy, Gerald Ford's energy advisor, William Simon, said, "We should be trying to get as many holes drilled as possible to get the proven oil reserves."[27] In other words, we need to exploit our resources even faster to keep up with our needs, and then deplete them sooner, not try to reduce our consumption so our supply might last longer. The late David Brower, longtime leader of the Sierra Club, called this strategy "Strength through exhaustion."[28]

In short, the eponymous cartoon opossum in Walt Kelly's classic comic strip, *Pogo,* said it best, "We have met the enemy and he is us." We humans are to blame for overpopulating the planet, and as we can now see the consequences of too many people and too much consumption, we are still blindly committed to the disastrous policies and thinking that got us into this mess in the first place. We are suffering from a dangerous cancer growing too fast, and we prescribe *more cancer* as a cure for the problem!

Yet there have long been voices that have argued a different viewpoint: there are limits to growth, and we are just passengers on this fragile lifeboat called earth, and must be careful about ruining it not only for ourselves but for all life on this planet. The Club of Rome, a group led by the late Donnella Meadows, started the most influential discussion with its 1972 book *The Limits to Growth.* Environmentalists and many other scientists have since pointed out over and over again that population growth and wasteful resource consumption cannot go on much longer; we need to change our habits. Although today the topic is not as trendy in the United States as it was in the environmentally conscious 1970s and 1980s, the spread of recycling, the emphasis on energy conservation, and the efforts to curb our oil use by changing our cars and building more mass transit are evidence that the message is gradually reaching a lot of people, and action is starting to occur. In most Western European countries, it has already happened as they have cut way back on their imported oil and wasteful energy practices, and aggressively adopted strategies for recycling and conservation and fighting acid rain and global warming in ways that the United States can only begin to imagine. Such changes are possible—it only takes people with the foresight to see the trouble coming, and political leaders who are willing to speak the truth and not repeat old platitudes about growing the economy.

THE GOOD NEWS AND THE BAD NEWS

Can you think of any problem in any area of human endeavor on any scale, from microscopic to global, whose long-term solution is in any demonstrable way aided, assisted, or advanced by further increases in population, locally, nationally, or globally?

Albert Bartlett

Thanks in large part to the environmentalists' 1970s message about population growth, many countries took notice and began to encourage birth control and other measures to slow the expansion of their populations. India pushed many forms of birth control, and China instituted a draconian policy of forbidding their people to have more than two children. Other factors came into play as well. Due to better health care, the people of the most affluent nations no longer need large families to offset the historic pattern of loss of children in infancy, nor do they need lot of kids to help with the family farms; most people in the developed world are residents of cities and suburbs. Consequently, these areas went through a *demographic transition,* and their birthrates dropped significantly. In countries such as Japan and some Eastern European countries, birthrates are now so low that they have declining populations, which some people regard as a problem. By the year 2000, the United Nations estimated that the worldwide population growth rate was only 1.16% (it was as high as 2.2% in 1963), or about 140 million new people born each year (in contrast to the 173 million born each year in the late 1990s, the peak of the world population boom).[29] These are all encouraging signs that the catastrophically high population growth rates of the past few decades are finally slowing down and might eventually level off, possibly at 9.2 billion around the year 2050, according to some United Nations estimates.[30] Other estimates place this number as high as 10.5 billion, but at least we are not hearing the alarming 15 to 20 billion people in this century that was forecast only decades ago.

Even in view of these optimistic projections, there are still alarmingly high birth rates in much of the impoverished and underdeveloped world, especially in Latin America, the Middle East, and sub-Saharan Africa.[31] According to the United Nations, the highest birthrates of all include 49.6 births per 1,000 people per year in the Democratic Republic of the Congo, Liberia, and Guinea-Bissau in central and western Africa; 49.0 per 1,000 in Niger; 48.0 per 1,000 in Mali; 47 per 1,000 in Angola, Burundi, and Uganda; and slightly lower numbers in Sierra Leone, Chad, Rwanda, Burkina-Faso, Somalia, Malawi, Nigeria, and Mozambique.[32] These rates work out to about six to eight children per couple in an average family, which means many are much larger. Only Afghanistan, with 48.2 births per 1,000 per year, breaks this long string of central African countries. By contrast, the U.S. birthrate is only 14.0 births per thousand per year (ranking 139th on the list), and most developed countries have birthrates in the single digits.

The saddest aspect of these high birthrates in Africa and other im- poverished, underdeveloped regions is that they are responses to the old problems of high infant mortality and deadly diseases that most developed countries seldom worry about, as well as the high death rates due to constant warfare and revolution in some of these countries. There is also famine due to bad droughts in the Sahel region south of the Sa- hara, as well as incompetent corrupt governments that fail to protect or feed their own people, or provide even minimal health care. Given that these problems have plagued the region for over a century, it seems unlikely that these countries will become affluent and developed in the near future and pass through the demographic transition, so we can expect their high birthrates (and therefore global population) to persist. Kenya is now going through a food shortage crisis, because its growth rate of 3% per year keeps producing a population that cannot be fed by its dwindling resources, especially since years of drought have drasti- cally reduced crop yields.[33] But as pointed out above, even countries with relatively low growth rates (anything significantly higher than zero population growth) are contributing to the problem as well. Remember, we have already produced a huge standing crop of seven billion people, so if each family has (on average) more than two children per couple, we will double in population in a short time and greatly exceed the capacity of the planet to hold us all.

THE LIMITS OF HUMAN POPULATION

The question of how many people the world can support is unanswerable in the finite sense. What do we want? Are there global limits, absolute limits, beyond which we cannot go without catastrophe or overwhelming costs? There are, most certainly.

George Woodwell, 1985

No species has ever been able to multiply without limit. There are two biological checks upon a rapid increase in number—a high mortality and a low fertility. Unlike other biological organisms, man can choose which of these checks shall be applied, but one of them must be.

Harold F. Dorn, 1962

A number of the cornucopians (who might better be described as "corn-utopians")—including Julian Simon, Bjørn Lomborg, and Matt Ridley—claim (with a straight face) that seven to fifteen billion people on the planet is no problem. They are convinced that somehow we will magically keep having more Green Revolutions to feed this huge doubling of the population—never mind that modern agriculture is largely dependent on cheap oil for pesticides and fertilizers. Somehow the laws of supply and demand will magically find a lot more oil and other natural resources to sustain this gigantic excess of humans—completely ignoring all the data described in chapter 11 that show resources are rapidly vanishing and no increase in prices will make more of them. They have a touching belief that somehow the net wealth of the world will keep growing and the richer nations will keep the poorer nations affluent as well—never mind the fact that so-called trickle-down global economics has never really transformed the underdeveloped nations. Never mind the facts of the exponential growth dynamic and the evidence we discussed earlier that cancerous growth is a bad thing, not a good thing.

Cornucopians seem to think that a planet with wall-to-wall people and cows and rats, and no room for nature or anything not in service of humans, is a good thing. Such abstract conceptions of planetary well-being may comfort the rich and powerful, snug in their mansions in the United States or Europe, who want to protect their place in society, but one only needs to travel to China or India or any number of African countries (as I have) to see the effects of population growth. Whether you visit the streets of Beijing, Mumbai, Cairo, or Mexico City, you realize how overpopulation has struck most of the habitable places of the world and many regions are already unbearably overcrowded. Countries such as China and India are throwing all caution to the wind to foster their economic growth and keep their huge populations somewhat happy with their lot in life, lest their governments face political turmoil. Of course, they have done so at the cost of environmental protections—often endangering a huge number of people (not to mention the natural world)—and they are committed to increasing their energy production through coal-fired power plants and other means that contribute huge amounts of greenhouse gases and would wipe out any gains made by developed countries in their efforts to curb greenhouse gas emissions.

In contrast to the cornucopians, there is another group of people that argues that the 7 billion people we have is already too much for the planet to sustain, let alone a population that gets much larger. This point of view argues that we must not just think about how many humans we can cram onto this planet with a poor quality of life, and without starving us all to death. Instead, we should think about the ecology of the entire planet, and what level of human population is actually sustainable for the future. A great number of studies have been conducted concerning the carrying capacity of the planet, and how big a human population the planet could support with reasonable affluence for most of the people of the world. Groups such as the Optimum Population Trust (www. optimumpopulation.org) maintain an online population clock that ticks off the total world population every second (this is unnerving to watch because several new people are born every second, greatly outstripping the death rate). The Optimum Population Trust is also a think tank and charity that deals with the problem of population growth and sustainability. You can click on their link to the "Living Planet Report," which talks about the method of ecological footprinting, and their techniques of estimating how much population is sustainable given the known resources of the planet.[34] Using their calculations (last updated in 2008), the carrying capacity of the planet with our present wasteful lifestyles and allowing for some biodiversity is only 4.6 billion people; with the forests, croplands, and grazing areas taken into account, the numbers are in the 4–9 billion range; and if the world keeps emitting greenhouse gases, the carrying capacity is only 2.7 billion! For the most part these numbers are significantly lower than the 7 billion people the planet now carries, and they call for populations about half of what we have now—a frightening prospect. If we compare this to natural examples, we are like the population of rabbits that multiplied beyond the capacity of their environment and then began starving to death until their population dropped and could fluctuate around the carrying capacity threshold (fig. 12.3).

Other approaches use the known quantities of various resources to calculate how much the planet has available, and how much of the terrestrial ecosystem has already been exploited. In 1986, Vitousek, Ehrlich, Ehrlich, and Matson did a simple calculation of the net primary productivity of terrestrial ecosystems around the world, and found that humans were already consuming about 40% of the total back in 1980.[35] Numerous other estimates have been generated since then. The most recent of them, published by a prestigious group of international experts in the preeminent scientific journal, *Proceedings of the National Academy of Sciences,* analyzed the "ecological overshoot" of the human economy, and also estimated that we had grown beyond the sustainable capacity of the earth back in the 1980s, and that we were already at 120% of the biosphere's capacity in 1999.[36] Joel Cohen's excellent *How Many People Can the Earth Support?* examined all the different estimates that have been produced (as of 1995), and found that these estimates vary widely since there are lots of different methods and assumptions.[37] However, most of them put the range between four and nine billion—and we are past seven billion right now. As Cohen shows, the solutions that have been proposed typically fall into three categories: the "bigger pie" approach (using technology to somehow find a way to support more people), the "fewer forks" approach (family planning and more vegetarianism to reduce the number of mouths to feed and how much they affect the planet), and the "better manners" approach (using shared governance and better public policy to more equitably distribute food, land, and wealth among the underdeveloped nations, while encouraging family planning). Of course, how we proceed with the problem is a matter of policy and political discussion, but (except for the cornucopians), there seems to be a widespread consensus among most scholars and scientists that seven billion people is already too many, and there is no chance that our planet can sustain fifteen to twenty billion people.

OUR FELLOW PLANETARY PASSENGERS: DO THEY COUNT?

Instead of controlling the environment for the benefit of the population, maybe we should control the population to ensure the survival of our environment.

Sir David Attenborough

One of the stories that my two young sons and I love the most is Dr. Seuss's ecological fable *The Lorax.* As the story demonstrates, there is an unspoken dialogue between humans and nature that is best captured in fiction (in this case, a child's story). Reacting to seeing the trees chopped down by the villain, known as the Once-ler, the Lorax says, "I am the Lorax. I speak for the trees. I speak for the trees, for the trees have no tongues. And I'm asking you sir, at the top of my lungs—he was very upset as he shouted and puffed—*What's that thing you've made out of my truffula tuft?*" Later in the story, the Once-ler reacts to the nagging of the Lorax by asserting the traditional human dominion over nature: "Now, listen here, Dad! All you do is yap-yap and say, 'Bad! Bad! Bad! Bad!' Well, I have my rights, sir, and I'm telling *you* I intend to go on doing just what I do! For your information, you Lorax, I'm figgering on biggering and biggering and biggering and BIGGERING."[38]

Throughout all these analyses and discussions, the writers we have cited (favorable to more population growth or not) always makes the assumption that humans are all that matter, and no other species or ecosystem is important except in the context of supporting us. Indeed, most human beings regard other species as inferior or something to be eaten or hunted for sport, or occasionally marveled at, but only rarely do you find people in the environmental literature asking this question: Don't all the other species on earth have a right to live as well? By what right does the human species subjugate nature and cause other species to suffer or vanish? Just because other species cannot speak for themselves, or fight back against our cancerous growth and expansion at their expense, does it mean that we have the right to take anything we want or need? Of course, the religious dogmas of the Western world (all the Abrahamic religions, and many others as well) give humans dominion over nature, but many people no longer use these ancient stories from Bronze Age goatherds as a guide for modern living in an overcrowded planet.

Even though we have no Lorax who speaks for the trees, many animal rights and environmental activists have commented that our exploitation of the planet is morally wrong, and that we have no right to drive other animals to extinction or wipe out their habitat. It is not easy to resolve an argument in which we try to balance human values and nature, because opinions are scattered over every possible point of view. So let us look instead on the human impact on the natural world, and see if we humans can continue as we are.

The answer to this is clear: humans have been an ecological disaster, especially in the past few centuries, and the natural world that sustains us is being destroyed at an alarming rate. As Niles Eldredge put it, "We are like loose cannons, able to wreak great damage on our own, and particularly dangerous if our effects happen to coincide with physically induced changes that are also causing extinctions."[39] The most vulnerable and important regions on land are the tropical rainforests, which support by far the bulk of the species on this planet (most of which have still not been discovered or documented by scientists yet, so we really do not know how many species live on this planet even now). *Each day 80 square miles* of rainforest are lost, mostly in tropical Africa, South America, and Southeast Asia. As this rainforest is cut down, dozens to hundreds of species vanish each year. Most of it is cut down by subsistence farmers who have no choice but to do so for their own survival (given the population pressure in their countries), and many countries actually encourage exploitation of their forests for economic reasons (although Brazil is now instituting policies to slow deforestation). Once the tropical rainforests are cut down by slash-and-burn methods, they leave a barren wasteland that supports farming for only a few years before it becomes infertile. Unlike soils in temperate regions, most of the nutrients in tropical soils are trapped in the vegetation, so when the trees vanish, all that is left are barren, highly leached laterite soils that bake into red bricks in the tropical sun. The poor farmer then must find another patch of the forest to cut down once his original land is worthless. Another effect of this deforestation is the release of huge amounts of greenhouse gases as the vegetation burns, whereas a growing forest is one of our best ways to pull carbon dioxide out of the atmosphere.

Wildlife biologist Peter Matthiessen documented the animal diversity of the tropics of Central Africa in *African Silences*.[40] He reports that most of the great forests of West Africa (home to most of Africa's tropical wildlife) are now so heavily hunted by starving people looking for bush meat that nearly all their vertebrate species are gone, from chimpanzees to most monkeys to small deer to even the rodents and lizards. Even the wildlife refuges do not have the resources to prevent poaching, so that no wild population is really safe. Larger animals are also vulnerable, since both the elephant and rhino have been poached to near extinction for their ivory and horn, respectively. As Matthiessen described it, "The great silence that resounds from the wild land without sign of human life, from which all the great animals are gone, is something ominous. Mile after mile, we stare down in disbelief."[41] The wonderful African jungles and savannas that kids and adults both love, and that have been the subject of so many nature documentaries, children's programs, Tarzan movies, and Disneyland rides, have nearly vanished—and are replaced by a sea of starving people.

The same can be said for the marine ecosystem as well. Fish species are vanishing worldwide, and the fishing industry is collapsing as it exhausts the ability of natural populations to recover from more and more efficient methods of catching them. Many biologists are sounding the alarm about the worldwide die-off of coral reefs, which are very sensitive to the changes in the ecosystem, and (like the rainforests) harbor a great diversity of species. Ask marine biologists who used to dive on reefs a few decades ago: they can see the dramatic changes as the reefs die off and bleach and nearly all the reef species of fish and invertebrates vanish as well. Biologists have shown that the biggest component of this die-off is the warming of the oceans as our climate changes, so that the ocean is now past the temperature tolerance of most reef corals. Even as human populations increase, we can count less and less on food from the sea to sustain us, and the dying of the oceans due to greenhouse warming will be as catastrophic as any mass extinction that happened in the distant past.

Indeed, biologists and paleontologists refer to the past few centuries as the Sixth Extinction, because the rate of extinction of species on this planet is thousands of times higher than occurred at any of the previous of the Big Five mass extinctions of the past 600 million years. Biologists now estimate that in the twentieth century alone, between 20,000 and 2 million species went extinct. Current estimates place the rate of extinction at about 140,000 species per year. This is enormously faster than any rate seen in the geologic past. Humans have turned out to be the most destructive force the earth has ever seen, wiping out more species than any mass extinction caused by a meteorite impact or gigantic volcanic eruption or major climate change.

But why should we care, especially if we think that humans have a right to do whatever it takes to survive? As I pointed out in *Catastrophes!* there are a number of possible answers to this question.[42] Some are practical, and others philosophical and moral. The practical answers remind us that nature provides huge benefits to us as humans. Many of our drugs come from rare tropical plants, and wild nature is essential to our food supply and providing other needs. If we wipe out biodiversity before it even has even been documented or studied, there is no way that the next undiscovered species of tropical plant will be analyzed and shown to provide us with a valuable medicine or important natural pesticide or other chemical we require. As James Leape, director general of the World Wildlife Federation, put it, "Reduced biodiversity means millions of people face a future where food supplies are more vulnerable to pests and disease and where water is in irregular or short supply. No one can escape the impact of biodiversity loss because reduced global diversity translates quite clearly into fewer new medicines, greater vulnerability to natural disasters and greater effects from global warming. The industrialised world needs to be supporting the global effort to achieve these targets, not just in their own territories where a lot of biodiversity has already been lost, but also globally."[43]

But what is the big deal if we wipe out a few species here and there? In their 1981 *Extinction,* Paul and Anne Ehrlich posed an interesting analogy.[44] Suppose you were flying in a jetliner and looked out the window to see a rivet pop out of the wing. Then you saw another rivet fall out, then another rivet—one after another. Perhaps one or two rivets would not damage the structural integrity of the aircraft and would be no cause for alarm. But how many rivets before the wing falls apart and you crash? Would you be willing to perform this experiment and take the chance that you would die? That is the unplanned experiment we are forcing on nature right now. Every species lost is another rivet holding the world's ecosystems together. One or two species may not make a big difference, but we really do not know how many can be lost before the entire planetary ecosystem collapses. In the case of the dying coral reefs, and the vanishing rainforest, we may already be past the point of no return.

FIGURE 12.4. The *moai,* giant
stone sculptures of heads from
Rano Raraku, from the extinct
culture of Easter Island. *Image
courtesy Wikimedia Commons.*

Nevertheless, there are many cornucopians and others who insist
that humans are too smart and inventive to ever wipe themselves out
with an ecological catastrophe. In answer to this, one need only look at
cultures documented by Jared Diamond (2004). Among the many cultures that vanished when their resources were outstripped by population
growth, Diamond points to the now extinct culture of tiny Easter Island
off the coast of Chile, source of the world-famous *moai* sculptures of giant heads (fig. 12.4). As archeologists discovered when they unearthed
the story of this civilization, at one time there was a thriving culture
on that island, but when the population got too large they wiped out
their food supply and deforested the island. Archeologists found signs
of stress as conditions got worse, skeletons showed the effects of starvation, brutal warfare, and many other bad consequences of too many
people and too little food. Finally, Easter Islanders died out completely
and left no descendants. When explorers first came to the islands and

discovered the *moai,* they could find no trace of the people who had once carved them.

As historian Will Durant wrote, "Civilization exists by geological consent, subject to change without notice."[45] It is a much more fragile thing than most of us realize, and history and archeology show us that humans can indeed make stupid decisions and outstrip their resources before their culture vanishes. As we look on a planet with seven billion people and with a huge standing population that has the potential to double to fourteen or twenty billion in a few decades, and see all the signs that the planet is already overpopulated beyond carrying capacity, we must ask ourselves if we are among those cultures who foolishly plunged forward with bad policies that ensured their extinction. In fact, 99% of all species on earth that have ever lived are now extinct. As a species we have only been around for about 100,000 years, much less than the average of several million years for most fossil species. As a biologist and paleontologist, I see no reason to be optimistic that humans are destined for survival, especially given our bad habits and propensity for creating things that threaten us even more than overpopulation, such as nuclear weapons and global warming. Sure, the planet and life will survive at some low level of diversity, populated by the ever-hardy rats and cockroaches—but it is doubtful that humans will be around to see it.

FOR FURTHER READING

Bartlett, A. A. 1978. Forgotten Fundamentals of the Energy Crisis. *American Journal of Physics* 46:876–888.

Brown, L., G. Gardner, and B. Halweil. 1999. *Beyond Malthus: Nineteen Dimensions of the Population Challenge.* New York: W. W. Norton.

Cohen, J. 1995. *How Many People Can the Earth Support?* New York: W. W. Norton.

Daly, H. 1997. *Beyond Growth: The Economics of Sustainable Development.* Boston: Beacon Press.

Diamond, J. 2004. *Collapse: How Societies Choose to Fail or Succeed.* New York: Viking.

Ehrlich, P. 1968. *The Population Bomb.* New York: Ballantine.

Ehrlich, P., and A. Ehrlich. 1991. *The Population Explosion.* New York: Touchstone.

Eldredge, N. 1991. *The Miner's Canary: Unraveling the Mysteries of Extinction.* New York: Prentice-Hall.

Jackson, T. 2009. *Prosperity without Growth: Economics for a Finite Planet.* New York: Earthscan.

Leakey, R., and R. Lewin. 1995. *The Sixth Extinction: Patterns of Life and the Future of Mankind*. New York: Doubleday.

Meadows, D. H. 1972. *The Limits to Growth*. New York: Macmillan.

Meadows, D. H., D. L. Meadows, and J. Randers. 1993. *Beyond the Limits: Confronting Global Collapse, Envisioning a Sustainable Future*. London: Chelsea Green.

Meadows, D. H., J. Randers, and D. L. Meadows. 2004. *The Limits to Growth: The 30-Year Update*. London: Chelsea Green.

Wackernagel, M., N. B. Schulz, D. Deumling, A. Callejas Linares, M. Jenkins, V. Kapos, C. Monfreda, J. Loh, N. Myers, R. Norgaard, and J. Randers. 2002. Tracking the Ecological Overshoot of the Human Economy. *Proceedings of the National Academy of Sciences* 99 (14): 9266–71.

13

The Rejection of Reality: How the Denial of Science Threatens Us All

Facts do not cease to exist because they are ignored.

Aldous Huxley

To treat your facts with imagination is one thing,
but to imagine your facts is another.

John Burroughs

Reality is that which, when you stop believing in it, doesn't go away.

Philip K. Dick

You are entitled to your own opinion, but you are not entitled to your own facts.

Daniel Patrick Moynihan, 2003

There are in fact two things, science and opinion; the
former begets knowledge, the latter ignorance

Hippocrates

UNSCIENTIFIC AMERICA

Blogger and author Farhad Manjoo (2008) paints a chilling (and savagely
funny) portrait of how our society has lost touch with reality. Only a
decade or two ago, there were just a few media outlets (major newspapers and TV news programs), and everyone got their news from these

mainstream, middle-of-the-road sources. Because they reported news to people of all political beliefs and were required by the FCC to follow the so-called Fairness Doctrine of balancing their coverage of controversial topics, the major media and their reporters strove for objectivity and made an effort to be truly "fair and balanced" when dealing with controversial issues—even to the point of giving someone who was clearly wrong at least some voice in the coverage. Newspapers, magazines, and the TV news all employed fact checkers who tracked down the details of a story and made sure there was some external corroboration before it was reported. When a story involved some expertise to evaluate, reporters routinely talked to numerous expert sources. Science reporting was often superficial and sometimes oversimplified, but at least the reporters talked to real scientists and strove to get a diversity of scientific expertise represented when they reported a scientific controversy.

Contrast that with the media landscape today. Thanks to the deregulation of the airwaves that the Reagan adminstration pushed through in the 1980s, the old Fairness Doctrine came to an end. No longer did TV news have to present both sides of a controversy in politics. A TV channel could be completely partisan with no effort to be fair and balanced. The deregulation of the airwaves also led to a huge number of new channels on TV and radio and elsewhere. The age of the old mainstream media giants in print and in broadcasting is slowly coming to an end, and the subscriptions for print news and the viewership for the major network news programs shrinks every year. In their place are hundreds of different TV channels, including Fox News, which shamelessly calls itself fair and balanced even though it is the mouthpiece for the right-wingers in the United States, and openly accepts money from right-wing causes and political organizations. The network was founded by Rupert Murdoch explicitly to push his conservative pro-business agenda, as he does in the many other media he owns. Fox News is run by Roger Ailes—former Republican media adviser to Nixon, Reagan, and George H. W. Bush— who has repeatedly stated his belief that their coverage should reflect the right-wing viewpoint. As Ailes himself said, "'The truth' is whatever people will believe."[1] We also have MSNBC, which is famous for its left-wing commentators. Only CNN was relatively centrist in their coverage, but then they took a hard right turn and tried to be a version of Fox, and now they are losing ground to Fox and MSNBC.

The talk radio world is dominated by Rush Limbaugh and the right-wingers, although for a brief time Air America tried to give progressive radio a chance at balancing the overwhelming conservative bias of the radio talk shows. Newspapers are fading so fast that there is little interest in the United States of owning newspapers, although Sun Myung Moon's conservative *Washington Times* and the business-friendly *Wall Street Journal* still maintain their editorial positions. The media world is even more fractured by the internet and the blogosphere, so there are thousands of websites where you can read or hear almost anything, without any fact checking, peer review, or quality control. Rumors, half-truths, and outright falsehoods gain equal status in cyberspace, where everyone has an equal voice, and there is no referee to declare something false or out of bounds. There are so many weird and probably false stories out there that political satirist Stephen Colbert coined the term "truthiness" to describe how we decide things "from the gut" without regard to evidence, logic, or the facts. We are told not to believe everything we hear or read, but it becomes harder and harder for the puzzled media consumer to tell the good stuff from the garbage.

As a consequence, we now have a media echo-chamber effect; anyone can get news that confirms all their biases and never encounter a different point of view or an outside reality that falsifies the stuff they read on the internet or hear on some news sources. We have a whole set of people in this country who hear only news filtered through what Bill Maher calls "the Fox bubble": the right-wing filter of Fox News, Rush Limbaugh, Glenn Beck, and the many reactionaries and religious fundamentalists in the blogosphere. That is how patently false ideas that President Obama is a Kenyan Muslim can be propagated for years, because no one on the right-wing media spectrum (or the GOP leadership) is willing to step up and point to the falsity of this rumor. Likewise, the conservative media almost exclusively propagate the lies about global warming, and most of their followers never encounter anyone who points out that these ideas are false. Walter Cronkite and Edward R. Murrow are probably rolling in their graves at the way the media have been distorted and disemboweled, so people are no longer exposed to balanced coverage, but only get confirmations of their preexisting biases.

Creationism is its own echo chamber, with large followings of loyal fundamentalists who view the world only as filtered by their ministry, and repeat the lies their leaders tell them, even though there is evidence right before their eyes that proves them wrong. Journalist Matt Taibbi went undercover in a fundamentalist megachurch in Texas, and tried to understand how these people think, and how they can be so misled.[2] He found that within their all-enveloping church community, they are discouraged from listening to any outside media or reading any books not recommended by the church. They are discouraged from talking to family or friends who are not part of the church, and their weekly calendars are so full of church events that they are as heavily indoctrinated as members of any of the old cults that used to be so controversial in the 1970s and 1980s. Taibbi documented the many lies that are preached from their pulpits; the collective brainwashing and peer pressure of the group is such that none of them protests or even dares ask questions. Their pulpits are openly political, campaigning against Democrats and for conservative Republicans, yet the IRS has not punished them for violating the tax laws that gives only nonpartisan churches their tax exemptions. This is the kind of insular, closed group that can be indoctrinated by groupthink into believing laughably false ideas such as Obama is a Kenyan Muslim, or the various lies told about evolution, and refuse to think about or even look at any evidence that contradicts their worldview.

This explains how creationists can be so blind to the evidence, no matter how clearly you present it to them. Even though my 2007 *Evolution* was a bestseller, it got almost no attention on the creationist websites or media, who studiously ignored it for over five years. I cannot recall how many times I have seen and read creationists robotically chanting, "There are no transitional fossils," as they been told to say—even though you can see these fossils in any natural history museum, and I wrote a whole book showing hundreds of examples. In a recent YouTube video, Richard Dawkins debates a particularly dogmatic creationist.[3] Over and over, he points to examples of transitional fossils, while the creationist refuses to acknowledge or understand what's right in front of her, and robotically repeats the lie, "There are no transitional fossils." No amount of evidence right in front of her would ever shake this deeply embedded false belief system.

THE SCIENCE TV WASTELAND

> This weekend on The History Channel, someone digs through old plastic junk ("It's a Dukes of Hazzard wastebasket!"), someone else tries to sell a doll head ("I used to take the heads off the bodies, and I kept the heads") ... and Larry the Cable Guy taste-tests Tabasco sauce ("I can't feel my dadgum tongue!"). The History Channel. What the hell happened to us?
>
> *Jimmy Kimmel Live*[4]

How about the role of the media in promoting scientific literacy? Let us look at TV first. If you flip through the various basic cable channels that are nominally science oriented (often grouped together on the dial if they feature scientific topics), you will come up with nothing but junk, pseudoscience, and worse. Reality television (whose programs, ironically, are totally unreal and phony in their content and staging) about subjects having little or no science content—ghosts, UFOs, Bigfoot, and creationism—all fill the airwaves on the Discovery Channel, The Learning Channel, the History Channel, and even the Science Channel and National Geographic Channel. We watch a few minutes of these and complain about them to anyone within earshot, then (usually) move on—or occasionally we get sucked in and watch the whole thing, like gawkers at a car crash.

I have seen it from both sides. I have appeared in prehistoric creature documentaries that have aired on all four channels (and keep reappearing years after I made them, so I feel like a reverse Dorian Gray, whose younger self is preserved in documentary limbo). Almost all these documentaries are made by small independent film outfits that are searching for any sexy topic that they can sell to the major cable networks, so they are under great pressure to come up with something flashy, noisy, scary, or mysterious. If I have any chance to review the script, I try my best to tone down the excessive hyperbole, but they usually ignore me. As I film segments with them, I try to be as dynamic and entertaining as a talking head can be, but they always push me to oversimplify and exaggerate to make the spiel more colorful (but less scientifically accurate). And then, when I see the final product, I discover that most of what I did ended up on the cutting room floor, the producers having used only a few seconds of material from many hours of filming. Even worse, I have put in many hours on projects that never got picked up at all. Documentary film-making is a high-risk, low-reward proposition—you have better odds of making big money in Vegas.

So we all complain about the changes in our basic cable channels, and wonder why such dreck can make it on the air, but seldom think much about the process. But the excellent TVTropes website does a very nice job of analyzing what happens to TV networks over time.[5] To no one's surprise, it comes down to one simple factor: ratings (and therefore money from advertisers), largely driven by the effort to woo those big-spending trendsetting male viewers aged eighteen to thirty-one, who already dictate the movie industry's bottom line (although movies aim even lower to reach teenage boys, the biggest-spending and most loyal movie audience). As TVTropes points out (and those of us old enough to remember can attest to), it was not always this bad on cable TV. When the Reagan administration deregulated the airwaves in the 1980s, two things happened: the end of the Fairness Doctrine and the new laws allowing the creation of hundreds of basic cable channels. These channels were initially set up to fill specific programming niches, from the Golf Channel to the Game Show Network and so on. In the early 1980s, all these new niche cable channels were very distinct and more or less true to their niche description. But since these are commercial channels that must sell ads based on numbers of viewers, the same factors that affect every other commercial enterprise came into play: keep tweaking it and give the customer whatever sells the most. (This dynamic does not apply to noncommercial stations such as PBS in the United States or the BBC in Britain, which can program whatever they feel is in the public interest.)

As TVTropes documents, nearly all these niche networks have undergone network decay since their establishment in the 1980s, as their programming shifts to find hit shows. Because they are nearly all chasing nearly the same demographic of eighteen- to thirty-one-year-old males, they end up programming a lot of the same kinds of things (or even the same shows). Their original mission and distinctive programming is lost in a sea of reality shows and junk that is supposed to keep you tuned in, whether it be explosions or dangerous occupations or what have you. Another factor has been the expansion of media conglomerates, so that multiple cable channels are owned by just a few corporations, and the CEO of each channel must answer to corporate bosses who are interested only in profits, not any abstract mission to air certain types of programming. So much for the high-minded idealism that drove the deregulation of the airwaves in the 1980s, with the intent of offering us dozens of distinct choices! Instead, they all decay to a lowest common denominator—a bottom-line "If it bleeds, it leads" mentality, negating whatever real advantages dozens of distinctive niche cable channels once offered. They could (and did) notice that professional wrestling is popular with the eighteen- to thirty-one-year-old male demographic, and see no problem with programming the WWE immediately before or after to a show about science.

TVTropes offers as a classic example the pioneering channel MTV, which singlehandedly changed the music business in the early 1980s and made telegenic pop artists (such as Michael Jackson and Madonna) into big stars while ending the careers of less telegenic musicians (such as Christopher Cross). But soon MTV found it was more profitable to offer reality shows, cartoons, and game shows—the music videos that it originally pioneered have vanished altogether. TVTropes analyzed the decay of the cable channels in various categories. Under "Total Abandonment" (of their original mission), they list MTV, A&E, G4, CMT, Biography, and The Learning Channel (TLC), among others. In their words,

> TLC, originally focusing around science and nature documentaries in the style of the Discovery Channel, drifted toward almost nothing but "home makeover"-style reality shows. In a somewhat confusing (in these days of internet porn) play at grabbing the all-important 18–30 male demographic, TLC acquired the rights to air the Miss America pageant. After sufficient decay, one would never guess that TLC used to be called The Learning Channel and was once co-owned by NASA.[6]

One need only check the Koi Koi Eleven website to see how far TLC has drifted away from "learning" and into the realm of bizarre sensationalism, or a TLC promo video that hilarious mocks their programming.[7]

Under the category "Slipped," we find The History Channel. As TVTrope comments,

> [Their] programming now consists of roughneck-focused reality shows (*Ice Road Truckers, Ax Men*) and conspiracy theory docu-soaps about UFOs, the Bible Code, ghosts, Atlantis, Nostradamus, and the end of the world, earning the network the derisive nickname "The Hysterical Channel." Heck, at least the "Hitler Channel," as they used to be known (back when everything was about either World War II, Nazis, or The American Civil War), was actual history.[8]

Their analysis of Discovery Channel is even more hilarious:

> The Discovery Channel still shows plenty of actual documentary material, despite having been decaying for almost as long as MTV has. In the late 80s the lineup was mostly serious documentaries, the most famous of which was *Wings* (no relation to the sitcom except for a focus on aircraft) but which also included classy repackaged BBC imports like *Making of a Continent*—and once a year there was Shark Week, which was just what you'd expect. By the mid-1990s, they showed an obscene amount of home improvement shows and cooking shows aimed at stay-at-home moms (enough to spawn the spin-off Discovery Home & Leisure Channel, now Planet Green) and *Wings* had proven so popular it had been farmed out to its own spin-off, Discovery Wings Channel (now Military Channel). Now, they're being swamped with "guys building and/or blowing things up" shows in the vein of *Mythbusters* and *Monster Garage*. And about four different shows about credulous idiots with no critical thinking skills ghost hunters. In 2005, Discovery debuted *Cash Cab*, a game show that takes place in the back of a cab, leaving one unsure whether it even has a theme beyond "non-fiction." It gets weird when you realize that they're knocking some of their own shows off, especially *Mythbusters* into *Smash Lab* (with a focus on safety measures) and *How It's Made* into *Some Assembly Required*. The latter has almost only done products featured in the former (though *How It's Made* has been on for just about ten years, so it's hard to find something they haven't done). The Discovery Channel also used to contain a lot of nature, which is where the now-classic Shark Week (which they still air regularly) originated from. But it seems that explosions have taken the place of tigers ripping stuff to pieces. Most of the nature shows have since been relegated to *Animal Planet*.[9]

Only the Science Channel and the National Geographic Channel still run mostly science documentaries with little junk, yet National Geographic still has *The Bounty Hunter, Is it Real?* and *The Dog Whisperer*. Science Channel has begun airing sci-fi programming, including *Firefly* and *Dark Matters: Twisted but True*, so they are running the sort of pop-pseudoscience garbage that now pollutes the History Channel.

Is it any wonder that it is almost impossible to get any science education from the public airwaves? Frankly, I see no light at the end of this tunnel. As long as these are commercial TV channels, they are driven by ratings and lowest common denominator programming aimed at eighteen- to thirty-one-year-old men. Only PBS and other noncommercial stations can escape this network decay—but then they more than compensate with annoying pledge drives that rerun old shows with sentimental value so that viewers will tune in and perhaps donate. The BBC, with its government support of top-quality science and drama programming (which the U.S. market then borrows or rips off) seems immune, although there are BBC channels that are lowbrow as well. After all, *The Benny Hill Show* reruns have done well in the United States for years.

THE LOOKING-GLASS WORLD OF DENIERS' BRAINS

"I can't believe that!" said Alice.

"Can't you?" the Queen said in a pitying tone. "Try again: draw a long breath, and shut your eyes."

Alice laughed. "There's not use trying," she said: "one can't believe impossible things."

"I daresay you haven't had much practice," said the Queen. "When I was your age, I always did it for half-an-hour a day. Why, sometimes I've believed as many as six impossible things before breakfast."

Lewis Carroll, Through the Looking-Glass

Reality has a well-known liberal bias.

Stephen Colbert

Anti-intellectualism has been a constant thread winding its way through our political and cultural life, nurtured by the false notion that democracy means that "my ignorance is just as good as your knowledge."

Isaac Asimov

If you tell a lie big enough, and keep repeating it, people will eventually come to believe it.

Josef Goebbels

For years, I have puzzled over why people can believe such weird things as creationism or other kinds of pseudoscience and science denial. In *Evolution* I devote an entire chapter to asking why creationists can so confidently believe patently false ideas, and refuse to look at any evidence placed in front of them. I have compared it to *Through the Looking-Glass*, in which Alice steps through the mirror and finds that the objects and the landscape look vaguely familiar—but all the rules of logic are reversed or turned inside out. How can people continue to believe things that are clearly wrong, and refuse to change their ideas or look at evidence?

It turns out that human brains are constructed very differently than we would like to believe. As described by Michael Shermer (2011) Chris Mooney (2012), our brains are not logical computers or unemotional Vulcans like Mister Spock, but organs in emotional animals who navigate the factual world according to our beliefs and biases. Mooney explains this by starting with an anecdote about the Marquis de Condorcet, an important figure in the French Enlightenment (he helped develop integral calculus and also wrote many important works on politics and philosophy). Condorcet believed in the Enlightenment ideal that humans would always be rational and guided by reason, and persuaded if logic and evidence were considered—and lost his life in 1794 during the irrational, emotional, highly political Reign of Terror. Even though Enlightenment philosophy and political science have long argued that humans are rational animals, modern psychology and neurobiology have shown this is not the case. Humans filter the world to see what fits their emotional and cultural biases, and easily neglect evidence and information that does not fit (confirmation bias). We are prone to what psychologists now call *motivated reasoning*—confirmation bias, reduction of cognitive dissonance, shifting the goalposts, ad hoc rationalization to salvage falsified beliefs, and other mental tricks cause us to constantly filter the world. Our minds do not objectively weigh all the evidence and listening to reason, but instead act as if we were lawyers seeking evidence to bolster our pre-existing beliefs. Counter to the Enlightenment ideal that humans would change their minds when the facts go against them, motivated reasoning explains why humans are adept at bending or ignoring facts to fit the world as we want to see it.

Our brains are governed by several different, often conflicting factors. What psychologists call System 1 are the rapid-fire emotions and reactions that date back to some of our earliest ancestors, and are controlled by the most fundamental animal parts of our brains, such as the limbic system and the amygdala, which are responsible for fear, feelings of pain, and our fight-or-flight response. Functional magnetic resonance imaging (fMRI) studies of the brain show that these regions are highly active in the brain of a politically conservative person when it is processing information. System 2 is the more rational, slow-moving, thoughtful conscious process of thinking things through and trying to arrive at reasoned decisions. The studies of people's brains by fMRI show that this type of rational, slow decision making is controlled by the anterior cingulated cortex (ACC), which tends to be active in the brains of liberals when they are confronted with new information. As psychologists have shown, most events in our life are first filtered through the emotional System 1, so that even if System 2 is working in a fully rational way, it is biased by what our emotional System 1 has told it. Our memories, too, work this way, so (as Mooney points out) if we have a set of associations of a certain person or concept, such as Sarah Palin, the associated memories with that stimulus (such as "woman," "Republican," "pregnant unwed teenage daughter," "death panels," or "wrong about Paul Revere") are immediately triggered and become part of the processing of any new information about that subject. As Mooney describes it,

> To see how it plays out in practice, consider a conservative Christian who has just heard about a new scientific discovery—a new hominid fossil, say, confirming our evolutionary origins—that deeply challenged something he or she believes ("human beings were created by God"; "the book of Genesis is literally true"). What happens next, explains Stony Brook University political scientist Charles Taber, is a subconcious negative (or "affective") response to the threatening new information—and that response, in turn, guides the types of memories and associations that are called into the conscious mind based on a network of emotionally laden associations and concepts. "They retrieve thoughts that are consistent with their previous beliefs," say Taber, "and that will lead them to construct or build an argument and challenge what they are hearing."[10]

We are all guilty of this to a greater or less extent, but Mooney explores some of the recent research that explains the psychological roots of these

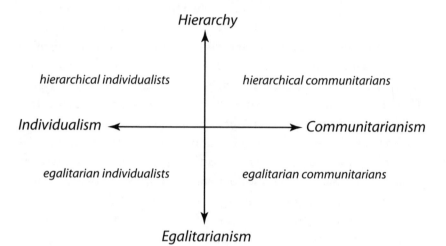

FIGURE 13.1. A plot of the cultural cognition worldviews, as recognized by psychologists. One axis contrasts hierarchy (support of the existing social order) vs. egalitarianism (support of a more equal society). The other axis plots individuals in their views on individualism (every man for himself, and everyone takes responsibility for their actions) vs. communitarianism (we must help our fellow humans). Most American conservatives fall in the hierarchy/individualism field, whereas liberals fall in the egalitarian/communitarian field.

beliefs. Individual belief systems, worldviews, and cultural biases have been categorized in research on cultural cognition by Yale law professor Daniel Kahan and his colleagues along two main axes (fig. 13.1).[11] Along one axis, humans range from individualists (people who value the individual over society, believe we are all responsible for our lot in life, and should be punished or rewarded for our choices or faults, and the government should not step in to change this) to communitarians (people who value the common good over individual welfare). On the other axis, the beliefs range from hierarchical (believing society should be highly structured, orderly, and stable, including rankings based on gender, class, and race) to egalitarian (believing everyone should strive for more equality and less hierarchy). When people are given a battery of psychological tests or respond to a series of polls, they tend to break out into discrete clusters along these axes. U.S. conservatives are—not surprisingly—very hierarchical and individualist, whereas liberals tend to be egalitarian and communitarian.

For whatever reasons we become liberal or conservative (largely upbringing in a conservative or liberal household, but other life events sometimes change this), most humans immediately identify with one or the other of these clusters, and this influences what they will think and what information they will absorb for the rest of their lives—unless a traumatic event changes them. These sets of beliefs or associations are strongly connected to one's sense of belonging to a community, and to our sense of well-being, and (for some) our sense of purpose or meaning in life, so they have strong emotional reinforcement that prevents them from being overturned by something as simple as facts or rational argument.

Consequently, one of the classic Enlightenment views—that rational arguments and evidence will eventually win out—turns out to be wrong in many cases. In people who have strong emotional and community connections to a belief system (whether it be a religion or a political party or what have you), their minds are preparing arguments against anything that weakens or challenges that belief (like a lawyer preparing a slanted case that defends one side of an argument), not listening to reason or evidence. Thus we have "smart idiots"—people who are actively engaged in an argument, well educated, and smart by any standard measure—but who have selectively biased what they have learned so they can argue against reality if it is important to defending their community and belief system. This is really discouraging to those of us battling irrationality. According to the Enlightenment view, truth, reason, and evidence should eventually persuade anyone, but what psychologists have shown is that the diehard creationists, climate deniers, and anti-vaxxers (along with other true believers) cannot be persuaded in this manner.

In fact, exposure to the facts can actually cause a backfire effect; people who are wedded to an ideology or religion react to such challenging information by clinging to their own beliefs even more firmly and becoming more entrenched in their worldview, using the tricks of motivated reasoning, such as ad hoc rationalization or moving the goalposts. Psychologists first documented this in the cases of religious sects that set a date for the Rapture and sold all their worldly possessions—only to have the prophecy fail them (as happened on May 21, and again on December 21, 2011, when evangelical minister Harold Camping falsely prophesied the end of the world and made worldwide news. The same scenario was repeated when the Mayan calendar apocalypse of December 21, 2012, did not come to pass). In each such case, we saw classic examples of motivated reasoning.[12] Instead of admitting the prophet was wrong, the followers (and sometimes the prophet) clung even more strongly to their failed belief system, and rationalized its failure by saying, "The Rapture did occur but it was invisible," or "Our prayers were so powerful that God spared the world." Thus, Mooney argues that all our efforts to educate doctrinaire conservatives in hopes of changing their minds are in vain, since evidence and reason do not work. Only some external factor that makes them change from their conservative stance and open their minds to other viewpoints have any chance at success.

Mooney (2012) also describes research which shows that conservatives (especially modern Republicans) are particularly prone to deny scientific realities such as evolution or climate change. Studies by John Jost and colleagues have shown a whole set of personality traits associated with the hierarchical-individualist conservatives versus the egalitarian-communitarian liberals.[13] Conservatism also tends to be associated with a variety of other personality traits, including dogmatism, intolerance of ambiguity and uncertainty, fear of death, fear of change, less openness to new experiences, less integrative complexity in their thinking, less nuanced thinking, more need for so-called closure, and so on. Liberals, on the other hand, are characterized by some of the opposite personality traits: rejection of dogma, tolerance of ambiguity and uncertainty, less fear of death or change, more openness to new experience, curiosity about the world, and more complex and nuanced thinking without the need for simplicity or closure. Thus, you find more artists and musicians and entertainers (so-called liberal Hollywood) have personalities that fit the open, experimental view of the world, whereas conservatives tend to be found in highly traditional institutions, such as churches, business, and the military. By their very nature academics and scholars tend to be more open; questioning; prone to complex, nuanced thinking; and comfortable with uncertainty—so it is no surprise that liberals tend to dominate universities (except, perhaps, in departments such as engineering or business, where the thinking is highly structured and the field is relatively unchanging).

Scientists also tend to be people with many of these liberal traits (especially open to challenges of dogma, curious, and comfortable with ambiguity or an absence of simple answers), because these are the traits that make a scientist successful and explain why science works. By contrast, someone with diehard conservative personality traits, or strong religious beliefs that bias their perspective on life, will view scientists with suspicion or scorn any time they discover an inconvenient truth that threatens or challenges their comforting belief systems about the world. Hence the widespread attempt to smear scientists and claim their research is motivated by greed for research grants in the global warming battle (chapter 5) and other battles over the environment (chapter 4); or the claim that all scientists are atheists in the creation/evolution wars (chapter 6); or the idea among the anti-vaxxers, AIDS deniers, and quacks that scientists and doctors are in cahoots with Big Pharma (chapters 7, 8, and 9).

To be fair to both sides, Mooney (2012) points out that science denial is not the exclusive province of the right wing. There are certain ideas, such as the vaccine fears of the anti-vaxxers, and fears of nuclear power, or of scary oil company practices such as fracking, that are predominately held by liberals and environmentalists. But there are important differences here. Adherence to pseudoscience and antiscience is not symmetrically distributed between the left and the right. Ideas such as anti-vaxx, antinukes, and anti-fracking are not held uniformly by the majority of liberals or progressives, but only a tiny subset, whereas studies show that the ideas of creationism and global climate change denial are virtually universal among conservatives in the United States. More importantly, *none of these ideas are held by the major leaders* of the Democratic Party, nor are they being actively written into law across the nation (with a few local exceptions). *By contrast, nearly all conservative politicians in the modern* GOP *must at least pay lip service to a litany of dogmas* such as lowering taxes; cutting spending on the poor; boosting military spending; opposing abortions, birth control, and stem-cell research; and homophobia—and, in recent years, they also have had to toe the line regarding global climate change denial, and throw at least a bone to creationism. These are important distinctions, and explain why the antiscientific attitudes of U.S. politics are not evenly or symmetrically distributed.

As Mooney (2012) reminds us, one personality trait that characterizes most liberals in the United States is an open, questioning attitude about their beliefs, and a respect for science. In the case of fracking, nuclear power, and anti-vaxxers, the scientific community has either spoken clearly (vaccines do not cause autism [see chapter 7]), or the scientific data are not in favor of the diehards and fearmongerers (nuclear power is not perfect, but all forms of energy have drawbacks, and we need power from somewhere; fracking might lead to a few cases of groundwater contamination, but so far there is no sign that it is always a problem). When scientists speak clearly and present their evidence, and liberal politicians respect their opinions, only minorities of liberals end up holding the unscientific or pseudoscientific views, and no policy changes occur. By contrast, when conservatives were or are in power (as in the Bush years of 2001 to 2009, or in many state governments now), laws are passed that either hinder scientific reseach or outright deny scientific reality.

All this research about our psychological dark sides is highly discouraging. It shows that reason, logic, and evidence cannot win when emotion or dogma clouds people's judgment, and even better education does not necessarily change people's false beliefs. It may mean only that people with these preexisting biases become dogmatic smart idiots, fighting scientific reality from a fortress of myths and misconceptions that no one can release them from. It explains why creationism has always held sway over about 40–45% of the U.S. population, despite decades of educational efforts and huge scientific advances and changes in our society. Apparently, the only way to change this is to change the culture, so that religion does not hold such a strong grip on us. In the case of environmental issues, climate change, and peak oil and population worries, it apparently takes really scary external demonstrations of reality (like the Cuyahoga River fire, or record deaths from heat waves or smog, or record oil prices) to get people to change their minds and come to realize what scientists have been saying for years. Apparently, the scary weather events of 2012—such as the record heat waves, droughts, and Hurricane Sandy—were just such a factor that shifted public opinion so that 80% of

Americans now agree that climate change is a serious problem.[14] Thus, for much of the antiscience we have discussed, the change will come, but only in the form of external events that force us to address reality.

OUR CANDLE IN THE DARKNESS

> It is far better to grasp the universe as it really is than to persist
> in delusion, however satisfying and reassuring.

> *Carl Sagan*

As we discussed in the first two chapters, the world of science has a built-in mechanism of rooting out bad ideas: peer review. Scientists are human, they are not perfect, and they can be misled by their own biases and ideologies, but in most cases, the harsh scrutiny of other scientists soon weeds out the bad data and gives us at least some basis on which to decide whether an idea has merit. Scientists are not immune to cultural forces, but by and large they are not openly ideological, either. Most of the ones I know are largely apolitical, and are appalled when they see other scientists bias their work to suit political or ideological ends. Most of them still follow the concept of a scientific reality that must be respected, no matter what your biases. There are, of course, examples of scientific bias and fraud, but pointing them out simply highlights how rare and unusual they really are—the exceptions to the rule. In most instances the scientific community does a good job of policing itself and trying to separate what is real from what is not. More to the point, scientists often discover things that go against our belief systems, but they must put aside their pet ideas and face this reality. When Copernicus and Galileo showed us that the earth (and us) are not in the center of the universe, it was not popular—but it was true. Everyone except a handful of religious nuts and the uneducated now look at the sun rising and setting and have come to terms with the counterintuitive notion that it is the earth turning instead. When Darwin showed that life had evolved and that we are closely related to other living things, not specially created, it offended many people (and still does)—but its truth was soon acknowledged by the entire scientific community and nearly all educated

FIGURE 13.2. Science tells us what we do not want to hear, and does not necessarily confirm our biases. The first panel portrays the Greek scientist and mathematician Archimedes, killed by a Roman soldier who did not realize he was beheading the smartest man in the ancient world. The LHC in the penultimate panel refers to the Large Hadron Collider and the silly idea that somehow it would cause an atomic catastrophe—which it did not. *abstrusegoose.com/47; courtesy abstrusegoose.com.*

Westerners who were not religiously biased, even before Darwin died. As the online comic "A Wise Man Once Said" puts it, "Science: if you ain't pissing people off, you ain't doin' it right" (fig. 13.2).

As discussed in chapters 1 and 2, science is our candle in the darkness (to use Carl Sagan's term). Our modern world, with all its benefits (discussed in chapter 2), is almost entirely the product of scientific advances. Most people respect and listen to science when it comes to practical issues such as engineering or medicine that improves their lives. In fact, in our gut we respect science so much that many people try to imitate it and put on its trappings and appearance, even pseudoscientists such as the creationists or quack doctors who are pushing garbage in the name of science. Science is practically the only form of thought or scholarship that is self-checking and relatively immune to intellectual fads, because ultimately scientific ideas have to prove their merit, and meet their own form of reality check when they are tested in the real world. The fact that science does not always tell us what we want to hear is further proof that we cannot impose our biases or ideologies on it and still be practicing scientists.

Mooney (2012) describes two case studies that reinforce this point. One involes Kerry Emanuel, a highly respected climate scientist from MIT who has published important research showing that hurricane strength will increase with global warming. Emanuel was a Republican (at least until the GOP swung to the extreme right), but he did not let his political views get in the way of doing good science. When the GOP-dominated House Committee of Science and Technology met on March 31, 2011, to discuss global climate change, Emanuel was the sole witness who was an actual climate scientist—and a Republican to boot—yet he was there only because the Democrats invited him. As discussed below, the rest of the witnesses were not climate scientists at all, but people brought in to enforce the GOP anti–climate change message. When Emanuel testified, he told his fellow Republicans the truth about climate change (contradicting their other witnesses). As he explained, even a freshman at MIT can do calculations on the back of an envelope that show why global warming is a serious concern. As Mooney put it, "You don't get an atmspheric sciences degree at MIT—with a climate focus, anyway—if you can't show on the back of an envelope what much of Congress now calls into question."[15] He also discussed the brouhaha over Climategate and the stolen e-mails that have been quote mined by the climate deniers. Emanuel was even assigned to one of the six committees that performed the review and investigation. In his words,

I am appalled at the energetic campaign of disinformation being waged in the climate arena. I have watched good, decent, hardworking scientists savaged and whole fields of scholarship attacked without merit. Consider as an example the issues surrounding the email messages stolen from some climate scientists. I know something about this as I served on a panel appointed by the Royal Society of Great Britain, under the direction of Lord Oxburgh, to investigate allegations of scientific misconduct by the scientists working at the Climate Research Unit of the University of East Anglia. Neither we nor several other investigative panels found any evidence of misconduct. To be sure, we confirmed what was by then well known, that a handful of scientists had exercised poor judgment in constructing a figure for a non peer-reviewed publication. Rather than omitting the entire record of a particularly dubious tree-ring-based proxy, the authors of the figure only omitted that part of it that was provably false. If this was a conspiracy to deceive, though, it was exceedingly poorly conceived as anyone with the slightest interest in the subject could (and did) immediately find the whole proxy record in the peer-reviewed literature.

The true scandal here is the enormously successful attempt to elevate this single lapse of judgment on the part of a small number of scientists into a sweeping condemnation of a whole scholarly endeavor. When the history of this event is written, the efforts of those seeking to discredit climate science will be seen for what they are; why many cannot see it now is a mystery to me.[16]

Mooney (2012) contrasts the upstanding behavior of Emanuel with a very different kind of Republican: conservative activist Andrew Schlafly. He is the son of antifeminist and homophobic activist Phyllis Schlafly, and the editor of Conservapedia, which is modeled on Wikipedia but has a strong conservative bias. Andrew Schlafly is not uneducated: he has an engineering degree from Princeton and a law degree from Harvard, and he has worked for Intel and Bell Labs. However, he has no training in any of the sciences that might challenge his conservative ideology, and it plainly shows when he writes about subjects beyond his expertise. His is a classic case of the smart idiot, someone educated enough to sound convincing but not educated in areas of expertise that would allow him to realize he is wrong.

Naturally, Conservapedia denies global climate change and evolution, denies much of astronomy (especially Big Bang cosmology) and geology, trashes environmentalism, lionizes capitalism and even the robber barons, trashes labor unions, claims that homosexuality is a mental disorder, claims that abortion leads to breast cancer, and makes many other demonstrably false claims about science and reality. It even obliquely supports the idea that the earth is the center of the universe, and implies that Copernicus and Galileo and modern astronomers are wrong![17] One would expect Conservapedia to push the idea that the earth is flat, but apparently those ideas are too retro even for Conservapedia. (Instead, it asserts that the "Flat Earth myth" about the past was cooked up by evolutionists to slander creationists, even though the idea is found in the Bible in many places![18])

Strangely, Conservapedia does go so far as to trash Einstein and relativity! It is hard to imagine why the century-old idea from physics is any threat to a conservative's view of the world. Apparently, like so many other people who do not understand relativity, Schlafly has confused the scientific idea with the philosophical notion of *relativism*, the notion that there are no absolute truths but only truths taken in relative context. This, he claims, has been used by liberal politicians to justify their agendas, and as a metaphor by Barack Obama in a law review article. (Which proves what, exactly?) Then Schlafly labors through six thousand words and many equations trying to debunk one of the best-tested ideas in all of science, making ridiculous claims that "relativity has been met with much resistance in the scientific world."[19] This may have been true when it was first proposed in 1905 and 1915, but it was widely accepted by nearly all physicists by the 1920s, when numerous experiments confirmed it. Schlafly points to the true but irrelevant fact that no one has received a Nobel Prize for relativity. Technically, Einstein received his 1921 Nobel "for his services to Theoretical Physics," without mentioning relativity directly—but it is implied in the citation that mentions his "services to Theoretical Physics." In addition, the Hulse-Taylor model of gravity waves, awarded a Nobel in Physics in 1993, depends on the notion of relativity. Even stranger is Schlafly's bizarre and highly irrelevant claim that "Virtually no one who is taught and believes Relativity continues to read the Bible, a book that outsells *New York Times* bestsellers by a hundred-fold." Finally, as proof that relativity is wrong, he cites examples of "action-at-a-distance by Jesus, described in John 4:46–54, Matthew 15:28, and Matthew 27:51."[20] Wow! That is the best way to debunk real science—quote Bible verses!

However, a whole spectrum of society will deny the inconvenient truths of science when facts gets in the way of their ideology. Conservative anti-environmentalists, and the business interests that fund them, have long tried to deny or discredit the evidence for acid rain, the ozone hole, global warming, the end of cheap oil, and the limits of human population on this planet, even though the evidence for all these facts is overwhelming and accepted by nearly all scientists, scholars, and educated individuals who are not biased by right-wing thinking. Religious extremists, both Christian Fundamentalists and radical Muslims, deny the reality of evolution, even though the scientific evidence has long been overwhelming, and nearly all the educated world (including most mainstream religions) has long ago accepted this reality. People from a wide spectrum of belief systems continually try to deny or subvert the realities of modern evidence-based medicine, whether it be denying that HIV causes AIDS, claiming that vaccinations cause autism, or resorting to quack cures like homeopathy and chiropractic. All of these alternative medical belief systems have been shown to be false, and in many cases they not only waste money and time, but they end up hurting or killing people as well.

Yet the human denial filter, motivated reasoning in the form of cognitive dissonance resolution and confirmation bias, and the tendency to believe what comforts us and fits our preconceptions still seems to trump all evidence from science and the real world. The truth is out there, in plain sight, and yet a significant portion of the American population cannot, or refuse to, accept it. Why is this? Is it just politics and ideology, or is it due to bad science education, or lack of training in critical thinking skills, or what?

SCIENCE POLICY DONE WRONG — AND RIGHT

Mr. Chairman, I rise in opposition to a bill that overturns the scientific finding that pollution is harming our people and our planet. However, I won't physically rise, because I'm worried that Republicans will overturn the law of gravity, sending us floating about the room. I won't call for the sunlight of additional hearings, for fear that Republicans might excommunicate the finding that the Earth revolves around the sun. Instead, I'll embody Newton's third law of motion and be an equal and opposing force against this attack on science and on laws that will reduce America's importation of foreign oil. This bill will live in the House while simultaneously being dead in the Senate. It will be a legislative Schrodinger's cat killed by the quantum mechanics of the legislative process! Arbitrary rejection of scientific fact will not cause us to rise from our seats today. But with this bill, pollution levels will rise. Oil imports will rise. Temperatures will rise. And with that, I yield back the balance of my time. That is, unless a rejection of Einstein's Special Theory of Relativity is somewhere in the chair's amendment pile.

Representative Edward Markey

Seeing how politicized scientific reality has become, one might despair of science itself functioning under such conditions. In this regard, there are both good signs and bad signs. Among the bad signs are the ridiculous attempt by the North Carolina legislature to redefine sea-level rise so that it is impossible to recognize if it is happening due to global warming.[21] Not to be outdone, a Virginia legislator refused to let the term "sea-level rise" be used in a $50,000 study of coastal flooding because it was a "liberal code word."[22] All of these things remind one of King Canute or Roman emperor Caligula, futilely commanding the sea to obey their wishes, only to become angry that they did not have such power. As Stephen Colbert stated in the June 4, 2012, edition of *Colbert Report:* "I think we should start applying this method to even more things we don't want to happen. For example, I don't want to die, but the actuaries at my insurance company are convinced that it will happen sometime in the next fifty years. However, if we only consider historical data, I've been alive my entire life, therefore I always will be."[23]

Even worse, when the GOP took over the House after the 2010 elections, the political climate surrounding controversial topics in science changed radically. The extremists who ran the House Energy and Commerce Committee did their best to challenge the enormous body of evidence supporting the reality of global climate change. On March 10, 2011, they set new lows by trying to redefine greenhouse gases to exclude carbon dioxide, methane, and all the other such gases science has recognized. The situation was so ludicrous that Representative Edward Markey (D-Massachusetts) mocked their ridiculous efforts to deny science by asking if they planned to repeal the laws of gravity, relativity, quantum mechanics, and heliocentrism (quoted in the deliciously sarcastic passage in the epigraph at the beginning of the section).[24]

On the positive side, there are scientists who might start out doubting a particular scientific idea, but still keep their integrity. For example, if the data about global climate change are indeed valid and robust, any qualified scientist should be able to look at them and see if the prevailing scientific interpretation holds up. Indeed, such a test took place. This occurred when the Republican leaders of the House Science and Technology Committee attacked the science of global warming. The agenda for their March 31, 2011, meeting (mentioned above) was explicitly slanted to challenge the climate science community and cast doubt on their data about global temperature change.[25] They openly stacked the deck with such "expert scientific witnesses" as an economist, a lawyer, and a profes-

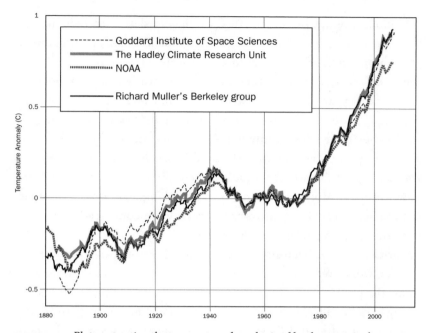

FIGURE 13.3. Plot contrasting the temperature data obtained by three original sources (NOAA, Goddard Institute of Space Sciences, and the Hadley Climate Research Unit of the University of East Anglia) with the data obtained by Richard Muller's Berkeley group, which originally attempted to deny the evidence of global warming, but found that in fact the original data were correct and the planet is getting warmer. *Redrawn from "Annual Land-Surface Average Temperature," Berkeley Earth Surface Temperature website, July 29, 2012, berkeleyearth.org/results-summary.*

sor of marketing, plus Kerry Emanuel—and Richard Muller, professor of physics at University of California Berkeley. Back in 2010, Muller's group had re-examined all the temperature data from the NOAA, the University of East Anglia Hadley Climate Research Unit, and the Goddard Institute of Space Science. Even though Muller started out skeptical of the temperature data, and was funded by the Koch brothers and other oil company sources, he carefully checked and rechecked the research himself. When the GOP leaders called him to testify before the House Science and Technology Committee on March 31, 2011, they expected him to discredit the temperature data that showed real change. Instead, Muller shocked his GOP sponsors by demonstrating his scientific integrity: the temperature increase was real, and the scientists who had demonstrated climate was changing were right (fig. 13.3). In fall 2011, his study was published, and the conclusions were clear: *global warming is real, even to a conservative skeptical scientist.*

Unlike the hired gun scientists who play political games, Muller did what a real scientist *should* do: if the data go against your biases and preconceptions, then do the right thing and admit it—even if you have been paid by sponsors who want to discredit global warming. Muller is a shining example of a scientist whose integrity and honesty came first, and did not sell out to the highest bidder but instead let the chips fall where they may.[26]

IS OUR CHILDREN LEARNING SCIENCE?

So try asking instead of telling. Ask people, "What is it about science that you don't like?" So often people think it's over their heads but it's not. Everybody knows what a magnet is in our society. Everybody knows wires. Everybody knows electricity. Everybody likes to drive their car. In short, everybody loves science and the way to reveal that is through the passion, beauty, and joy of science.

Bill Nye the Science Guy

We've arranged a global civilization in which the most critical elements profoundly depend on science and technology. We have also arranged things so that almost no one understands science and technology. This is a prescription for disaster.

Carl Sagan

Scientific literacy may likely determine whether or not democratic society will survive into the 21st century.

Leon M. Lederman, Nobel Laureate

For many years, the *Tonight Show with Jay Leno* has produced short comedy segments called "Jay Walking." Jay and his small camera crew would stroll the streets of Hollywood or Universal Citywalk or Burbank, and ask the man on the street simple questions about current events, culture, history, government, science, and so on. Invariably the interviewees would respond with astounding demonstrations of their ignorance of basic facts about the world, most of which they should have learned in high school or much earlier. Interviewees ranged from people who thought that Abraham Lincoln was the first president, or did not know the color of the White House or where the Panama Canal is located. (Groucho Marx used to tweak his contestants on *You Bet Your Life* with questions such as "Who is buried in Grant's Tomb?" but these examples are not comic exaggerations.) The displays of misinformation and lack of knowledge were so appalling they made the both the T V viewer and the studio audience laugh with scorn (and experience a bit of uncomfortable self-recognition). Of course, Jay and the camera crew taped plenty of people that *did* know the correct answers, and they edited out all but the funniest displays of ignorance. In fact, my wife and I witnessed Jay and his crew taping a segment at the Americana on Brand in Glendale, California, in early 2011. Very little of what we heard ended up on the show that night.

Even though "Jay Walking" is entertainment and not a scientific poll, many rigorous studies confirm the general ignorance and lack of cultural knowledge of the American public, despite the fact that 85% of Americans complete high school (up from only 25% in 1940), and almost 30% get a college education. A recent poll was conducted by the American Revolution Center of 1,001 American adults. Over 89% were confident they could pass it, but 83% actually failed. They uncovered the following appalling results:

· More Americans could identify Michael Jackson as the composer of "Beat It" and "Billie Jean" than could identify the Bill of Rights as a body of amendments to the Constitution.
· More than 50 percent of respondents attributed the quote "From each according to his ability to each according to his needs" to Thomas Paine, George Washington, or President Obama. The quote is from Karl Marx, author of "The Communist Manifesto."
· More than a third did not know the century in which the American Revolution took place, and half of respondents believed than the Civil War, the Emancipation Proclamation, or the War of 1812 occurred before the American Revolution.
· With a political movement now claiming the mantle of the Revolutionary-era Tea Party, more than half of respondents

misidentified the outcome of the eighteenth-century agitation as a repeal of taxes, rather than as a key mobilization of popular resistance to British colonial rule.

· A third mistakenly believed that the Bill of Rights does not guarantee a right to a trial by jury, while 40% mistakenly thought that it did secure the right to vote.

· More than half misidentified the system of government established in the Constitution as a direct democracy, rather than a republic—a question that must be answered correctly by immigrants qualifying for U.S. citizenship.[27]

Yet another survey found that over 80% of Americans could not name a single Supreme Court justice, and some of the people they named (including Sandra Day O'Connor and David Souter) had left the court.[28] In 2011, *Newsweek* gave 1,000 Americans the standard test that immigrants must pass to earn U.S. citizenship. Over 38% of native-born Americans failed a simple test about American history and civics covering things they were taught in eighth grade and again in high school. [29] Among the respondents, 29% were unable to name the vice president (and Joe Biden has not really been hiding in the shadows like some VPs); 73% were unable to explain why we fought the Cold War; 44% could not define the Bill of Rights; 6% were unable even to identify the date of Independence Day on the calendar. Even more alarming was the general level of ignorance about world events compared to that of citizens of just about any other developed nation, who scored far higher than U.S. citizens. For example, Europeans were far more literate about the world: 68% of Danes, 75% of Brits, and 76% of Finns could identify the Taliban, but only 58% of Americans can—even though we are fighting them right now in Afghanistan (and the other nations are not). A century ago, perhaps such ignorance of the outside world and isolationism might have not been a problem, but now the United States is the sole remaining military superpower, and we are constantly facing threats from not only the wars Bush dragged us into (in Iraq and Afghanistan) but just about nearly every other conflict (such as Libya in 2011).

Other polls show that American ignorance of their own government and its processes leads to all sorts of myths that politicians can manipu-

late. A 2010 survey that most voters have no clue what the federal government actually spends money on.[30] We hear one party constantly crying,
"Cut federal spending!" but nearly all the federal budget is tied up in
categories (servicing our debts; military expenses in the time of war;
and Social Security, Medicaid, and Medicare) that no politician dares to
touch (the same poll indicated that 81% opposed cuts to Medicare, 78%
opposed cuts to Social Security, and 70% opposed cuts to Medicaid).
Instead, the Republicans attack budget categories such as NPR or the
NEA or Planned Parenthood, which are a minuscule fraction of 1% of
the total federal budget. The poll showed that Americans wanted to cut
foreign aid spending from the 27% they thought it represented to 13%; it is
actually less than 1% of the Federal budget. A study done by Stanford professor James Fishkin showed that people, when polled about the issues
blind and then given the facts of the situation, tended to make rational
choices on budget issues. The problem, as he sees it, is not that Americans
are stupid about budgetary issues, but simply ignorant or misinformed,
so that they are easily misled by politicians.[31]

Such news stories pop up every few months, further underlining not
only the general factual ignorance of Americans, especially their lack
of curiosity about the world around them. The reporters telling these
stories typically wring their hands in shame and shock that more people know who (*pop star or actor*) is than (*important political figure, like
Speaker of the House or Supreme Court Justice*). The general American
ignorance of political and important cultural matters is indeed appalling.
It explains why much of the current political debate about obeying the
Constitution (which Teabaggers claim to believe in) is followed by false
and ignorant statements about the Constitution (such as claiming that
the Constitution eliminated slavery, or that the Founding Fathers tried
to establish a Christian nation) or by cries for actions that are blatantly
unconstitutional (such as their frequent attempts to eliminate the separation of church and state).[32]

If the general ignorance of Americans is not shocking enough, their
ignorance of science is even more staggering. Study after study over the
years shows a virtually unchanging and an abysmally poor understanding of how the world really works.[33] The studies describe such howlers
as these:

- Only 53% of adults know how long it takes for the earth to revolve around the sun;
- Only 59% of adults know that dinosaurs and humans never coexisted (the "Flintstones model of prehistory");
- Only 47% of adults can guess correctly the percentage of the earth's surface that is covered by water;
- Only 21% of adults answered all three of these questions correctly.

And a surprisingly large number of American adults still think the sun revolves around the earth!

This is not just the crackpot fanatics from Galileo Was Wrong, or people who think this out of pure ignorance.[34] No one knows how many American adults even think the earth is flat, but it is probably a lot more than just the crazies who are part of the flat earth creationist movement.

There are shockingly large numbers of adults do not know which is larger, an electron or an atom. Most adults cannot give simple definitions of concepts such as the cell, the molecule, or DNA. Only about 33% of adults agree with the notion that more than half of human genes are identical to those of mice, and only 38% of adults realize that humans share 98% of their DNA with chimpanzees.[35] Only 35% think the Big Bang describes the early history of our universe. Carl Sagan (1997) estimated that 95% of American adults were scientifically illiterate. Sagan was thinking of a far higher level of science literacy than these simple middle-school level science knowledge questions we have just mentioned, and judging from numbers we have just cited, he is not far off.

If American adults are so appallingly illiterate in science, what about teenagers who are still supposed to be taking science classes in school? Sadly, the numbers are just as depressing. Most high school students know about the same amount of science or less than adults who have not sat in a science class for years. According to a study by Jon Miller of Northwestern University (an expert on scientific literacy who has studied it for years), U.S. high school students are "below average and below most European countries" on virtually every academic achievement test administered in the past thirty years.[36] Miller found that exposure to a college science course, the other hand, made significant improvements on science literacy, but only as measured against a baseline of almost total ignorance. Currently, scholars are studying the concept of civic science literacy, which involves more than just knowledge of science facts, but a sufficient understanding of science to apply to their everyday lives. Here again, the results are equally depressing. Although the numbers are slowly rising, Miller found that the "civic science literacy" of Americans was still less than 30%. As Miller put it, "We should take no pride in a finding that 70 percent of Americans cannot read and understand the science section of the *New York Times*."[37]

Another way to frame the question is to ask how we stack up against other countries. Study after study has shown that the United States is near the bottom of industrialized nations in scientific literacy. One recent study found that among fifteen-year-olds, the United States ranked twenty-ninth among the nations of the world.[38] At the top of the list was Finland, followed by a number of other northern European countries (the other Scandinavian countries, Germany, France, and the United Kingdom), and developed or developing Asian countries such as Japan, South Korea, and China. Nearly every other ranking in recent years gives similar results, although the exact order of the top ten countries might be shuffled a bit—but the United States always comes out near the bottom, along with countries such as Turkey and Cyprus, which enjoy a fraction of our wealth and spending on education. That alone is a mark of disgrace for our society—we spend so much money per child, and yet end up with such miserable results, and nearly every other industrialized country does far better. What does that say for our future economic well-being when we are near the bottom in crucial matters?

It does not help when politicians actively prevent good science from being taught in our schools. Such efforts originate in the 1920s and the Scopes Monkey Trial; creationists have since tried to prevent evolution being taught, or failing that, tried to insert creationism (in the various guises described in chapter 6) into the curriculum instead. Kentucky, Tennessee, and Louisiana have passed laws pushed by creationists that allow "balanced treatment" of the "strengths and weaknesses" in scientific theories to be taught in their schools. They do not attack all scientific theories, just the ones that the right-wingers hate: evolution, global climate change, sex education, and stem-cell research. If the political goals of this unscientific policy were not already apparent, one need only follow the paper trail of the people who pushed these bills through—without exception, this was done entirely with a fundamentalist or right-wing ideological motive, not balanced science education.

Perhaps the nadir of this kind of political interference in science education was just articulated by the platform of the Texas State GOP in June 2012.[39] The document is complete laundry list of pro-business, social conservative, and extremist right-wing talking points: abstinence-only sex education (which has been a miserable failure so far); corporal punishment of children; trying juveniles as adults; opposing the UN Convention of the Rights of the Child (which forbids child slavery); faith-based drug rehabilitation; flat-rate income tax; repeal of the income tax and return of the sales tax (which hits poor people most and favors the rich); return to the gold standard (despite the warnings of economists, both liberal and conservative, that it will not work); privatizing Social Security (as if the lessons of the 2008 Wall Street meltdown were not enough); repealing the minimum wage; opposition to homosexuality in any form; opposition to the Employment Non-Discrimination Act (so that employers can go to the bad old days of racial and sexual discrimination and harassment); continued opposition to ACORN (which has not existed since 2010); opposition to states' rights or voting rights for the District of Columbia; unyielding support of Israel (because in the minds of fundamentalists, Israel must exist so that biblical prophesies can come true); ending multicultural education (so that the white supremacist viewpoint would not be challenged by the histories of minorities or women); a decree that controversial theories, such as evolution and climate change, should be taught in a manner that can be challenged and their content debated or minimized; reducing or ending public education and switching to voucher-supported private schools; repealing the Civil Rights Act, Affirmative Action, and the Voting Rights Act (because they do not want minorities to challenge their white-run society). But the most shocking statement in the entire document was the following: "Knowledge-Based Education—We oppose the teaching of Higher Order Thinking Skills (HOTS) (values clarification), critical thinking skills and similar programs that are simply a relabeling of Outcome-Based Education (OBE) (mastery learning) which focus on behavior modification and have the purpose of challenging the student's fixed beliefs and undermining parental authority."[40]

You have not misread things. This is not a satirical lampoon, a Poe (a parody so broad that it could be mistaken for some crazy creationist screed), or an article from the *Onion*. No, it is right there in black and white: the Texas State GOP is *officially opposed to critical thinking!* What we suspect they whisper among themselves is now a public proclamation of policy. And the last phrase is almost too bizarre for belief: they explicitly do not want their kids to think critically because their ideas might challenge "the student's fixed beliefs and [undermine] parental authority." That is it in a nutshell—the classic dogma of fundamentalist churches and authoritarian states. Do not think for yourself, do not ask questions, and do not rock the boat—because the Church, the GOP, and your parents are the ultimate repository of truth, and we do not want to have the lies we told you as children undermined by anything like education.

So why are we so scientifically illiterate? Everyone has a favorite culprit. Certainly the media share a lot of the blame, filling the airwaves and print and internet with mountains of empty reality television and pseudoscience and celebrity gossip. Even the science they do present is watered down and oversimplified, often to the point of being distorted or just plain wrong. This is apparently where most scientists feel the blame lies.[41] There are many who blame our educational system and argue that students need to be turned on to science early and provided with hands-on experiments and active learning. This is probably also true, but unrealistic in an age when education budgets are being slashed to meet politicians' needs to cut costs without raising taxes. I know many high school science teachers personally, and they are at their wits' end. To them, the issue is not just the problem of small budgets, inadequate supplies and equipment, and huge classes. They battle an almost impossible uphill struggle to hold the interest and attention of the average American teenager—filled with raging hormones and interests in cars, pop culture, video games, and the opposite sex—and to get them to pay much attention to science classes, no matter how wonderful and inspiring the teachers try to make them.

Certainly, we cannot underestimate the effects of raging hormones and social pressures as children develop. As I pointed out in *Greenhouse of the Dinosaurs,* one need only watch the transformation in children's programming. Many of the shows for preschoolers and preteens are highly educational and filled with dinosaurs and astronomy and other real science. Science is clearly cool. But switch channels to the programs that cater to tweens and teens and you will see it is all about boy-girl relationships and getting along and being cool among your peers, along with lots of teen celebrity gossip and pop music marketed just for teeny-boppers. Science is no longer cool but nerdy; the popular kids try to avoid looking like they might enjoy it, even if they do. (Although teenagers do love computers and technology, if only to better communicate with their friends and catch the latest music or video or movie or game.) About the only factor that explains this change is adolescence. However, most of the other industrialized countries have no such problems with scientific literacy, and their teenagers certainly go through the same stresses and influences when they reach adolescence.

But there is another likely culprit—the elephant in the room that no one wants to mention. Consider the graph in figure 13.4. It shows the relationship between the acceptance of evolution (here defined as "humans beings, as we know them, evolved from earlier species of animals," a reasonably good metric of true acceptance of evolution) in various countries around the world versus their relative wealth (as measured by GDP adjusted for purchasing power parity). The main trend across countries forms a well-defined cloud with a reasonably curvilinear fit. At the top is a well-defined cluster of northern and western European nations (and Japan), with the southern European nations just behind them. Near the bottom are the former Soviet Bloc countries of Eastern Europe, which still suffer the effects of decades of backward Soviet educational and economic policies. (China, South Korea, and Singapore are not shown, but on other surveys, they all rate high on the acceptance of evolution scale, so they would plot high on the ordinate or *y* axis, no matter what their GDP.)

The same relationship could be shown if you consider any of the recent surveys that measure science literacy on an international scale. The

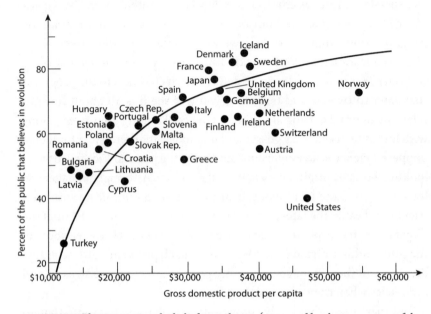

FIGURE 13.4. Plot comparing the belief in evolution (measured by the acceptance of the idea that "human beings, as we know them, evolved from earlier species of animals") vs. national wealth (as measured by GDP per capita). *Modified from www.calamitiesofnature .com/archive/?c=559; by permission of calamitiesofnature.com.*

northern and western European nations (especially Germany and the Scandinavian countries plus Iceland) nearly always come out near the top, along with Japan, Singapore, South Korea, and sometimes China. The exact order differs from survey to survey, but they only shuffle within the top ten or top fifteen. In other words, the acceptance of evolution in these countries is a very strong predictor of overall science literacy.

Now look at the position of the United States. It is a striking outlier on the graph shown here, because its low rate of acceptance of evolution relative to its national wealth (and the same would be true if you plotted it against the money spent on education per student). It falls down near the bottom of the curve on evolution acceptance along with Islamic nations such as Turkey, which spend much less per student. What is this telling us?

We have just reviewed all sorts of arguments about why our U.S. students are so illiterate in science despite all the money spent on their education. No doubt all of these things are true to some extent, but they all miss the elephant in the room that is apparent in these data: the stultifying influence of creationism in U.S. science education. Most of the examples of scientific illiteracy revealed in common survey questions, such as the mistaken notions about the age of the earth and Big Bang, or whether humans lived with dinosaurs, or whether we share a lot of DNA with chimps, are clearly so out of line with reality because they are part of the creationist dogma. No matter what kids learn in school about these subjects, their religious training at home overcomes the best efforts of their teachers—and their ideas rarely change as they become scientifically illiterate adults.

The single biggest predictor of national success in science literacy is the degree to which a country is not dominated by dogmatic religious beliefs, whether it be fundamentalist Christianity or conservative Islam. As Jon Miller documented, most of these industrialized European and Asian countries have no such strong forces of religious dogmatism in their politics and culture, and their schools teach evolution, climate change, and other scientific topics with almost no interference by religious zealots.[42]

CONSEQUENCES

An educated citizenry is the only safe repository for democratic values.

Thomas Jefferson

In a democracy, it is very important that the public have a basic understanding of science so that they can control the way that science and technology increasingly affect our lives.

Stephen Hawking

If you are scientifically literate the world looks very different to you. It's not just a lot of mysterious things happening. There is a lot we understand out there. And that understanding empowers you to, first, not be taken advantage of by others who do understand it. And second there are issues that confront society that have science as their foundation.

If you are scientifically illiterate, in a way, you are disenfranchising
yourself from the democratic process, and you don't even know it.

Neil deGrasse Tyson

Item: Scientists around the world are excited in July 2012 when the
Large Hadron Collider (LHC) in the CERN lab near Geneva, Switzerland, finally discovers the Higgs boson, one of the most important particles in nuclear physics, nicknamed (improperly) the God particle.[43]
Accolades come from around the world for the team of European scientists that made the discovery just a year or two after the LHC had finally
become operational. Forgotten in all the fuss was that the United States
was once the leader in high-energy physics, and its several large colliders
had discovered most of the major types of subatomic particles that were
the new foundation of the science. This march of progress abruptly ended
in 1993, when the next generation of instruments, the Superconducting Supercollider, scheduled to be built in Waxahachie, Texas, was canceled due to budget cuts by a shortsighted Congress.[44] Several scientists
pointed out, however, that if the Texas supercollider had been built, the
Higgs boson, along with many other subatomic particles that the LHC
at CERN is not big enough to produce, would have been discovered by
U.S. scientists much earlier than 2012.

Item: In the 1960s, the United States led the world in the space race,
going from simple rockets in the early 1960s to a moon landing on July
20, 1969, and then several more moon missions—and then Apollo was
canceled. During the 1970s through the 1990s, the space shuttle and the
International Space Station were among the crowning achievements of
NASA. But now the shuttle program has ended without a replacement,
Russian rockets and crews service the International Space Station, and
the only missions that NASA has in front of it are small ones to send ro-
botic rovers to Mars. All around the United States, astronomers bemoan
the fact that the United States once led in space science and technology
(with many important inventions as side benefits), but no longer sees
fit to spend money on one of the most exciting and uplifting areas of
science. When a reporter asks America's most famous astronomer, Neil

deGrasse Tyson, why space exploration is essential, he replies that without it, we have stopped dreaming, and that we need it to inspire future generations of kids to become scientists:

> Most people who don't agree say, "We have problems here on Earth. Let's focus on them." Well, we are focusing on them. The budget of social programs in the federal tax base is fifty times greater for social programs than it is for NASA. We're already focused in ways that many people who are NASA naysayers would rather it become. NASA is getting half a penny on a dollar—I say let's double it. A penny on a dollar would be enough to have a real Mars mission in the near future.[45]

This is an important perspective to keep in mind. The United States is not a poor country, but it has to make informed decisions about what is important to us. The amount we spend on science ($32 billion in 2012, including all of NASA, NOAA, and all our health research at the NIH) is minuscule. It is less than a twentieth of the $737 billion we spend on defense[46]—and that figure does not count the wars in Afghanistan and Iraq, which cost us many billions more. Our defense budget is larger than that of the next twenty largest defense budgets of other countries *combined!*[47] *Each one* of the trouble-plagued and controversial F-22 Raptor fighter aircraft cost about $150 million to build,[48] ten times the $14 million that was allocated to build the Superconducting Supercollider— which was then canceled for being too expensive. For the cost of another $4.5 billion *Nimitz*-class aircraft carrier,[49] we could pay for more than half of the National Science Foundation budget ($7.7 billion in 2012), which funds nearly of all the U.S. federal science research except for health science (which is covered by the NIH) and space research (covered by NASA).

Why do we need such huge defense expenditures that are bigger than those of the next 20 countries' defense budgets combined? The Cold War ended twenty-three years ago, and many of our large, expensive conventional ships and planes meant to fight the Soviet military sit barely used, because there is no longer a superpower opposing us. We have over three hundred thousand troops and hundreds of bases scattered in 192 countries worldwide, only a fraction of which are fighting in our only war zone in Afghanistan.[50] What do we need them all for? Most of those countries are our allies or are at least friendly nations. As recent decades have shown, the modern battlefield is not a Cold War showdown of armies and navies, but small cadres of terrorists and guerrillas, who are best fought with good intelligence and small groups of elite forces and drone aircraft, not giant battleships. Surely we could cut back a lot of this excess obsolete hardware and focus the money on improving our competiveness in science and technology. Meanwhile, China, Japan, and the European nations—with a fraction of our defense spending—are now in the forefront of particle physics and many other areas of science, sending their own space missions out to do things we used to do, and investing in science and technology to win the economic struggles of the future.

Item: The headline from CNN reads "China Shoots Up Rankings as Science Power, Study Finds."[51] As the article summarizes, a recent study by the Royal Society of London, the world's foremost and oldest scientific organization, found that although the United States was still the dominant scientific power in terms of scientific publications, the Chinese scientific had experienced a "meteoric rise" in scientific publications and new research.[52] Back in 2003, fewer than 5% of scientific articles came out of China. By 2008, 10% were Chinese authored, putting China second only the United States. Meanwhile, the American share of scientific publications dropped from 26% to 21%. Sir Chris Llewellyn Smith FRS, Chair of the Advisory Group for the study, commented, "The scientific world is changing and new players are fast appearing. Beyond the emergence of China, we see the rise of South-East Asian, Middle Eastern, North African and other nations. The increase in scientific research and collaboration, which can help us to find solutions to the global challenges we now face, is very welcome. However, no historically dominant nation can afford to rest on its laurels if it wants to retain the competitive economic advantage that being a scientific leader brings."[53]

China is climbing in many rankings; it is the second-largest economic power as well. Unencumbered by global warming deniers or stem-cell research antagonists or creationists who interfere with science policy, China is making huge investments in new technologies for a world with global warming and limited oil, while the United States and United Kingdom slip down the rankings of countries investing in green technology.[54] Germany and several Scandinavian countries have long led the world in their investments in green technology and their societal commitment to low energy use and reducing greenhouse gases—yet their economies are stronger than ours or than most of those in southern Europe or elsewhere.[55] Not surprisingly, these northern European and East Asian countries also rank at the top of science literacy rankings, and we already saw the correlation between acceptance of evolution and science literacy and other factors.

The United States still has the largest number of Nobel prizes in science, and has since 1956, when the effect of Germany's experiment with Hitler, antisemitism, and World War II caused a "brain drain" from Germany to the United States and other countries, and ended German supremacy in science.[56] But how long can this U.S. supremacy in science last when our population is less scientifically literate than those of most Asian or northern European nations? How long can it last when political and religious ideologues and zealots interfere with stem-cell research, deny evolution, and try to stifle American awareness of—and preparedness for—issues of global warming, population growth, and the limits of our resources? The answer is already apparent when you look at the next generation of scientists being trained as graduate students at most U.S. university science labs: they are populated largely by foreign students (especially Chinese and Indian students) who are better trained as undergraduates and tend to work harder.[57] Some may stay in the United States to teach another generation of American students, but many are going back to their home countries—and taking their training and intelligence and creativity with them.

Some people say, "It can't happen here. The United States has been the number-one power ever since World War II, and now we're the only superpower left." But as historians have pointed out, many other powerful societies with enormous economic reach and flourishing sciences and the arts have also declined in the past. Only 150 years ago, the British Empire spanned the entire globe, but now Britain is a relatively minor player among global powers, as it lost most of its economic strength and its colonial empire during and after the two world wars. The once-mighty Soviet Union fell in just a matter of years during the 1990s. The United States has long been embroiled in two different unfunded wars in the

Middle East, costing billions of dollars and thousands of American lives, while running up huge economic deficits in a time of recession. We like to think of ourselves as exceptional and bulletproof, but that is not the lesson history teaches us.

Imagine a society where a great flowering of science and technology spanned several centuries—and then, due to dogmatism, it throws away all this progress and recedes into the Dark Ages. Is it hard to imagine our own society sliding back into darkness and preindustrial conditions? Well, it has happened before. The Greeks made huge advances in mathematics, geometry, engineering, philosophy, arts, and literature—especially during the golden ages of Periclean Athens, and again during the Hellenistic period, where the descendants of Alexander's conquest of the known world flourished in Alexandria (where Ptolemy set up the famous model of the geocentric world) or in Syracuse (where Archimedes made great intellectual leaps in geometry, mathematics, and engineering). But this all vanished when the Greeks were conquered by the Roman Empire. A Roman soldier killed Archimedes during the conquest of Syracuse (fig. 13.2). The soldier did not recognize him, or realize the genius of the man he had killed, even though there were orders from the Roman generals to capture him alive. Archimedes, completely absorbed in geometry, allegedly said, "Do not disturb my circles" before he was killed. And the Roman conquest of the other great centers of learning, such as Alexandria, did much to set back science and philosophy, although the Romans did great feats of engineering and spread the benefits of Greek mathematics and engineering and civilized life to almost all of Europe.

The last remnants of the Roman Empire fell in 456 CE, and the Western world slipped into the Dark Ages. The advances in science and mathematics and engineering were lost for almost a thousand years, and would not return to the European world until the Renaissance. The ancient texts of the great Greek and Roman authors were largely destroyed by the Catholic Church for being heretical, or (since papyrus and parchment were rare and precious), reused by medieval monks to write religious documents right over the ancient texts (a palimpsest). In fact, most of our copies of the works of classical Greek and Roman authors come from palimpsests, because medieval monk copyists placed no value on ancient learning, but only saw the ancient parchments as vehicles for their own religious ideas. Only centuries later did scholars realize that these palimpsests contained the key documents of the ancient Greeks and Romans.

Few people realize that there was more science and scholarship going on in Baghdad about 1000 CE than in any European city at the time. In the Arabic golden age from about 800 to 1100 CE, Baghdad and many other Arabic cities experienced their own Renaissance, and made incredible scientific discoveries.[58] These scholars made advances in agriculture, the arts, economics, industry, law, literature, navigation, philosophy, sciences, sociology, and technology, both by preserving earlier traditions and by adding inventions and innovations of their own.[59] There was a long period of religious tolerance in Baghdad and elsewhere, allowing Jews, Christians, and even nonbelievers to live in peace in a predominantly Muslim world. Thanks to them, we all use Arabic numerals rather than clumsy Roman numerals for most mathematical tasks. Arabic scholars invented the concept of zero, and invented algebra (derived from an Arabic word, as is the word "algorithm"). Many of the stars in the sky have Arabic names, and Arabic astronomers made huge advances, mostly in navigation and in service of their large seagoing trade networks. Some of the discoveries, inventions, concepts, and cultural advances they made include the camera obscura, coffee, bar soap, toothpaste, shampoo, distilled alcohol, uric acid, nitric acid, alembic, valves, reciprocating suction piston pumps, mechanized water clocks, quilting, surgical catgut, vertical-axle windmills, inoculation, cryptanalysis, frequency analysis, three-course meals, stained glass and quartz glass, Persian rugs, and the celestial globe.

For three centuries, advances continued under this period of relatively benign rule and religious tolerance. Then, during the twelfth century and later, their Renaissance collapsed in a spectacular fashion. According to George Sarton, "The achievements of the Arabic speaking peoples between the ninth and twelfth centuries are so great as to baffle our understanding. The decadence of Islam and of Arabic is almost as puzzling in its speed and completeness as their phenomenal rise."[60] Although there is much debate among historians as to the cause of this spectacular decline, much of it can be attributed to religious dogmatism and intolerance.[61] By the thirteenth and fourteenth centuries, extremism dominated the Muslim world, and still does today. In the past few decades, the Muslim world has been so dominated by extremists that it is hard for us to think of Muslims as being tolerant of other religions or concerned with concepts in science or philosophy that might threaten their concept of Islam. Indeed, most of the Muslim countries are highly resistant to the notion of evolution, and have their own virulent form of creationism that borrows heavily from the U.S. fundamentalist version.

There are other examples of religious or political intolerance and oppression of science when it conflicts with the established powers. Take the infamous case of Trofim Lysenko, Stalin's favorite scientist. Lysenko held almost absolute power over Soviet science from 1927 until 1964. Most modern historians of science consider him a mediocre geneticist who promoted ideas of how Lamarckian inheritance might improve Soviet crop yields and prevent famine. His experimental results were inconclusive or outright fraudulent, yet he told Stalin that he could produce incredible bounties of food. As a result, he became the most powerful figure in the Soviet scientific establishment, and conspired with Stalin to suppress Mendelian geneticists, who really did understand how inheritance worked. Most of them were killed outright, sent to concentration camps, or driven into exile—forever destroying the vitality and strength of Soviet genetics and biology. Soviet genetics fell decades behind that of the rest of the world until the 1960s, when Lysenko was finally denounced, his work discredited, and he died in disgrace. Millions of people died in frequent famines when his nonsensical ideas were applied to agriculture.

No matter whether it is Classical Greek science, Arabic science, or Soviet science, the conclusion is clear: science cannot be subservient to ideology, and scientists cannot be forced to distort their message or results in order to please the political or religious powers that be. Lysenko and Stalin did not believe in Mendelian genetics or Darwinian biology, and they murdered hundreds of legitimate scientists who had the temerity to disagree with them. Other regimes (such as the Nazis or the devout Muslims after 1100) have distorted science to support their ideas, but ultimately scientific reality must win.

It is true that we do not live in the Soviet Union of Stalin, and that the United States has some safeguards against such oppression of scientific ideas. But as Mooney (2005) and Shulman (2007) showed, the Bush administration actively interfered with legitimate scientists, rewriting reports by federal scientists that disagree with their right-wing ideology, encouraging fringe scientists and non-experts to testify as legitimate equals with well-regarded scientists in order to cancel out their politically inconvenient message, and generally ignoring the conclusions of scientists who disagree with them. As we saw in previous posts, the current Republican House majority brings global warming deniers and other fringe scientists to testify in front of Congress, and passes bills denying obvious scientific facts. Stem-cell research in the United States has been set back compared to that in other countries, as our best scientists go to countries with less political oppression. Likewise, the foot-dragging and denials of global warming by the Bush administration and the flunkies of the oil industry in Congress may have cost the world valuable time in addressing this serious crisis.

When the prophet Cassandra told the Trojans what they did not want to hear, they ignored her and were eventually destroyed. If science tells us that we have evolved from the animal kingdom, or that microbes are evolving resistances to all our medicines, or that our wasteful society is destroying our planet, we had better learn from it, rather than shooting the messenger—and letting our children pay the ultimate price for our folly.

The late, great Carl Sagan said it best:

> There's another reason I think popularizing science is important, why I try to do it. It's a foreboding I have—maybe ill-placed—of an America in my children's generation, or my grandchildren's generation, when all the manufacturing industries have slipped away to other countries; when we're a service and information-processing economy; when awesome technological powers are in the hands of a very few, and no one representing the public interest even grasps the issues; when the people (by "the people" I mean the broad population in a democracy) have lost the ability to set their own agendas, or even to knowledgeably question those who do set the agendas; when there is no practice in questioning those in authority; when, clutching our crystals and religiously consulting our horoscopes, our critical faculties in steep decline, unable to distinguish between what's true and what feels good, we slide, almost without noticing, into superstition and darkness.[62]

FOR FURTHER READING

Brock, D. 2012. *The Fox Effect: How Roger Ailes Turned a Network into a Propaganda Machine*. New York: Anchor.

Faircloth, S. 2012. *Attack of the Theocrats: How the Religious Right Harms Us All*. New York: Pitchstone.

Frazier, K. 2009. *Science Under Siege: Defending Science, Exposing Pseudoscience*. Buffalo, N.Y.: Prometheus.

Gilovich, T. 1993. *How We Know What Isn't So: The Fallibility of Human Reason in Everyday Life*. New York: Free Press.

Grant, J. 2011. *Denying Science: Conspiracy Theories, Media Distortions, and the War against Reality*. Buffalo, N.Y.: Prometheus.

Harrison, G. P. 2011. *50 Popular Beliefs People Think Are True*. Buffalo, N.Y.: Prometheus.

Hofstadter, R. 1966. *Anti-Intellectualism in American Life*. New York: Vintage.

Law, S. 2011. *Believing Bullshit: How Not to Get Sucked into an Intellectual Black Hole*. Buffalo, N.Y.: Prometheus.

Manjoo, F. 2008. *True Enough: Learning to Live in a Post-Fact Society*. New York: Wiley.

Mooney, C. 2005. *The Republican War on Science*. New York: Basic.

———. 2012. *The Republican Brain: The Science of Why They Deny Science—and Reality*. New York: John Wiley.

Mooney, C., and S. Kirshenbaum. 2010. *Unscientific America: How Scientific Illiteracy Threatens Our Future*. New York: Basic.

Otto, S. L. 2011. *Fool Me Twice: Fighting the Assault on Science in America*. New York: Rodale Press.

Pierce, C. P. 2010. *Idiot America: How Stupidity Became a Virtue in the Land of the Free*. New York: Anchor.

Pigliucci, M. 2010. *Nonsense on Stilts: How to Tell Science from Bunk*. Chicago: University of Chicago Press.

Prothero, D. R. 2007. *Evolution: What the Fossils Say and Why It Matters*. New York: Columbia University Press.

———. 2009. *Greenhouse of the Dinosaurs*. New York: Columbia University Press.

Sagan, C. 1996. *The Demon-Haunted World: Science as a Candle in the Darkness*. New York: Ballantine.

Shermer, M. 2002. *Why People Believe Weird Things: Pseudoscience, Superstition, and Other Confusions of Our Time*. New York: Holt.

———. 2011. *The Believing Brain: From Ghosts to Gods to Politics and Conspiracies—How We Construct Beliefs and Reinforce Them as Truths*. New York: Times.

Shulman, S. 2007. *Undermining Science: Suppression and Distortion in the Bush Administration*. Berkeley: University of California Press.

Specter, M. 2009. *Denialism: How Irrational Thinking Hinders Scientific Progress, Harms the Planet, and Threatens Our Lives*. New York: Penguin.

Taibbi, M. 2009. *The Great Derangement: A Terrifying True Story of War, Politics, and Religion*. New York: Spiegel and Grau.

NOTES

1. Reality Check

1. K. Goldstein, "Creationists and Climate Deniers Take On Teaching Climate Science in Schools," *Huffington Post,* May 4, 2010, www.huffingtonpost.com/2010/03/04/creationists-and-climate_n_485572.html.

2. P. Plait, "I'm Skeptical of Denialism," *Discover,* June 9, 2009, blogs.discover magazine.com/badastronomy/2009/06/09/im-skeptical-of-denialism.

3. "Galileo Was Wrong: The Church Was Right," *Galileo Was Wrong,* www.galileo waswrong.com/galileowaswrong.

4. R. Hood, "Mack Wolford's Death a Reminder That Serpent Handlers Should Be Lauded for Their Faith," *Washington Post,* June 5, 2012, articles.washingtonpost .com/2012-06-05/local/35460981_1_serpent-handlers-religious-liberty.

2. Science, Our Candle in the Darkness

1. As many philosophers of science have shown, Popper's falsifiability criterion does not work in every aspect of science. Still, they agree that science must be testable, and the most rigorous method of doing so is the criterion of falsifiability.

2. R. Carroll, "Pope Says Sorry for Sins of Church," *Guardian,* March 13, 2000, www .guardian.co.uk/world/2000/mar/13/catholicism.religion.

3. T. H. Huxley, Presidential Address to the British Association, "Biogenesis and Abiogenesis" (1870), alepho.clarku.edu/huxley/CE8/B-Ab.html.

4. M. W. Browne, "Physicists Debunk Claim of a New Kind of Fusion," *New York Times,* May 3, 1989, partners.nytimes.com/library/national/science/050399sci-cold-fusion.html.

5. "Ricky Gervais: Why I'm an Atheist," *Wall Street Journal,* December 19, 2010, blogs.wsj.com/speakeasy/2010/12/19/a-holiday-message-from-ricky-gervais-why-im-an-atheist.

6. Sagan (1996, 10; M. Shermer, *Why People Believe Weird Things: Pseudoscience, Superstitions and Other Confusions of Our Times* (New York: Holt, 2002), 48; D. R. Prothero, *Evolution: What the Fossils Say and Why It Matters* (New York: Columbia University Press, 2007), 14–17.

7. M. Shermer, "Show Me the Body," *The Work of Michael Shermer,* May 2003, www .michaelshermer.com/2003/05/show-me-the-body.

3. Betrayers of the Truth

1. *The Global Burden of Disease: 2004 Update,* World Health Organization website, www.who.int/healthinfo/global_burden_disease/GBD_report_2004update_full.pdf.

2. "Appendix III: Internationally Comparable Prevalence Estimates," World Health Organization website, www.who.int/tobacco/mpower/mpower_report_prevalence_data_2008.pdf.

3. Figure, *Cigarette Smoking Among Adults: United States, 2006,* Centers for Disease Control website, www.cdc.gov/mmwr/preview/mmwrhtml/mm5644a2.htm#fig.

4. N. Wolchover, "Want to Live Longer? Move to NYC," *Yahoo! News,* June 13, 2012, news.yahoo.com/want-live-longer-move-nyc-204217030.html.

5. See N. Oreskes and E. M. Conway, *Merchants of Doubt: How a Handful of Scientists Obscured the Truth on Issues from Tobacco Smoke to Global Warming* (New York: Bloomsbury, 2010); D. Michaels, *Doubt Is Their Product: How Industry's Assault on Science Threatens Your Health* (Oxford: Oxford University Press, 2008); and T. O. McGarity, *Bending Science: How Special Interests Corrupt Public Health* (Cambridge, Mass.: Harvard University Press, 2010).

6. S. Glantz, J. Slade, L. A. Bero, P. Hanauer, and D. E. Barnes, *The Cigarette Papers* (Berkeley: University of California Press, 1996).

7. Oreskes and Conway 2010, 14–16.

8. Glantz et al. 1996.

9. Oreskes and Conway 2010, 16.

10. Oreskes and Conway 2010, 34.

11. Glantz et al. 1996.

12. Glantz et al. 1996.

13. Glantz et al. 1996.

14. Glantz et al. 1996.

15. Glantz et al. 1996.

16. Oreskes and Conway 2010, 129.

17. Oreskes and Conway 2010, chapter 2.

18. Oreskes and Conway 2010, 47.

19. W. Burr and S. Savranskaya, eds., "Previously Classified Interviews with Former Soviet Officials Reveal U.S. Strategic Intelligence Failure Over Decades," National Security Archive website, September 11, 2009, www.gwu.edu/~nsarchiv/nukevault/ebb285.

20. Oreskes and Conway 2010, 43.

21. A. H. Kahn, *Killing Détente: The Right Attacks the CIA* (College Station: Pennsylvania State University Press, 1998), 167.

22. For the current scientific acceptance of this idea, see D. R. Prothero, *Greenhouse of the Dinosaurs: Evolution, Extinction, and the Future of Our Planet* (New York: Columbia University Press, 2009), chapter 5.

23. R. P. Turco, O. B. Toon, T. P. Ackerman, J. B. Pollack, and C. Sagan, "Nuclear Winter: Global Consequences of Multiple Nuclear Explosions," *Science* 222, no. 4630 (December 23, 1983): 1283–1292.

24. Oreskes and Conway 2010, 38.

25. Oreskes and Conway 2010, 63–65.

26. "President Barack Obama's Inaugural Address," White House blog, January 21, 2009, www.whitehouse.gov/blog/inaugural-address.

4. Making the Environment the Enemy

1. "Cuyahoga River Fire," Ohio History Central website, www.ohiohistorycentral .org/entry.php?rec=1642.

2. J. Nielsen, "The Killer Fire of '52: Thousands Died as Poisonous Air Smothered London," National Public Radio website, December 11, 2002, www.npr.org/templates/ story/story.php?storyId=873954.

3. G. Hardin, "The Tragedy of the Commons," *Science* 162 (1968): 1243–1248.

4. G. Likens, "Acid Precipitation," *Chemical and Engineering News* 54, no. 48 (Nov. 22, 1976): 29–44.

5. N. Oreskes and E. M. Conway, *Merchants of Doubt: How a Handful of Scientists Obscured the Truth on Issues from Tobacco Smoke to Global Warming* (New York: Blooms- bury, 2010), 74.

6. Royal Society of Canada, *Acid Deposition in North America: A Review of Docu- ments Prepared under the Memorandum of Intent between Canada and the United States of America, 1980, on Transboundary Air Pollution* (n.p.: Royal Society of Canada, 1983), sec. II-9.

7. Oreskes and Conway 2010, 77.

8. Oreskes and Conway 2010, 77.

9. National Research Council (U.S.) Committee on the Atmosphere and the Bio- sphere. *Atmosphere-Biosphere Interaction: Toward a Better Understanding of the Ecological Consequences of Fossil Fuel Combustion* (Washington, D.C.: National Academy of Sci- ences, 1981).

10. Oreskes and Conway 2010, chapter 3.

11. "Sulfur Dioxide," United States Environmental Protection Agency website, www .epa.gov/airtrends/sulfur.html.

12. "Acid Rain Program 2007 Progress Report," United States Environmental Pro- tection Agency website, January 2009, www.epa.gov/airmarkt/progress/arp07.html.

13. Z. Coile, "'Cap-and-Trade' Model Eyed for Cutting Greenhouse Gases," *San Francisco Chronicle*, www.sfgate.com/green/article/Cap-and-trade-model-eyed-for- cutting-greenhouse-3300270.php.

14. See Dotto and Schiff 1978, 68–70.

15. M. J. Molina and F. S. Rowland, "Stratospheric Sink for Chlorofluoromethanes," *Nature* 249 (June 28, 1974): 810–812.

16. Oreskes and Conway 2010, chapter 4.

17. Dotto and Schiff 1978, 225.

18. See the discussion of Singer in Oreskes and Conway 2010, chapters 3 and 4.

19. F. S. Singer, "My Adventures in the Ozone Layer." *National Review* 41, no. 12 (June 10, 1989): 34–38.

20. R. W. Scheuering, *Shapers of the Great Debate on Conservation: A Biographical Dictionary* (Westport, Conn.: Greenwood Press, 2004).

21. As documented by Oreskes and Conway 2010, chapter 4, n. 95, from SEPP IRS form 990 for 2007, lines 8d and 21, dated May 15, 2008.

22. Oreskes and Conway 2010, 134.

23. Oreskes and Conway 2010, chapter 7.

24. Oreskes and Conway 2010, chapter 7.

5. Hot Enough for You?

1. S. C. Webster, "Palin: America Does Not Need 'This Snake Oil Science Stuff,'" *Raw Story,* April 10, 2010, www.rawstory.com/rs/2010/04/10/palin-this-snake-oil-science-stuff.

2. Webster 2010.

3. J. Romm, "Major Analysis Finds 'Less Ice Covers the Arctic Today than at Any Time in Recent Geologic History,'" Think Progress website, September 8, 2010, thinkprogress.org/climate/2010/09/08/206152/arctic-sea-ice-history-paleoclimate-polar-amplification.

4. J. Amos, "Arctic Summers Ice-Free 'by 2013,'" BBC News website, December 12, 2007, news.bbc.co.uk/2/hi/7139797.stm.

5. J. Bryner, "Greenland Ice Sheet Melting Breaks 30-Year Record," *Huffington Post,* August 15, 2012, www.huffingtonpost.com/2012/08/15/greenland-ice-sheet-melting_n_1783063.html.

6. "Pacific Island Villagers Become Climate Change Refugees," Environment News Service website, December 6, 2005, www.ens-newswire.com/ens/dec2005/2005-12-06-02.asp.

7. A. D. Barnosky, *Heatstroke: Nature in an Age of Global Warming* (Washington, D.C.: Island Press, 2009).

8. Shifting Baselines website, www.shiftingbaselines.org/index.php.

9. "Team," Shifting Baselines website, www.shiftingbaselines.org/team/index.html.

10. "Team."

11. D. Zabarenko, "Oceans' Acidic Shift May Be Fastest in 300 Million Years," Reuters website, March 1, 2012, www.reuters.com/article/2012/03/01/us-climate-oceans-acid-idUSTRE82025S20120301.

12. K. Gammon, "Sea Creatures in a Warming World: Winners and Losers," Live Science website, October 1, 2012, www.livescience.com/23601-ocean-acidification-sea-creatures-survival.html.

13. S. Johnson, "Acidic Oceans Are Dissolving Sea Creatures' Shells Leaving Them Defenceless Against Predators," *Daily Mail,* November 25, 2012, www.dailymail.co.uk/sciencetech/article-2238302/Acidic-oceans-dissolving-sea-creatures-shells-leaving-defenceless-predators.html.

14. Zabarenko 2012.

15. B. Hönisch, A. Ridgwell, D. N. Schmidt, E. Thomas, S. J. Gibbs, A. Sluijs, R. Zeebe, et al., "The Geological Record of Ocean Acidification," *Science* 335, no. 6072 (March 2, 2012): 1058–1063; www.sciencemag.org/content/335/6072/1058.abstract.

16. V. Masson, F. Vimeux, J. Jouzel, V. Morgan, M. Delmotte, P. Ciais, C. Hammer, et al., "Holocene Climate Variability in Antarctica Based on 11 Ice-Core Isotopic Records," *Quaternary Research* 54 (2000): 348–358; D. S. Kaufman, T. A. Ager, N. J. Anderson, P. M. Anderson, J. T. Andrews, P. J. Bartlein, L. B. Brubaker, et al., "Holocene Thermal Maximum in the Western Arctic (0–180 W)," *Quaternary Science Reviews* 23 (2004): 529–560.

17. "What Ended the Little Ice Age?" Skeptical Science website, www.skeptical science.com/coming-out-of-little-ice-age.htm.

18. "Solar Activity & Climate: Is the Sun Causing Global Warming?" Skeptical Science website, www.skepticalscience.com/solar-activity-sunspots-global-warming.htm.

19. "What's the Link between Cosmic Rays and Climate Change?" Skeptical Science website, www.skepticalscience.com/cosmic-rays-and-global-warming.htm.

20. "What's the Link between Cosmic Rays and Climate Change?"

21. "Do Volcanoes Emit More Co2 Than Humans?" Skeptical Science website, www .skepticalscience.com/volcanoes-and-global-warming.htm.

22. "What is Methane's Contribution to Global Warming?" Skeptical Science website, www.skepticalscience.com/methane-and-global-warming.htm.

23. "What Has Global Warming Done since 1998?" Skeptical Science website, www .skepticalscience.com/global-warming-stopped-in-1998.htm.

24. "1934 Is the Hottest Year on Record," Skeptical Science website, www .skepticalscience.com/1934-hottest-year-on-record.htm.

25. B. Chameides, "Statistically Speaking: 2008's Place in Warmest Years Sweepstakes," *The Green Grok,* December 22, 2008, www.nicholas.duke.edu/ thegreengrok/2008temps.

26. "Does Cold Weather Disprove Global Warming?" Skeptical Science website, www.skepticalscience.com/global-warming-cold-weather.htm.

27. "FACTSHEET: Competitive Enterprise Institute, CEI," Exxon Secrets website, www.exxonsecrets.org/html/orgfactsheet.php.

28. "Is Co2 a Pollutant?" Skeptical Science website, www.skepticalscience.com/ co2-pollutant.htm.

29. "Why Ocean Heat Cannot Drive Climate Change, Only Chase It," Skeptical Science website, www.skepticalscience.com/ocean-and-global-warming.htm; "Does Ocean Cooling Prove Global Warming Has Ended?" Skeptical Science website, www .skepticalscience.com/cooling-oceans.htm; J. M. Lyman, S. A. Good, V. V. Gouretski, M. Ishii, G. C. Johnson, M. D. Palmer, D. M. Smith, and J. K. Willis, "Robust Warming of the Global Upper Ocean," *Nature* 465 (2010): 334–337.

30. A. H. Knoll, R. K. Bambach, J. P. Grotzinger, and D. Canfield, "Comparative Earth History and Late Permian Mass Extinction," *Science* 273 (1996): 452–457.

31. A. J. Bloom, M. Burger, J. S. R. Asensio, and A. B. Cousins, "Carbon Dioxide Enrichment Inhibits Nitrate Assimilation in Wheat and *Arabidopsis,*" *Science* 328 (2010): 899–903.

32. "The Human Fingerprint in Global Warming [Basic]," Skeptical Science website, www.skepticalscience.com/its-not-us.htm.

33. "The Human Fingerprint in Global Warming [Intermediate]," Skeptical Science website, www.skepticalscience.com/its-not-us-intermediate.htm.

34. "The Human Fingerprint in Global Warming [Intermediate]"; "Stratospheric Cooling and Tropospheric Warming: Revised," Skeptical Science website, skepti-calscience.com/Stratospheric_Cooling.html.

35. "Americans' Global Warming Concerns Continue to Drop," Gallup website, March 11, 2010, www.gallup.com/poll/126560/americans-global-warming-concerns-continue-drop.aspx.

36. N. Oreskes, "Beyond the Ivory Tower: The Scientific Consensus on Climatic Change," *Science* 306 (2004): 1686.

37. P. Doran and M. K. Zimmerman, "Examining the Scientific Consensus on Climatic Change," EOS 90, no. 3 (2009): 22.

38. W. R. L. Anderegg, J. W. Prall, J. Harold, and S. H. Schneider, "Expert Credibility on Climate Change," *Proceedings of the National Academy of Sciences (USA)* 107 (2010): 12107–12109.

39. J. L. Powell, "Why Climate Deniers Have No Scientific Credibility: In One Pie Chart," Desmog Blog, November 15, 2012, www.desmogblog.com/2012/11/15/why-climate-deniers-have-no-credibility-science-one-pie-chart.

40. A. Glikson, "Climate 'Skeptics' Renew War Against Science While Rome Burns," Todays Alternative News website, April 4, 2009, www.todaysalternativenews.com/index2.php?event=link,news_view2&values%5B0%5D=5987.

41. J. Gillis and L. Kaufman, "Leak Offers Glimpse of Campaign Against Climate Science," *New York Times,* February 15, 2012, www.nytimes.com/2012/02/16/science/earth/in-heartland-institute-leak-a-plan-to-discredit-climate-teaching.html.

42. A. M. McCright and R. E. Dunlap, "Defeating Kyoto: The Conservative Movement's Impact on U.S. Climate Change Policy," *Social Problems* 50, no. 3(2003): 348–373; J. A. Curry, P. J. Webster, and G. J. Holland, "Mixing Politics and Science in Testing the Hypothesis That Greenhouse Warming Is Causing a Global Increase in Hurricane Intensity," *Bulletin of the American Meteorological Society* 87, no. 8 (2006): 1025–1037; N. Williams, "Heavyweight Attack on Climate-Change Denial," *Current Biology* 15, no. 4 (2005): R109–R110; C. Mooney, *The Republican War on Science* (New York: Basic, 2006); C. Mooney, *Storm World: Hurricanes, Politics, and the Battle over Global Warming* (New York: Harcourt, 2009); J. Hoggan, *Climate Cover-Up: The Crusade to Deny Global Warming* (Vancouver, B.C.: Greystone, 2009); N. Oreskes and E. M. Conway, *Merchants of Doubt: How a Handful of Scientists Obscured the Truth on the Issues from Tobacco Smoke to Global Warming* (New York: Bloomsbury Press, 2010).

43. I. Sample, "Scientists Offered Cash to Dispute Climate Study," *Guardian,* February 2, 2007.

44. "James M. Inhofe," Open Secrets website, www.opensecrets.org/politicians/summary.php?cid=N00005582.

45. "Ranking Member's Senate Minority Report on Global Warming Not Credible, Says Center for Inquiry," Center for Inquiry website, July 17, 2009, www.centerforinquiry.net/opp/news/senate_minority_report_on_global_warming_not_credible.

46. "Oregon Petition," Debunking website, debunking.pbworks.com/w/page/17102969/Oregon%20Petition.

47. B. Lomborg, *The Skeptical Environmentalist: Measuring the Real State of the World* (Cambridge: Cambridge University Press, 2001); B. Lomborg, *Cool It: The Skeptical Environmentalist's Response to Global Warming* (New York: Knopf, 2007).

48. "UCS Examines 'The Skeptical Environmentalist,'" Union of Concerned Scientists website, www.ucsusa.org/global_warming/science_and_impacts/global_warming_contrarians/ucs-examines-the-skeptical.html.

49. H. Friel, *The Lomborg Deception: Setting the Record Straight about Global Warming* (New Haven, Conn.: Yale University Press, 2010).

50. J. Jowit, "Bjørn Lomborg: The Dissenting Dlimate Change Voice Who Changed His Tune," *Guardian,* August 30, 2010, www.guardian.co.uk/environment/2010/aug/30/bjorn-lomborg-climate-change-profile.

51. I. Plimer, *Heaven and Earth: Global Warming, the Missing Science* (Boulder, CO: Taylor, 2009).

52. "Plimer's Homework Assignment," RealClimate website, August 24, 2009, www .realclimate.org/index.php/archives/2009/08/plimers-homework-assignment/; "The Science Is Missing from Ian Plimer's "Heaven and Earth," *Deltoid,* April 23, 2009, sci-enceblogs.com/deltoid/2009/04/the_science_is_missing_from_ia.php; I. G. Enting, "Ian Plimer's 'Heaven + Earth': Checking the Claims," Australian Research Council Centre of Excellence for Mathematics and Statistics of Complex Systems website, www .complex.org.au/tiki-download_file.php?fileId=91.

53. "The CRU Hack," RealClimate website, November 20, 2009, www.realclimate .org/index.php/archives/2009/11/the-cru-hack.

54. K. Adam and J. Eilperin, "Academic Experts Clear Ccientists in 'Climate-Gate,'" *Washington Post,* April 15, 2010, www.washingtonpost.com/wp-dyn/content/ article/2010/04/14/AR2010041404001.html.

55. J. L. Powell, *The Inquisition of Climate Science* (New York: Columbia University Press, 2011), 187.

56. "Discovery Institute Praises Global Warming Deniers," *The Sensuous Curmudgeon,* March 19, 2010, sensuouscurmudgeon.wordpress.com/2010/03/19/ discovery-institute-praises-global-warming-deniers.

57. "30-Country Poll Finds Worldwide Consensus that Climate Change Is a Serious Problem," World Public Opinion website, April 25, 2006, www.worldpublicopinion.org/ pipa/articles/btenvironmentra/187.php?nid=&id=&pnt=187.

58. "Man Causing Climate Change: Poll," BBC News website, September 25, 2007, news.bbc.co.uk/2/hi/in_depth/7010522.stm.

59. "Factsheet: Heatland Institute, Heartland," Exxon Secrets website, http://www .exxonsecrets.org/html/orgfactsheet.php?id=41.

60. R. Harrabin, "Climate Sceptics Rally to Expose 'Myth,'" BBC News website, May 21, 2010, news.bbc.co.uk/2/hi/science/nature/8694544.stm.

61. D. Harris and C. Brouwer, "Climate Scientists Claim 'McCarthy-Like Threats,' Say They Face Intimidation, Ominous E-Mails," *ABC World News,* May 23, 2010, abcnews.go.com/WN/Media/climate-scientists-threat-global-warming-proponents -face-intimidation/story?id=10723932.

62. D. Safier, "Palin: Democrats in the Crosshairs," *Blog for Arizona,* www .blogforarizona.com/blog/2010/03/palin-democrats-in-the-crosshairs.html.

63. R. Piltz, "Sen. Inhofe Inquisition Seeking Ways to Criminalize and Prosecute 17 Leading Climate Scientists," Climate Science Watch website, February 24, 2010, www.climatesciencewatch.org/index.php/csw/details/ sen._inhofe_inquisition_seeking_to_criminalize_climate_scientists.

64. C. Stuart, "Oh, Mann: Cuccinelli targets UVA papers in Climategate salvo," *The Hook,* April 29, 2010, www.readthehook.com/blog/index.php/2010/04/29/ oh-mann-cuccinelli-targets-uva-papers-in-climategate-salvo.

65. "Poll Question," HamptonRoads website, May 1, 2010, hamptonroads.com/polls/ what-do-you-think-attorney-general-ken-cuccinelli%25E2%2580%2599s-decision-issue-lapel-pins-state-seal-cover?t=1355867417.

66. "The American Clean Energy and Security Act (Waxman-Markey Bill)," Center for Climate and Energy Solutions website, www.c2es.org/federal/congress/111/acesa.

67. A. Horowitz, "Global Warming Poll: Climate Change A 'Serious Problem' To 68% Of Americans," *Huffington Post,* November 9, 2012, www.huffingtonpost .com/2012/11/09/global-warming-poll-climate-change_n_2105600.html.

68. S. Borenstein, "AP-GfK Poll: Science doubters say world is warming," Yahoo! News website, December 14, 2012, news.yahoo.com/ap-gfk-poll-science-doubters-world-warming-080143113.html.

69. P. R. Epstein, "Two Outta Three Ain't a Bad Start: Lurching Toward Global Governance in Copenhagen," *Solutions,* December 2009, www.thesolutionsjournal.com/ node/526.

70. A. Landman, "BP's 'Beyond Petroleum' Campaign Losing Its Sheen," *PRWatch,* May 3, 2010, www.prwatch.org/node/9038.

71. J. M. Broder, "Climate Change Seen as Threat to U.S. Security," *New York Times,* August 8, 2009, www.nytimes.com/2009/08/09/science/earth/09climate.html?hp.

72. P. Schwartz and D. Randall, *An Abrupt Climate Change Scenario and Its Implications for United States National Security* (Washington, D.C.: U.S. Departmentt of Defense, 2003), 14, www.gbn.com/articles/pdfs/Abrupt%20Climate%20Change%20 February%202004.pdf.

73. Schwartz and Randall 2003, 21.

74. P. R. Epstein, and E. Mills, *Climate Change Futures: Health, Ecological and Economic Dimensions, Center for Health and the Global Environment* (Boston: Harvard Medical School, 2005).

6. Gimme That Old Time Religion

1. K. Q. Seelye, "At G.O.P. Debate, Candidates Played to Conservatives," *New York Times,* May 5, 2007, www.nytimes.com/2007/05/05/us/politics/05repubs.html.

2. D. Amira, "Which GOP Presidential Candidates Believe in Evolution?" *New York Magazine,* August 18, 2011, nymag.com/daily/intel/2011/08/evolution_gop_candidates .html.

3. D. Amira, "Evolution Is Just a 'Theory That's Out There,' According to Rick Perry," *New York Magazine,* August 18, 2011, nymag.com/daily/intel/2011/08/rick _perry_evolution_theory.html.

4. Jon Huntsman, Twitter feed, August 18, 2011, twitter.com/jonhuntsman/ status/104250677051654144.

5. M. Cooper, "Fanning the Controversy Over 'Intelligent Design,'" *Time,* August 3, 2005, www.time.com/time/nation/article/0,8599,1089733,00.html.

6. J. Linkins, "Bradley Byrne, Alabama Gubernatorial Candidate, Attacked For Supporting Evolution," *Huffington Post,* May 11, 2010, www.huffingtonpost.com/2010/05/11/ bradley-byrne-alabama-gub_n_572223.html.

7. N. Wing, "Marco Rubio: Actual Age Of Earth Is 'One Of The Great Mysteries,'" *Huffington Post,* November 19, 2012, www.huffingtonpost.com/2012/11/19/marco-rubio-earth-age_n_2158555.html.

8. J. Rayfield, "Quote of the Day: Lies from 'the Pit of Hell,'" *Salon,* Ocotber 5, 2102, www.salon.com/2012/10/05/quote_of_the_day_lies_from_the_pit_of_hell.

9. M. Pearce, "U.S. Rep. Paul Broun: Evolution a Lie 'from the Pit of Hell,'" *Los Angeles Times,* October 7, 2012, articles.latimes.com/2012/oct/07/nation/ la-na-nn-paul-broun-evolution-hell-20121007.

10. A. P. Beadle, "Todd Akin: No 'Science' behind Evolution," Think Progress website, October 12, 2012, thinkprogress.org/election/2012/10/12/1003121/todd-akin-no-science-behind-evolution.

11. "By one count there are some 700 scientists with respectable academic credentials (out of a total of 480,000 U.S. earth and life scientists) who give credence to creation-science, the general theory that complex life forms did not evolve but appeared 'abruptly'" (L. Martz and A. McDaniel, "Keeping God Out of Class [Washington and Bureau Reports]," *Newsweek,* June 29, 1987, 23).

12. C. Delgado, "Finding the Evolution in Medicine," NIH *Record* 58, no. 15 (July 28, 2006), nihrecord.od.nih.gov/newsletters/2006/07_28_2006/story03.htm.

13. "Few Biologists but Many Evangelicals Sign Anti-Evolution Petition," *Panda's Thumb,* February 21, 2006, pandasthumb.org/archives/2006/02/few-biologists.html.

14. J. F. Ashton, *In Six Days: Why Fifty Scientists Choose to Believe in Creation,* (Green Forest, Ark.: Master, 2001).

15. National Center for Science Education, "Project Steve," NCSE website, October 17, 2008, ncse.com/taking-action/project-steve.

16. Clergy Letter Project website, www.theclergyletterproject.org.

17. M. Matsumura, "What Do Christians Really Believe About Evolution?" *Reports of the National Center About Evolution* 18, no. 2 (1998): 8–9.

18. John Paul II, "Message to the Pontifical Academy of Sciences: On Evolution," Global Catholic Network website, October 22, 1996,"www.ewtn.com/library/papaldoc/jp961022.htm

19. J. D. Bales, *Forty-Two Years on the Firing Line,* (Shreveport, La.: Lambert, 1977), 71–72.

20. See D. R. Prothero, *Evolution: What the Fossils Say and Why It Matters* (New York: Columbia University Press, 2007), 107–109.

21. National Center for Science Education, "Icon 4: Haeckel's Embryos," NCSE website, November 23, 2006, ncse.com/creationism/analysis/icon-4-haeckels-embryos.

22. Quoted in J. Weiner, *The Beak of the Finch: A Story of Evolution in Our Own Time* (New York: Knopf, 1994), 255.

23. See Prothero 2007, 115–118.

24. J. Weiner, "Evolution in Action," *Natural History* 115, no.9 (2005): 47–51.

25. T. Dobzhansky, "Nothing in Biology Makes Sense Except in the Light of Evolution," *The American Biology Teacher* 35 (March 1973): 125–129.

26. Prothero 2007, chapter 2.

27. R. Friedman, *Who Wrote the Bible?* (New York: Harper and Row, 1987), 54.

28. R. Numbers, *The Creationists: The Evolution of Scientific Creationism* (New York: Knopf, 1992), 39–40.

29. Prothero 2007, 32–35.

30. McLean v. Arkansas Board of Education, TalkOrigins Archive, updated January 30, 1996, www.talkorigins.org/faqs/mclean-v-arkansas.html.

31. Prothero 2007, 35–43.

32. Documented in Prothero 2007, 41–42.

33. Center for the Renewal of Science and Culture, "The Wedge Strategy," AntiEvolution website, www.antievolution.org/features/wedge.html.

34. M. Shermer, "The Cowardice and Calumny of Creationism," *Huffington Post,* December 3, 2009, www.huffingtonpost.com/michael-shermer/the-cowardice-and-calumny_b_379529.html

35. Shermer 2009.

36. Shermer 2009.

37. Shermer 2009.

38. J. Wells, "Darwinism: Why I Went for a Second Ph.D.," Tparents website, www.tparents.org/library/unification/talks/wells/darwin.htm.

39. "Cdesign Proponentsists," National Center for Science Education website, September 25, 2008, ncse.com/creationism/legal/cdesign-proponentsists.

40. *Tammy Kitzmiller v. Dover Area School District,* National Center for Science Education website, December 20, 2005, ncse.com/files/pub/legal/kitzmiller/highlights/2005-12-20_Kitzmiller_decision.pdf.

41. M. Guarino, "Teaching Creationism: Louisiana Law That Skirts US Ban Survives Challenge," *Christian Science Monitor,* June 2, 2011, www.csmonitor.com/USA/Education/2011/0602/Teaching-creationism-Louisiana-law-that-skirts-US-ban-survives-challenge.

42. National Center for Science Education website, ncse.com.

43. J. Arthur, "Creationism: Bad Science or Immoral Pseudoscience?" *Skeptic* 4, no. 4 (1996): 88–93, www.holysmoke.org/gish.htm.

44. K. Miller, "Answers to the Standard Creationist Arguments," *Creative Evolution Journal* 3, no. 1 (1982): 1–13, ncse.com/cej/3/1/answers-to-standard-creationist-arguments.

45. M. Shermer, "25 Creationists' Arguments and 25 Evolutionists' Answers," Geological Society of America website, www.geosociety.org/criticalissues/ev_shermer.htm.

46. See Prothero 2007, 74–78.

47. G. B. Dalrymple, *Ancient Earth, Ancient Skies: The Age of the Earth and Its Cosmic Surroundings* (Palo Alto, Calif.: Stanford University Press, 2004).

48. "The Ancient Bristlecone Pine," www.sonic.net/bristlecone/intro.html; "World's Oldest Living Tree Discovered in Sweden," Umeå University website, April 16, 2008, info.adm.umu.se/NYHETER/PressmeddelandeEng.aspx?id=3061.

49. "Sumerians Look On in Confusion as God Creates World," *The Onion,* December 15, 2009, www.theonion.com/articles/sumerians-look-on-in-confusion-as-god-creates-wor1,2879.

50. See Prothero 2007, chapter 7.

51. See Prothero 2007, chapter 15.

52. See this excellent video on the topic":The Light of Evolution: What Would Be Lost," YouTube video, 10:55, posted by "Concordance," January 6, 2010, www.youtube.com/watch?v=rt5CfQvaYSM.

53. D. Prothero, "Denial of Evolution Can Be Hazardous to Your Health . . . ," Skeptic Blog, August 10, 2011, www.skepticblog.org/2011/08/10/denial-of-evolution-can-kill-you.

54. Prothero 2007, 353–354.

7. Jenny's Body Count

1. J. Browne, *Charles Darwin: A Biography, Vol. 1: Voyaging,* (New York: Knopf, 1995), 499.

2. Quoted in Browne 1995, 501.

3. F. E. W. Harper, "Thank God for Little Children," Read Book Online website, www.readbookonline.net/readOnLine/28161/

4. "Infant Mortality," Wikipedia, en.wikipedia.org/wiki/Infant_mortality.

5. R. Aylward, "Eradicating Polio: Today's Challenges and Tomorrow's Legacy," *Ann Trop Med Parasitol* 100, nos. 5–6 (2006): 401–413.

6. F. Fenner, D. A. Henderson, I. Arita, Z. Jeek, and I. D. Ladnyi, *Smallpox and Its Eradication* (Geneva: World Health Organization, 1988).

7. R. W. Sutter and C. Maher, "Mass Vaccination Campaigns for Polio Eradication: An Essential Strategy for Success," *Curr Top Microbiol Immunol* 304 (2006): 195–220.

8. Centers for Disease Control and Prevention, "Progress toward Elimination of *Haemophilus influenzae* Type B Invasive Disease among Infants and Children: United States, 1998–2000," *MMWR Morb Mortal Wkly Rep* 51, no. 11 (2002): 234–237.

9. "Cholera Vaccines: A Brief Summary of the March 2010 Position Paper," World Health Organization website, March 10, 2010, www.who.int/immunization/Cholera_PP_Accomp_letter__Mar_10_2010.pdf.

10. W. Atkinson, J. Hamborsky, L. McIntyre, and S. Wolfe, "Diphtheria," in *Epidemiology and Prevention of Vaccine-Preventable Diseases (The Pink Book),* 10th ed. (Washington D.C.: Public Health Foundation, 2007), 59–70.

11. World Health Organization, *Weekly Epidemiological Record* 84, no. 49 (December 4, 2009): 505–516, www.who.int/wer/2009/wer8449/en/index.html.

12. D. L. Kasper, E. Braunwald, A. S. Fauci, S. L. Hauser, D. L. Longo, J. L. Jameson, and K. J. Isselbacher, eds,*Harrison's Principles of Internal Medicine,* 16th ed. (New York: McGraw-Hill Professional, 2004).

13. A. Park, "How Safe Are Vaccines?" *Time,* June 2, 2008, www.time.com/time/printout/0,8816,1808438,00.html.

14. A. Wakefield, S. Murch, A. Anthony, et al., "Ileal-Lymphoid-Nodular Hyperplasia, Non-specific Colitis, and Pervasive Developmental Disorder in Children," *Lancet* 351, no. 9103 (1998): 637–641, quotation on 641.

15. B. Carey, "Father's Age Is Linked to Risk of Autism and Schizophrenia," *New York Times,* August 22, 2012, www.nytimes.com/2012/08/23/health/fathers-age-is-linked-to-risk-of-autism-and-schizophrenia.html.

16. M. J. Dougherty, "The Genetics of Autism," ActionBioScience website, September 2000, www.actionbioscience.org/genomic/dougherty.html.

17. P. McIntyre, and J. Leask, "Improving Uptake of MMR Vaccine," BMJ 336, no. 7647 (2008): 729–730; M. B. Pepys, "Science and Serendipity," *Clin Med* 7, no. 6 (2007): 562–567.

18. Jenny McCarthy Body Count website, www.jennymccarthybodycount.com/Jenny_McCarthy_Body_Count/Home.html.

19. idoubtit, " Congressional Autism Hearing Goes down the Toilet," Doubtful News website, December 3, 2012, doubtfulnews.com/2012/12/congressional-autism-hearing-goes-down-the-toilet.

20. M. Specter, *Denialism* (New York: Penguin, 2009), 62.

21. B. Deer, "Exposed: Andrew Wakefield and the MMR-autism fraud," Brian Deer website, briandeer.com/mmr/lancet-summary.htm.

22. Deer, "Exposed."

23. Deer, "Exposed."

24. N. Triggle, "MMR Scare Doctor 'Acted Unethically,' Panel Finds," BBC News website, January 28, 2010, news.bbc.co.uk/2/hi/health/8483865.stm.

25. T. H. Maugh II, "Journal Retracts Study That Linked Autism to Vaccine," *Statesman,* February 2, 2010, www.statesman.com/news/world/journal-retracts-study-that-linked-autism-to-vaccine-210033.html.

26. A. J. Wakefield, A. Anthony, S. H. Murch, M. Thomson, S. M. Montgomery, S. Davies, J. J. O'Leary, M. Berelowitz and J. A. Walker-Smith, "Retraction: Enterocolitis in Children with Developmental Disorders," *American Journal of Gastroenterology* 95 (2000): 2285–2295, www.nature.com/ajg/journal/v105/n5/full/ajg2010149a.html.

27. J. Meikle and S. Boseley, "MMR Row Doctor Andrew Wakefield Struck off Register," *Guardian,* May 24, 2010, www.guardian.co.uk/society/2010/may/24/mmr-doctor-andrew-wakefield-struck-off.

28. T. Jefferson, D. Price, V. Demicheli, and E. Bianco, "Unintended Events Following Immunization with MMR: A Systematic Review," *Vaccine* 21, nos. 25–26 (2003): 3954–3960.

29. "Statement on Thiomersal," World Health Organization website, July 2006, www.who.int/vaccine_safety/committee/topics/thiomersal/statement_ju12006.

30. Centers for Disease Control, "Thimerosol," Centers for Disease Control website, www.cdc.gov/vaccinesafety/Concerns/thimerosal/index.html; U.S. Food and Drug Administration, "Thimerosal in Vaccines," U.S. Food and Drug Administration website, www.fda.gov/BiologicsBloodVaccines/SafetyAvailability/VaccineSafety/UCM096228.

31. P. A. Offit, "Thimerosal and Vaccines: A—Cautionary Tale," *The New England Journal of Medicine* 357, no. 13 (2007): 1278–1279.

32. Dan Burton, "The Status of Research into Vaccine Safety and Autism," Whale website, June 19, 2012, www.whale.to/a/a.html.

33. U.S. Food and Drug Administration, "FDA Warns Marketers of Unapproved 'Chelation' Drugs," U.S. Food and Drug Administration website, www.fda.gov/ForConsumers/ConsumerUpdates/ucm229358.htm.

34. W. A. Thompson, C. Price, B. Goodson, D. K. Shay, P. Benson, V. L. Hinrichsen, E. Lewis, et al., " Early Thimerosal Exposure Neuropsychological Outcomes at 7 to 10 Years," *The New England Journal of Medicine* 357, no. 13 (2007): 1281–90, www.nejm.org/doi/full/10.1056/NEJMoa071434.

35. Dougherty 2000.

36. Carey 2012.

37. G. Laden, "Are You Are Real Skeptic, or Are You Just Faithing It?" Greg Laden's blog, June 30, 2010, networkedblogs.com/519vG.

38. "Misconceptions about Immunization," Quackwatch website, http://www.quackwatch.org/03HealthPromotion/immu/autism.html.

39. Steven Novella, "Is the Rise in Autism Rates Real?" Neurologica blog, January 8, 2009, theness.com/neurologicablog/?p=454.

40. Laden 2010.

41. D. Gorski, "Welcome Back, My Friends, to the Show That Never Ends: The Jenny and Jim Antivaccine Propaganda Tour Has Begun," Science-Based

Medicine website, April 6, 2009, www.sciencebasedmedicine.org/index.php/
the-jenny-and-jim-antivaccine-propaganda-tour-has-begun.

42. S. Murch, "Separating Inflammation from Speculation in Autism," *Lancet* 362,
no. 9394 (2003): 1498–1499.

43. P. Asaria and E. MacMahon, "Measles in the United Kingdom: Can We Eradicate
It by 2010?" BMJ 333 (2006): 890, www.bmj.com/cgi/content/full/333/7574/890.

44. "England and Wales in Grip of Mumps Epidemic," *New Zealand Herald,* May 13,
2005, www.nzherald.co.nz/world/news/article.cfm?c_id=2&objectid=10125382.

45. European Centre for Disease Prevention and Control, "Measles Once Again Endemic in the United Kingdom," *Eurosurveillance* 13, no. 27 (July 3, 2008), www
.eurosurveillance.org/ViewArticle.aspx?ArticleId=18919.

46. California Department of Public Health, "Whooping Cough Epidemic May Be
Worst in 50 Years," California Department of Public Health website, June 23, 2010, www
.cdph.ca.gov/Pages/NR10-041.aspx

47. Park 2008.

48. "The Vaccine War," episode of *Frontline,* PBS website, April 27, 2010, www.pbs.org/wgbh/pages/frontline/vaccines/
view/?utm_campaign=viewpage&utm_medium=grid&utm_source=grid.

49. P. Nguyen, "Parents Who Don't Vaccinate Their Kids Put Us All at Risk," *Los Angeles Times,* June 1, 2010.

8. Victims of Modern Witch Doctors

1. P. M. Sharp, E. Bailes, R. R. Chaudhuri, C. M. Rodenburg, M. O. Santiago, and
B. H. Hahn, "The Origins of Acquired Immune Deficiency Syndrome Viruses: Where
and When?" *Philos. Trans. R. Soc. Lond., B, Biol. Sci.* 356, no. 1410 (2001): 867–876.

2. F. Gao, E. Bailes, D. L. Robertson, Y. Chen, C. M. Rodenburg, S. F. Michael, L.B.
Cummins, et al., "Origin of HIV-1 in the Chimpanzee Pan troglodytes troglodytes."
Nature 397, no. 6718 (1999): 436–441; B. F. Keele, F. van Heuverswyn, Y. Y. Li, E. Bailes,
J. Takehisa, M. L. Santiago, F. Bibollet-Ruche, et al., "Chimpanzee Reservoirs of Pan-
demic and Nonpandemic HIV-1." *Science Online* 5786 (May 25, 2006): 523.

3. M. Worobey, M. Gemmel, MD. E. Teuwen, et al., "Direct Evidence of Extensive
Diversity of HIV-1 in Kinshasa by 1960." *Nature* 455, no. 7213 (October 2008): 661–664.

4. P. A. Marx, P. G. Alcabes, and E. Drucker, "Serial Human Passage of Simian Im-
munodeficiency Virus by Unsterile Injections and the Emergence of Epidemic Human
Immunodeficiency Virus in Africa," *Philos. Trans. R. Soc. Lond., B, Biol. Sci.* 356, no. 1410
(June 2001): 911–920.

5. M. Schindler, J. Münch, O. Kutsch, et al., "Nef-mediated Suppression of T Cell
Activation Was Lost in a Lentiviral Lineage That Gave Rise to HIV-1," *Cell* 125, no. 6
(2006): 1055–1067.

6. T. Zhu, B. T. Korber, A. J. Nahmias, E. Hooper, P. M. Sharp, and D. D. Ho, "An
African HIV-1 Sequence from 1959 and Implications for the Origin of the Epidemic,"
Nature 391, no. 6667 (1998): 594–597.

7. R. F. Garry, M. H. Witte, A. A. Gottlieb, et al., "Documentation of an AIDS Virus
Infection in the United States in 1968," JAMA 260, no. 14 (October 1988): 2085–2087.

8. M. T. Gilbert, A. Rambaut, G.Wlasiuk, T. J. Spira, A. E. Pitchenik, and M. Worobey, "The Emergence of HIV/AIDS in the Americas and Beyond," *Proc Natl Acad Sci USA* 104, no. 47 (2009): 18566–18570.

9. R. Shilts, *And the Band Played On: Politics, People and the AIDS Epidemic* (New York: St. Martin's Press, 1987).

10. H. W. Jaffe, W. W. Darrow, D. F. Echenberg, P. M. O'Malley, J. P. Getchell, V. S. Kalyanaraman, R. H. Byers, D. P. Drennan, E. H. Braff, J. W. Curran, et al., "The Acquired Immunodeficiency Syndrome in a Cohort of Homosexual Men: A Six-Year Follow-Up Study," *Annals of Internal Medicine* 103, no. 2 (1985): 210–214; N. Bowdler, "Key HIV Strain 'Came from Haiti,'" BBC News website, October 30, 2007, news.bbc .co.uk/2/hi/health/7068574.stm.

11. D. Cummings, "On This Day: Magic Johnson Reveals That He Has HIV," Finding Dulcinea website, November 7, 2011, www.findingdulcinea.com/news/on-this-day/ November/Magic-Johnson-Reveals-That-He-He-Has-HIV.html.

12. J. Mcgeary, "Death Stalks a Continent," *Time*, February 12, 2001, www.time.com/ time/magazine/article/0,9171,999190–1,00.html.

13. UNAIDS, "Know Your Epidemic," UNAIDS website, www.unaids.org/en/ KnowledgeCentre/HIVData/GlobalReport/2008/2008_Global_report.asp.

14. "Swaziland," note 6, Wikipedia, en.wikipedia.org/wiki/ Swaziland#cite_note-UNDP-6.

15. S. Robson and K. B. Sylvester, "Orphaned and Vulnerable Children in Zambia: The Impact of the HIV/AIDS Epidemic on Basic Education for Children at Risk," *Educational Research* 49, no. 3 (2007): 259–272.

16. Mcgeary 2001.

17. "About PEPFAR," United States President's Emergency Plan for AIDS Relief website, www.pepfar.gov/about.

18. UNAIDS, "Know Your Epidemic."

19. T. Leonard, "Scientists Rip S. African AIDS Policies," *Washington Post*, September 6, 2006, www.washingtonpost.com/wp-dyn/content/article/2006/09/06/ AR2006090600586.html.

20. AVERT, "HIV and AIDS in Zimbabwe," AVERT website, www.avert.org/aids-zimbabwe.htm.

21. M. Schoofs, "Debating the Obvious: Inside the South African Government's Controversial AIDS Panel," *Village Voice*, July 4, 2000, www.villagevoice .com/2000-07-04/news/debating-the-obvious.

22. "Controversy Dogs Aids Forum," BBC News website, July 10, 2000, news.bbc .co.uk/2/hi/africa/826742.stm.

23. "The Durban Declaration," *Nature* 406, nos. 15–16 (July 6, 2000), www.nature .com/nature/journal/v406/n6791/full/406015a0.html.

24. C. W. Dugger, "Study Cites Toll of AIDS Policy in South Africa," *New York Times*, November 25, 2008, www.nytimes.com/2008/11/26/world/africa/26aids.html.

25. S. Robinson, "No Place for Denial," *Time*, January 9, 2005, www.time.com/time/ magazine/article/0,9171,1015842,00.html.

26. Schoofs 2000.

27. N. Nattrass, "Estimating the Lost Benefits of Antiretroviral Drug Use in South Africa," *African Affairs* 107, no. 427 (February 2008): 157–176.

28. Dugger 2008.

29. D. France, "The HIV Disbelievers," *Newsweek*, August 19, 2000.

30. "A Letter to the Community," ACT UP New York website, March 21, 2000, www .actupny.org/indexfolder/actupgg.html.

31. ACT UP San Francisco website, www.actupsf.com/aids/index.htm.

32. C. Ornstein and D. Costello, "A Mother's Denial, a Daughter's Death," *Los Angeles Times*, September 24, 2005, articles.latimes.com/2005/sep/24/local/me-eliza24.

33. Ornstein and Costello 2005.

34. "Alive & Well AIDS Alternatives," Wikipedia, en.wikipedia.org/wiki/ Alive_%26_Well_AIDS_Alternatives.

35. A. Gorman and A. Zavis, "Christine Maggiore, Vocal Skeptic of AIDS Research, Dies at 52," *Los Angeles Times*, December 30, 2008, www.latimes.com/news/local/ la-me-christine-maggiore30-2008dec30,0,7407966.story.

36. Toxi-Health International website, www.toxi-health.com.

37. R. A. Weiss and H. W. Jaffe, "Duesberg, HIV and AIDS," *Nature* 345 (June 21, 1990): 659–660, www.nature.com/nature/journal/v345/n6277/pdf/345659a0.pdf.

38. J. Cohen, "The Controversy over HIV and AIDS," *Science* 266 (December 9, 1994), www.sciencemag.org/feature/data/cohen/cohen.dtl.

39. J. Cartwright, "AIDS Contrarian Ignored Warnings of Scientific Misconduct," *Nature*, May 4, 2010, http://www.nature.com/news/2010/100504/full/news.2010.210 .html.

40. Cartwright 2010.

41. T. Bethell, *The Politically Incorrect Guide to Science* (Washington, D.C.: Regnery, 2005).

42. "Statement on CDC Fraud," Virusmyth website, www.virusmyth.com/aids/hiv/ pjcdc.htm.

43. M. J. Brauer, B. Forrest, and S. G. Gey, "Is It Science Yet? Intelligent Design Creationism And The Constitution," *Washington University Law Quarterly* 83, no. 1 (2005), www.canada.com/vancouversun/columnists/story. html?id=b0cb194b-51d3-4140-88f7-e4099445c554.

44. S. Epstein, *Impure Science: AIDS, Activism, and the Politics of Knowledge* (Berkeley: University of California Press, 1996).

45. "AIDS Denialists Who Have Died," AIDSTruth website, www.aidstruth.org/ denialism/denialists/dead_denialists.

9. If It Quacks like a Quack

1. For an excellent review of premodern medicine, see S. Singh and E. Ernst, *Trick or Treatment: The Undeniable Facts about Alternative Medicine* (New York: W. W. Norton, 2008), chapter 1.

2. B. Martin, "Archived Famous Health Quotes," Dr. Bob Martin website, www .doctorbob.com/famoushealthquotes.html.

3. Martin, "Archived Famous Health Quotes."

4. J. H. Young, *The Toadstool Millionaires: A Social History of Patent Medicines in America before Federal Regulation* (Princeton, N.J.: Princeton University Press, 1961).

5. S. Anderson and P. Homan, "'Best for Me, Best For You': A History of Beecham's Pills 1842–1998." *The Pharmaceutical Journal* 269 (2004): 921–924.

6. F. Benedetti, *Placebo Effects: Understanding the Mechanisms in Health and Disease* (New York: Oxford University Press, 2009).

7. Singh and Ernst 2008, 118.

8. Singh and Ernst 2008, 120.

9. Singh and Ernst 2008, 93.

10. "CBC Marketplace: Homeopathy: Cure or Con?" YouTube video, 9:34, posted by "jonnyeh," January 15, 2011, www.youtube.com/watch?v=kFKojcTknbU.

11. K. F. Schulz, I. Chalmers, R. J. Hayes, and D. G. Altman, "Empirical Evidence of Bias: Dimensions of Methodological Quality Associated With Estimates of Treatment Effects in Controlled Trials," *JAMA*, 273 (February 1995): 408–412; D. Moher, B. Pham, A. Jones, D. J. Cook, A. R. Jadad, M. Moher, P. Tugwell, and T. P. Klassen, "Does Quality of Reports of Randomised Trials Affect Estimates of Intervention Efficacy Reported in Meta-analyses?" *Lancet*352, no. 9128 (August 1998): 609–613.

12. E. Davenas, F. Beauvais, J. Amara, M. Oberbaum, B. Robinzon, A. Miadonna, A.Tedeschi, et al., "Human Basophil Degranulation Triggered by Very Dilute Antiserum against IgE," *Nature* 333, no. 6176 (June 1988): 816–818.

13. J. Maddox, J. Randi, and W. W. Stewart, "'High-Dilution' Experiments a Delusion," *Nature* 334 (July 28, 1988): 287–290.

14. "The 1998 Ig* Nobel Prize Ceremony," Improbablty Research website, improbable.com/ig/miscellaneous/ig-98.htm.

15. E. Ernst, "A Systematic Review of Systematic Reviews of Homeopathy," *Br J Clin Pharmacol* 54, no. 6 (2002): 577–582.

16. A. Shang, K. Huwiler-Müntener, L. Nartey, P. Jüni, S. Dörig, J. A. Sterne, D. Pewsner, and M. Egger, "Are the Clinical Effects of Homoeopathy Placebo Effects? Comparative Study of Placebo-Controlled Trials of Homoeopathy and Allopathy," *Lancet* 366, no. 9487 (August 27–September 2, 2005): 726–732, National Center for Biotechnology Information website, www.ncbi.nlm.nih.gov/sites/entrez?cmd=Retrieve&db=PubMed&list_uids=16125589.

17. S. Milazzo, N. Russell, and E. Ernst, "Efficacy of Homeopathic Therapy in Cancer Treatment," *Eur J Cancer* 42, no. 3 (February 2006): 282–289.

18. J. Jacobs, B. Guthrie, G. Montes, L. Jacobs, N. Colman, A. Wilson, and R. DiGiacomo, "Homeopathic Combination Remedy in the Treatment of Acute Childhood Diarrhea in Honduras," *The Journal of Alternative and Complementary Medicine* 12, no. 8 (October 2006): 723–732.

19. Singh and Ernst 2008, 136.

20. "Sceptics Stage Homeopathy 'Overdose,'" BBC News website, January 30, 2010, news.bbc.co.uk/2/hi/8489019.stm.

21. World Health Organization, "Legal Status of Traditional Medicine and Complementary/Alternative Medicine: A Worldwide Review," WHO website, whqlibdoc.who .int/hq/2001/WHO_EDM_TRM_2001.2.pdf.

22. "Homeopathy Prescriptions Falling," BBC News website, July 24, 2008, news .bbc.co.uk/2/hi/health/7523302.stm.

23. D. Colquhon, "Complementary Medicine Courses inUuniversities: How I Beat the Varsity Quacks," *Telegraph*, January 31, 2012, www.telegraph.co.uk/science/science-news/9051103/Complementary-medicine-courses-in-universities-how-I-beat-the -varsity-quacks.html.

24. Science and Technology Committee, "Evidence Check 2: Homeopathy," U.K. Parliament website, Fevruary 22, 2010, www.parliament.uk/business/committees/ committees-archive/science-technology/s-t-homeopathy-inquiry/.

25. "Government Response to the Science and Technology Committee Report 'Evidence Check 2: Homeopathy,'" Department of Health website, July 2010, www.dh.gov .uk/prod_consum_dh/groups/dh_digitalassets/@dh/@en/@ps/documents /digitalasset/dh_117811.pdf.

26. National Center for Complementary and Alternative Medicine, "The Use of Complementary and Alternative Medicine in the United States: Cost Data," National Institutes of Health website, July 2009, nccam.nih.gov/news/camstats/costs/costdatafs .htm.

27. Singh and Ernst 2008, 147.

28. H. Hall, "Kaiser Rejects Neck Manipulation," Science-Based Medicine website, August 31, 2010, www.sciencebasedmedicine.org/?p=6681.

29. C. Johnson, R. Baird, P. E. Dougherty, et al., "Chiropractic and Public Health: Current State and Future Vision," *J Manipulative Physiol Ther* 31, no. 6 (2008): 397–410.

30. Singh and Ernst 2008, chapter 4; E. Ernst, "Chiropractic: A Critical Evaluation," *J Pain Symptom Manage* 35, no. 5 (2008): 544–562.

31. E. Ernst, "Adverse Effects of Spinal Manipulation: A Systematic Review," *J R Soc Med* 100, no. 7 (2007): 330–338; L. O. Gouveia, P. Castanho, and J. J. Ferreira, "Safety of Chiropractic Interventions: A Systematic Review," *Spine* 34, no. 11 (2009): E405–413; S. Vohra, B. C. Johnston, K. Cramer, and K. Humphreys, "Adverse Events Associated with Pediatric Spinal Manipulation: A Systematic Review," *Pediatrics* 119, no. 1 (2007): e275–283; Ernst 2008; E. Ernst, "Deaths after Chiropractic: A Review of Published Cases," *Int J Clinical Practice* 64 (2010): 1162–1165.

32. Ernst 2008.

33. E. Ernst and P. D. Canter, "A Systematic Review of the Systematic Reviews of Spinal Manipulation," *JR. Soc. Med.* 9 (2006): 192–196.

34. E. Ernst, "Chiropractic Manipulation, with a Deliberate 'Double Entendre,'" *Arch Dis Child* 94, no. 6 (2009): 411; B. F. Walker, S. D. French, W. Grant, and S. Green, "Combined Chiropractic Interventions for Low-Back Pain," *Cochrane Database Syst Rev* 4, no. 4 (2010): CD005427; Singh and Ernst 2008, chapter 4; Ernst 2008.

35. Singh and Ernst 2008, 171.

36. Singh and Ernst 2008, chapter 4.

37. Ernst 2010.

38. Ernst and Canter 2006; Ernst 2010.

39. "Gallup Poll: Americans Have Low Opinion of Chiropractors' Honesty and Ethics," *Dynamic Chiropractic* January 29, 2007, www.dynamicchiropractic.com/mpacms/ dc/article.php?id=52038.

40. H. A. Tindle, R. B. Davis, R. S. Phillips, and D. M. Eisenberg, "Trends in Use of Complementary and Alternative Medicine by US Adults: 1997–2002," *Altern Ther Health Med* 11, no. 1 (2005): 42–49.

41. R. Eden, "Doctors Take Simon Singh to Court," *Telegraph*, August 16, 2008, www.telegraph.co.uk/news/newstopics/mandrake/2570744/Doctors-take-Simon- Singh-to-court.html; S. Singh, "Beware the Spinal Trap: Some Practitioners Claim It Is a Cure-All but Research Suggests Chiropractic Therapy Can Be Lethal," *Guardian,* April

19, 2008, cached on Svetlana Pertsovich's website, svetlana14s.narod.ru/Simon_Singhs_silenced_paper.html.

42. M. Robbins, "Furious Backlash from Simon Singh Libel Case Puts Chiropractors on Ropes," *Guardian,* March 1, 2010, www.guardian.co.uk/science/2010/mar/01/simon-singh-libel-case-chiropractors.

43. Robbins 2010.

10. What's Your Sign?

1. C. Huygens, *Cosmotheoros* (London: Timothy Childe, 1698), 68; available online at www.staff.science.uu.nl/~gento113/huygens/huygens_ct_en.htm.

2. L. Lyons, "Paranormal Beliefs Come (Super)Naturally to Some," Gallup website, November 1, 2005, www.gallup.com/poll/19558/paranormal-beliefs-come-supernaturally-some.aspx; "More Believe in God Than Heaven," Fox News website, June 18, 2004, www.foxnews.com/story/0,2933,99945,00.html.

3. North Texas Skeptics, "Astrology Fact Sheet," North Texas Skeptics website, www.ntskeptics.org/factsheets/astrolog.htm.

4. "American Public Media Promotes Astrology," *Discover,* January 1, 2009, blogs.discovermagazine.com/badastronomy/2009/01/01/npr-promotes-astrology.

5. G. Dell'orto, "Ariz. Astrology School Accredited," *Washington Post,* August 27, 2001, www.washingtonpost.com/wp-srv/aponline/20010827/aponline135357_000.htm.

6. P. Plait, "Astrology," Bad Astronomy blog, www.badastronomy.com/bad/misc/astrology.html.

7. "James Randi on Astrology," YouTube video, 1:36, posted by "hocobo," June 16, 2006, www.youtube.com/watch?v=3Dp2Zqk8vHw.

8. G. Dean and I. W. Kelly, "Is Astrology Relevant to Consciousness and Psi?" Imprint Academic website, www.imprint.co.uk/pdf/Dean.pdf.

9. P. Kruger, "European Researchers Debunk Astrology," *The World Today,* April 28, 2006, www.abc.net.au/worldtoday/content/2006/s1626391.htm.

10. "Skeptical Studies in Astrology," Psychic Investigator website, psychicinvestigator.com/demo/AstroSkc.htm; S. Carlson, "A Double-Blind Test of Astrology," *Nature* 318 (December 5, 1985): 419–425.

11. Carlson 1985.

12. J. Ashmun, "Astrology on the Internet: Quality of Discussion," *Correlation,* 15 (1996): 35–51, quotation on 41–43.

13. Astrology and Science website, www.astrology-and-science.com/hpage.htm.

14. Plait, "Astrology."

15. Plait, "Astrology."

16. Plait, "Astrology."

17. J. Wadler, A. Blessing, D. Mathison, and M. B. Sellinger, "The President's Astrologers," *People,* May 23, 1988, www.people.com/people/archive/article/0,,20099022,00.html.

11. Down the Slope of Hubbert's Curve

1. B. Trumbore, "The Arab Oil Embargo of 1973–74," Buy and Hold website, www.buyandhold.com/bh/en/education/history/2002/arab.html.

2. J. Leggett, *Half Gone: Oil, Gas, Hot Air and the Global Energy Crisis* (London, Portobello, 2005), 150, lines 12–13.

3. R. Harris, "Scientists Find Thick Layer Of Oil On Seafloor,"National Public Radio website, September 10, 2010, www.npr.org/templates/story/story.php?storyId=1297 82098&sc=17&f=1001.

4. For an excellent short video clip of the seventy-three-year-old Hubbert explaining his ideas, see "1976 Hubbert Clip," YouTube video, 1:58, posted by "David Room," March 26, 2007, www.youtube.com/watch?v=ImV1voi41YY&feature=channel_page.

5. E. Cook, "The Depletion of Geologic Resources," *Technology Review* 77 (1975): 15–27.

6. R. A. Hagerman, "U.S. Reliance on Africa for Strategic Minerals," April 6, 1984, Global Security website, www.globalsecurity.org/military/library/report/1984/HRA .htm.

7. A. Aston, "China's Rare-Earth Monopoly," MIT *Technology Review,* Ocotober 15, 2010, www.technologyreview.com/energy/26538/?p1=A2.

8. G. Palast, "Secret US Plans for Iraq's Oil," BBC News website, March 17, 2005, news.bbc.co.uk/2/hi/programmes/newsnight/4354269.stm; A. Cockburn, "Bush, Oil and Iraq: Some Truth at Last," *Counterpunch,* January 14, 2004, www.counterpunch .org/cockburn01142004.html. http://georgewashington2.blogspot.com/2008/07/ cheney-amd-oil-bigs-planned-us-war.html.

9. "Price Rise and New Deep-Water Technology Opened Up Offshore Drilling," Boston.com website, December 11, 2005, www.boston.com/news/world/articles/2005 /12/11/price_rise_and_new_deep_water_technology_opened_up_offshore_drilling.

10. P. Wheatcroft, "The Next Crisis: Prepare for Peak Oil," *Wall Street Journal,* February 11, 2010, online.wsj.com/article/SB10001424052748704140104575057260398292350 .html.

11. "Alaska Crude Oil Production," Wikipedia, en.wikipedia.org/wiki/File:Alaska_ Crude_Oil_Production.PNG.

12. *Arctic National Wildlife Refuge, 1002 Area, Petroleum Assessment, 1998, Including Economic Analysis,* U.S. Geological Survey website, pubs.usgs.gov/fs/fs-0028-01/fs-0028-01.pdf.

13. M. T. Klare, "World Energy Report 2012," *Nation of Change,* November 28, 2012, www.nationofchange.org/world-energy-report-2012-1354118005.

14. S. Foucher, "Application of the Dispersive Discovery Model," The Oil Drum website, November 27, 2007, www.theoildrum.com/node/3287.

15. K. Aleklett, "International Energy Agency Accepts Peak Oil," Association for the Study of Peak Oil and Gas website, www.peakoil.net/uhdsg/we02004 /TheUppsalaCode.html.

16. Aleklett, "International Energy Agency Accepts Peak Oil"; K. S. Deffeyes, *Hubbert's Peak: The Impending World Oil Shortage* (Princeton, N.J.: Princeton University Press, 2001), 146.

17. *S. A. Korpela,* "Oil Depletion in the United States and the World," Peak Oil and Sustainability website, May 1, 2002, www.greatchange.org/ov-korpela,US_and_world_ depletion.html; Deffeyes 2001, 147; C. J. Campbell and J. H. Laherrère, "The End of Cheap Oil," *Scientific American,* March 1998, archived at www.hubbertpeak.com/ _archive/ScientificAmerican199803/EndOfCheapOil.htm.

18. G. Monbiot, "When Will the Oil Run Out?" *Guardian,* December 14, 2008, www .guardian.co.uk/business/2008/dec/15/oil-peak-energy-iea.

19. "World Oil Capacity to Peak in 2010 Says Petrobras CEO," The Oil Drum website, February 4, 2010, www.theoildrum.com/node/6169.

20. R. L. Hirsch, R. Bezdek, and R. Wendling, "Peaking of World Oil Production: Impacts, Mitigation, and Risk Management," National Energy Technology website, February 2005, www.netl.doe.gov/publications/others/pdf/Oil_Peaking_NETL.pdf.

21. C. Campbell and J. Laherrère, "The End of Cheap Oil," Scientific American 278, no. 3, (March 1998): 78–83, doi:10.1038/scientificamerican0398-78.

22. The View from the Peak website, www.theviewfromthepeak.net; cached at www.zoominfo.com/CachedPage/?archive_id=0&page_id=1997335973&page_url=//www.theviewfromthepeak.net/&page_last_updated=2007-11-14To3:31:57&firstName=Oil&lastName=Shale.

23. A. Landman, "BP's 'Beyond Petroleum' Campaign Losing Its Sheen," PRWatch, May 3, 2010, www.prwatch.org/node/9038.

24. M. Grunwald, "The Clean Energy Scam," Time, March 27, 2008, www.time.com/time/magazine/article/0,9171,1725975,00.html.

25. "Excuse Me, I Am Going to Need This to Run My Car," China Daily, June 26, 2008, bbs.chinadaily.com.cn/thread-607981-1-1.html.

26. P. Roberts, The End of Oil (New York: Houghton Mifflin, 2004), 290–295.

12. Far from the Madding Crowd

1. "Paul Ehrlich, Famed Ecologist, Answers Questions," Grist website, August 10, 2004, http://grist.org/article/ehrlich/full.

2. J. Ziegler, l'Empire de la honte (n.p.: Fayard, 2007), 130.

3. P. Angela and A. Angela, The Extraordinary Story of Human Origins (New York: Prometheus, 1993).

4. Q. D. Atkinson, R. D. Gray, and A. J. Drummond, "Bayesian Coalescent Inference of Major Human Mitochondrial DNA Haplogroup Expansions in Africa," Proceedings of the Royal Society B 276 (2009): 367–373.

5. Gary Stix, "Traces of a Distant Past," Scientific American, July 2008, 56–63.

6. C. K. Brain, "A Taphonomic Overview of the Swartkrans Fossil Assemblages," In C. K. Brain, ed., Swartkrans, a Cave's Chronicle of Early Man, Transvaal Museum Monographs 8 (Pretoria: Transvaal Museum, 1993), 1–295.

7. M. R. Rampino and S. Self, "Volcanic Winter and Accelerated Glaciation Following the Toba Super-Eruption," Nature 359 (September 2, 1992). 50–52; M. R. Rampino and S. Self, "Climate–Volcanism Feedback and the Toba Eruption of ~74,000 Years Ago," Quaternary Research 40 (1993): 269–280.

8. D. Whitehouse, "When Humans Faced Extinction," BBC News website, June 9, 2003, news.bbc.co.uk/2/hi/science/nature/2975862.stm.

9. "Of Lice And Men: Parasite Genes Reveal Modern and Archaic Humans Made Contact," Science Daily website, October 5, 2004, www.sciencedaily.com/releases/2004/10/041005075751.htm.

10. B. Linz, F. Ballou, Y. Moodley, A. Manica, H. Liu, P. Roumagnac, D. Falush, et al., "An African Origin for the Intimate Association between Humans and Helicobacter pylori," Nature 445, no. 7130 (February 2007): 915–918.

11. S. H. Ambrose, "Late Pleistocene Human Population Bottlenecks, Volcanic Winter, and Differentiation of Modern Humans," Journal of Human Evolution 34, no.

6 (1998): 623–651; M. R. Rampino and S. Ambrose, "Volcanic Winter in the Garden of Eden: The Toba Supereruption and the Late Pleistocene Human Population Crash," *Geological Society of America Special Paper* 345 (2000): 71–82.

12. T. L. Goldberg, "Genetics and Biogeography of East African Chimpanzees (*Pan troglodytes schweinfurthii*)." Ph.D. diss., Harvard University.

13. M. E. Steiper, "Population History, Biogeography, and Taxonomy of Orangutans (Genus: *Pongo*) Based on a Population Genetic Meta-analysis of Multiple Loci," *Journal of Human Evolution* 50 (2006): 509–522.

14. R. D. Hernandez, M. J. Hubisz, D. A. Wheeler, D. G. Smith, B. Ferguson, D. Ryan, J. Rogers, et al., "Demographic Histories and Patterns of Linkage Disequilibrium in Chinese and Indian *Rhesus* Macaques," *Science* 316 (2007): 240–243.

15. O. Thalman, A. Fisher, F. Lankester, S. Pääbo, and L. Vigilant, "The Complex History of Gorillas: Insights from Genomic Data." *Molecular Biology and Evolution* 24 (2007): 146–158.

16. S.-J. Luo, J.-H. Kim, W. E. Johnson, J. Van der Walt, J. Martenson, N. Yuhid, D. G. Miquelle, et al., "Phylogeography and Genetic Ancestry of Tigers (*Panthera tigris*)," *PLoS Biology* 2 (2004): 2275–2293.

17. G. Stix, "Traces of a Distant Past," *Scientific American*, July 2008, 56–63.

18. J. Diamond, *Guns, Germs, and Steel* (New York: W. W. Norton, 1997).

19. "Population Distribution Over Time," United States Census Bureau website, http://www.census.gov/history/www/reference/maps/population_distribution_over_time.html.

20. "Early Medieval and Byzantine Civilization: Constantine to Crusades," Tulane University website, www.tulane.edu/~august/H303/handouts/Population.htm.

21. "Demographic and Agricultural Growth," *Encyclopedia Britannica*, www.britannica.com/EBchecked/topic/195896/history-of-Europe/276190/Demographic-and-agricultural-growth#ref=ref994290.

22. "Population Distribution Over Time."

23. M. C. Buer, *Health, Wealth and Population in the Early Days of the Industrial Revolution*, (London: Routledge, 1926), 30.

24. A. A. Bartlett, "Forgotten Fundamentals of the Energy Crisis," *American Journal of Physics* 46 (1978): 876–888; reprinted as "Forgotten Fundamentals of the Energy Crisis," *Journal of Geological Education* 28, no. 1 (1980): 4–35.

25. J. Carter, "Proposed Energy Policy," *American Experience*, April 18, 1977, PBS website, www.pbs.org/wgbh/americanexperience/features/primary-resources/carter-energy.

26. "Energy: Curbing the Strippers," *Time*, May 19, 1975, www.time.com/time/magazine/article/0,9171,945407,00.html.

27. A. Bartlett, " Forgotten Fundamentals of the Energy Crisis: Part 7," Al Bartlett website, www.albartlett.org/articles/art_forgotten_fundamentals_part_7.html.

28. D. Brower, "Not Man Alone," *Friends of the Earth* 6, no. 20 (November 1976).

29. United Nations Department of Economic and Social Affairs, *Population Newsletter* 87 (June 2009), United Nations website, www.un.org/esa/population/publications/popnews/Newsltr_87.pdf.

30. United Nations Department of Economic and Social Affairs, *World Population Prospects: The 2006 Revision*, United Nations website, www.un.org/esa/population/publications/wpp2006/WPP2006_Highlights_rev.pdf.

31. R. Nielsen, *The Little Green Handbook* (New York: Picador, 2006).

32. United Nations Department of Economic and Social Affairs, "Population Estimates," United Nations website, esa.un.org/wpp/Sorting-Tables/tab-sorting_fertility.htm.

33. J. Otieno, "More Kenyans to Face Chronic Food Shortages, *East African,* November 30, 2010, www.theeastafrican.co.ke/news/More%20Kenyans%20to%20face%20chronic%20food%20shortages/-/2558/1063324/-/item/0/-/137y691z/-/index.html.

34. R. McDougall, "Too Many People: Earth's Population Problem," Population Matters website, 2010, www.populationmatters.org/wp-content/uploads/population_problem.pdf.

35. P. M. Vitousek, P. R. Ehrlich, A. Ehlich, and P. A. Matson, "Human Appropriation of the Products of Photosynthesis," *BioScience* 34 (1986): 368–373, biology.duke.edu/wilson/EcoSysServices/papers/VitousekEta11986.pdf.

36. M. Wackernagel, N. B. Schulz, D. Deumling, A. Callejas Linares, M. Jenkins, V. Kapos, C. Monfreda, et al., "Tracking the Ecological Overshoot of the Human Economy," *Proceedings of the National Academy of Sciences* 99, no. 14 (2002): 9266–9271.

37. J. Cohen, *How Many People Can the Earth Support?* (New York: W. W. Norton, 1995).

38. Dr. Seuss, *The Lorax* (New York: Random House, 1971), n.p.

39. N. Eldredge, *The Miner's Canary: Unraveling the Mysteries of Extinction* (New York: Prentice-Hall, 1991).

40. P. Matthiessen, *African Silences* (New York: Random House, 1992).

41. Matthiessen 1992, 120.

42. D. R. Prothero, *Catastrophes! Earthquakes, Tsunamis, Tornadoes, and other Earth-Shattering Disasters* (New York: Columbia University Press, 2009).

43. J. Godoy, "Environment: Not Enough Done to Protect Biodiversity," Inter Press Service website, May 21, 2008, ipsnews.net/news.asp?idnews=42441.

44. P. Ehrlich and A. Ehrlich, *Extinction: The Causes and Consequences of the Disappearnce of Species* (New York: Random House, 1981).

45. "Quote Details: Will Durant," Quotations Page website, www.quotationspage.com/quote/21280.html.

13. The Rejection of Reality

1. "Fox News Channel Controversies," Wikipedia, en.wikipedia.org/wiki/Fox_News_Channel_controversies.

2. M. Taibbi, *The Great Derangement: A Terrifying True Story of War, Politics, and Religion* (New York: Spiegel and Grau, 2009).

3. "Richard Dawkins Interviews Creationist Wendy Wright (Part 1/7)," YouTube video, 10:04, posted by "AtheistPlanet2," February 19, 2010, www.youtube.com/watch?v=YFj0EgYOgRo.

4. "Creator: The History Channel," TV Tropes website, tvtropes.org/pmwiki/pmwiki.php/Creator/TheHistoryChannel.

5. "Network Decay," TV Tropes website, tvtropes.org/pmwiki/pmwiki.php/Main/NetworkDecay.

6. "Network Decay: Total Abandonment," TV Tropes website, http://tvtropes.org/pmwiki/pmwiki.php/NetworkDecay/TotalAbandonment.

7. "Educational Programming on The Learning Channel," Koi Koi Eleven website, July 17, 2008, koikoi11.blogspot.com/2008/07/education-programming-on-learning. html; "TLC Promo," YouTube video, 0:53, posted by "jayismostlywater," January 28, 2011, www.youtube.com/watch?v=_8jeuYMHX9Y&feature=autofb.

8. "Network Decay: Total Abandonment."

9. "Network Decay: Slipped," TV Tropes website, tvtropes.org/pmwiki/pmwiki .php/Main/NetworkDecay/Slipped.

10. C. Mooney, *The Republican Brain: The Science of Why They Deny Science—and Reality* (New York: John Wiley, 2012), 32.

11. D. Kahan, "Cultural vs. Ideological Cognition, Part 1," Cultural Cognition blog, December 20. 2011, www.culturalcognition.net/blog/2011/12/20/cultural-vs-ideological-cognition-part-1.html.

12. "Harold Camping's 21st May Doomsday Prediction Fails: No Earthquake in New Zealand," *International Business Times,* May 21, 2011, www.ibtimes.com /articles/149529/20110521/may-21st-doomsday-prediction-end-of-the-world-failed-earthquake-new-zealand-twitter-geological-surve.htm.

13. J. T. Jost, C. M. Federico, and J. L. Napier, "Political Ideology: Its Structure, Functions, and Elective Affinities," New York University Department of Psychology website, www.psych.nyu.edu/jost/Political%20Ideology__Its%20structure,%20 functions,%20and%20elective%20a.pdf.

14. S. Borenstein, "AP-GfK Poll: Science doubters say world is warming,' Yahoo! News website, December 14, 2012, news.yahoo.com/ap-gfk-poll-science-doubters-world-warming-080143113.html.

15. Mooney 2012, 129–130.

16. C. Mooney, "Kerry Emanuel's Powerful Testimony on Climate," *Discover,* March 31, 2011, blogs.discovermagazine.com/intersection/2011/03/31/ kerry-emanuels-powerful-testimony-on-climate.

17. "Geocentric Theory," Conservapedia website, www.conservapedia.com/ Geocentric_theory.

18. "Flat Earth," Conservapedia website, www.conservapedia.com/Flat_Earth.

19. "Theory of Relativity," Conservapedia website, www.conservapedia.com/ Theory_of_relativity.

20. "Theory of Relativity."

21. R. Schiffman, "North Carolina Legislature Prepares to Ban Sea From Rising," *Huffington Post,* June 6, 2012, www.huffingtonpost.com/richard-schiffman/north-carolina-legislature-sea-level-rising_b_1567213.html.

22. N. Wing, "Chris Stolle, Virginia Lawmaker: 'Sea Level Rise' In Climate Change Study Is 'Left-Wing' Term," *Huffington Post,* June 11, 2012, www.huffingtonpost .com/2012/06/11/chris-stolle-virginia-gop-climate-change_n_1586128.html.

23. See "Colbert: North Carolina Sea Level Rise Findings Can Simply Be Made Illegal," *Huffington Post,* June 5, 2012, www.huffingtonpost.com/2012/06/05/colbert-north-carolina-sea-level_n_1571329.html.

24. B. Johnson, "During Climate Hearing, Markey Asks If Anti-Science GOP Will Repeal Gravity, Heliocentrism, Relativity," Think Progress website, March 10, 2011, thinkprogress.org/climate/2011/03/10/174942/markey-flat-earthers.

25. "Full Committee Hearing: Climate Change: Examining the Processes Used to Create Science and Policy," Committee on Science, Space, and Technology website, March 31, 2011, science.house.gov/hearing/full-committee-hearing-climate-change.

26. M. Roosevelt, http://articles.latimes.com/keyword/news"Critics' Review Unexpectedly Supports Scientific Consensus on Global Warming," *Los Angeles Times,* April 4, 2011, articles.latimes.com/2011/apr/04/local/la-me-climate-berkeley-20110404.

27. S. Cavanaugh, "The American Revolution: Flunking the Test," Free Republic website, December 3, 2009, www.freerepublic.com/focus/f-news/2400379/posts.

28. "Two-Thirds of Americans Can't Name Any U.S. Supreme Court Justices, Says New FindLaw.com Survey," PR Newswire website, June 1, 2012, www.prnewswire.com/news-releases/two-thirds-of-americans-cant-name-any-us-supreme-court-justices-says-new-findlawcom-survey-95298909.html.

29. A. Romano, "How Dumb Are We?"Daily Beast website, March 20, 2011, www.thedailybeast.com/newsweek/2011/03/20/how-dumb-are-we.html.

30. "American Public Vastly Overestimates Amount of U.S. Foreign Aid," World Public Opinion website, November 29, 2010, worldpublicopinion.org/pipa/articles/brunitedstatescanadara/670.php.

31. "American Public Vastly Overestimates Amount of U.S. Foreign Aid."

32. J. Linkins, "What Does Michele Bachmann's Grasp Of History Mean For The Children?" *Huffington Post,* March 14, 2011, www.huffingtonpost.com/2011/03/14/what-does-michele-bachman_n_835689.html.

33. "Science Literacy: American Adults 'Flunk' Basic Science, Says Survey," Science 2.0 website, March 12, 2009, www.science20.com/news_releases/science_literacy_american_adults_flunk_basic_science_says_survey; L. Gross, "Scientific Illiteracy and the Partisan Takeover of Biology," Curious Cat Science and Engineering blog, June 17, 2006, engineering.curiouscatblog.net/2006/06/17/scientific-illiteracy.

34. Galileo Was Wrong blog, galileowaswrong.blogspot.com.

35. T. Oswald, "Americans Less Likely to Accept Evolution than Europeans," Michigan State University website, August 10, 2006, news.msu.edu/story/1087.

36. J. Raloff, "Science Literacy: U.S. College Courses Really Count, *ScienceNews,* March 13, 2010, www.sciencenews.org/view/generic/id/56517/title/Science_%2B_the_Public__Science_literacy_U.S._college_courses_really_count.

37. "Scientific Literacy: How Do Americans Stack Up?" Science Daily website, Feburary 27, 2007, www.sciencedaily.com/releases/2007/02/070218134322.htm.

38. A. Paulson, "New Report Ranks U.S. Teens 29th in Science Worldwide," *Christian Science Monitor,* December 5, 2007, www.csmonitor.com/2007/1205/p02s01-usgn.html.

39. R. Whittaker, "GOP Opposes Critical Thinking," *Austin Chronicle,* June 27, 2012, www.austinchronicle.com/blogs/news/2012-06-27/gop-opposes-critical-thinking.

40. "Texas GOP Platform Opposes Teaching 'Critical Thinking Skills' in Schools: The Stupid, it burns!" Daily Kos website, June 27, 2012, www.dailykos.com/story/2012/06/27/1101959/-Texas-GOP-Platform-to-ban-teaching-Critical-Thinking-Skills-in-schools-The-stupid-IT-BURNS.

41. "Public Praises Science; Scientists Fault Public, Media," Pew Research Center website, July 9, 2009, www.people-press.org/2009/07/09/public-praises-science-scientists-fault-public-media.

42. J. Owen, "Evolution Less Accepted in U.S. than Other Western Countries, Study Finds," *National Geographic,* August 10, 2006, news.nationalgeographic.com/news/2006/08/060810-evolution.html.

43. L. Davies, "Higgs Boson Announcement: Cern Scientists Discover Subatomic Particle," *Guardian,* July 4, 2012, www.guardian.co.uk/science/blog/2012/jul/04/higgs-boson-discovered-live-coverage-cern.

44. S. Weinberg, "The Crisis of Big Science," *New York Review of Books,* May 10, 2012, www.nybooks.com/articles/archives/2012/may/10/crisis-big-science.

45. C. Heller, "Neil deGrasse Tyson: How Space Exploration Can Make America Great Again," *Atlantic,* March 5, 2012, www.theatlantic.com/technology/archive/2012/03/neil-degrasse-tyson-how-space-exploration-can-make-america-great-again/253989.

46. "2012 United States Federal Budget," Wikipedia, en.wikipedia.org/wiki/2012_United_States_federal_budget.

47. W. Wheeler, "The Military Imbalance: How the U.S. Outspends the World," AOL Defense website, March 16, 2012, defense.aol.com/2012/03/16/the-military-imbalance-how-the-u-s-outspends-the-world.

48. D. Axe, "Buyers Remorse: How Much Has the F-22 Really Cost?" *Wired,* December 14, 2011, www.wired.com/dangerroom/2011/12/f-22-real-cost.

49. "*Nimitz*-Class Aircraft Carrier," note 19, Wikipedia, en.wikipedia.org/wiki/Nimitz-class_aircraft_carrier#cite_note-19.

50. J. Dufour, "The Worldwide Network of US Military Bases: The Global Deployment of US Military Personnel," Centre for Research on Globalization website, July 1, 2007, www.globalresearch.ca/index.php?context=va&aid=5564.

51. R. A. Greene, "China Shoots Up Rankings as Science Power, Study Finds," CNN website, March 29, 2011, articles.cnn.com/2011-03-29/world/china.world.science_1_china-output-papers?_s=PM:WORLD.

52. " Knowledge, Networks and Nations: Final Report," Royal Society website, March 28, 2011, royalsociety.org/policy/projects/knowledge-networks-nations/report.

53. "New Countries Emerge as Major Players in Scientific World," Royal Society website, March 28, 2011, royalsociety.org/news/new-science-countries.

54. R. Matthews, "China's Green Innovation and the Challenge for America," Global Warming Is Real website, February 23, 2011, globalwarmingisreal.com/2011/02/23/chinas-green-innovation-and-the-challenge-for-america; F. Harvey, "UK Slips Down Global Green Investment Rankings," *Guardian,* March 29, 2011, www.guardian.co.uk/environment/2011/mar/29/uk-global-green-investment-rankings.

55. "China and Germany Seize Lead in Green Technologies," DW website, November 21, 2011, www.dw.de/china-and-germany-seize-lead-in-green-technologies/a-15531139.

56. "List of Nobel Laureates by Country," Wikipedia, en.wikipedia.org/wiki/List_of_Nobel_laureates_by_country.

57. C. Gross, "U.S. Sees Large Increase in Foreign Graduate Students," *U.S. News and World Report,* November 10, 2011, www.usnewsuniversitydirectory.com/articles/us-sees-large-increase-in-foreign-graduate-student_11910.aspx.

58. H. R. Turner, *Science in Medieval Islam* (Austin: University of Texas Press, 1997).

59. "How Islamic Inventors Changed the World," *Independent,* March 11, 2006, www .independent.co.uk/news/science/how-islamic-inventors-changed-the-world-469452. html.

60. G. Sarton, *The Incubation of Western Culture in the Middle East: A George C. Keiser Foundation Lecture Delivered in the Coolidge Auditorium of the Library of Congress, March 29, 1950* (Washington, D.C.: Library of Congress, 1951).

61. "The Erosion of Progress by Religions," YouTube video, 10:19, posted by "Reason-ableMe," September 3, 2010, www.youtube.com/watch?v=6oxTMUTOzow&feature=re lated.

62. C. Sagan, "Wonder and Skepticism," *Skeptical Enquirer* 19, no. 1 (January–February 1995), available online at Positive Atheism website, www.positiveatheism.org/writ/ saganws.htm.

INDEX

DONALD R. PROTHERO has taught college geology for thirty-three years, most recently as Professor of Geology at Occidental College and Lecturer in Geobiology at the California Institute of Technology. He has published thirty-two books, including *Rhinoceros Giants: The Paleobiology of Indricotheres* (Indiana University Press, 2013); *Earth: Portrait of a Planet; The Evolution of Earth; Evolution: What the Fossils Say and Why It Matters; Catastrophes!* and *After the Dinosaurs: The Age of Mammals* (Indiana University Press, 2006). He lives in La Crescenta, California.